The Killing of Karen Silkwood

The Story Behind the Kerr-McGee Plutonium Case

The Killing
of Karen Silkwood

Richard Rashke

HOUGHTON MIFFLIN COMPANY   BOSTON

Library of Congress Cataloging in Publication Data

Rashke, Richard L
    The killing of Karen Silkwood.

    1. Silkwood, Bill.   2. Kerr-McGee Nuclear Corporation.
3. Silkwood, Karen, 1946–1974.   4. Negligence — United States.
5. Plutonium — Safety regulations — United States.   I. Title.
KF228.S54R37   346.7303'2   80-27051
ISBN 0-395-30233-1

Printed in the United States of America

V 10 9 8 7 6 5 4 3 2

Interview by Barbara Newman, copyright © 1976 by National Public Radio, provided courtesy of National Public Radio, was originally broadcast April 6, 1976, on "All Things Considered." Interview by Barbara Newman, copyright © 1975 by National Public Radio, provided courtesy of National Public Radio, was originally broadcast March 23, 1975 on "Options." The author is grateful for permission to quote from *Wheeling and Dealing,* by Bobby Baker, published by W. W. Norton & Co., 1978. The photographs of Jacque Srouji, James V. Smith, and Peter Stockton were taken by Howard Kohn.

EKA 1

To Angeline

Contents

In the laboratory, we've got eighteen- and nineteen-year-old boys . . . And they didn't have schooling, so they don't understand what radiation is. They don't understand, Steve. They don't understand.

The Killing

James Mullins sat high in the cab of a one-ton flatbed welding truck. Behind him sprawled Oklahoma City, where Shell Oil and Phillips Petroleum pumped black gold from pools deep under the state capital. Thirty-five miles ahead of him, north on Highway 74, lay the tiny town of Crescent, with the Hub Cafe and Ted Sebring's and the Kerr-McGee plutonium plant.

It was around 7:30 P.M. The air was crisp and dry and cold. A light wind kicked the red dust across the road. Mullins' boss, John Trindle, was a quarter of a mile ahead in a pickup truck, and Mullins' fourteen-year-old brother-in-law, Dalton Ervin, was dozing in the flatbed cab.

Mullins had been trucking down Highway 74 every day for two weeks. He used to be a wrecker driver in Hydro, so he saw more on the lonely Oklahoma highway than most. He knew the farmscape well between Oklahoma City and Crescent, the pastures and pines, the wells pumping oil from the red clay, the barns stripped gray by sand and wind.

As Mullins sped over a culvert 7.3 miles south of Crescent, his headlights caught a glimmer in the ditch on the right side of the road. He checked his rearview mirror for traffic, pressed his airbrakes, and lumbered onto the shoulder along the bar ditch. Then he flashed Trindle and backed up his red rig.

The flatbed's headlights beamed over the ditch into the pasture, spilling just enough light to outline a white Honda Civic trapped at the entrance of the concrete culvert.

3

Ervin crawled out of the cab and called down into the ditch. No one answered. He picked up a flashlight and sidestepped down the bank. The left front of the car was squashed, and there was an arm sticking out the window.

She was dead. Her purse rested against the concrete wingwall, as if she had dropped it there. Her Kerr-McGee paycheck lay in the soft mud.

A *New York Times* reporter and a Washington union official were waiting for her at the Northwest Holiday Inn. She had promised them important documents that would prove that Kerr-McGee was making defective plutonium fuel rods.

She was a wisp of a woman, just under 100 pounds, twenty-eight years old, with thick black hair that tumbled over her shoulders, and bangs that almost hid her tiny face. Her eyes were dark, mostly suspicious and frightened the past few weeks. A Texan, independent and stubborn and bright. She had wanted to go back to Texas, but not before she'd finished her assignment, for she was a fighter who wouldn't give up. Her boyfriend had told her to stop struggling. It was burning her out, consuming her, he'd said. But she hung on. Now she was dead.

There was plutonium in her lungs and clinging to her bones. She had eaten and breathed it just one week ago. Scientists would argue about how many nanocuries she actually had or how much of a full-body burden she carried. They would bisect her right lung and cut her liver in pieces. In the end, some would say the plutonium in her wouldn't have hurt her; others would claim she was married to cancer.

Her family would have to bury her in a new dress, because the clothes in her apartment, contaminated with plutonium, had been sealed in drums. No representative of Kerr-McGee, for whom she had worked for more than two years, or of the Atomic Energy Commission, which had followed her case, would be at the funeral.

But first, police would watch as they cut her out of the car, dried blood on her face. They would find two marijuana cigarettes and a Quaalude in her purse and assume they knew what had happened. And when someone would find strange dents on the

4

left rear side of her car, they would spend two months trying to explain them away.

The FBI would investigate the bizarre events that had begun three months earlier. So would the United States Congress. Both would summarize her case in one word — *mystery.*

But she was no mystery. In life, she was an ordinary woman who stuck her neck out. In death, she became a nuclear martyr, a symbol to the feminists, the environmentalists, and the labor movement.

And her enemies, all the men and the powerful institutions they run — the police, the FBI, the Kerr-McGee Corporation, the AEC — would label her emotionally disturbed, sexually promiscuous, and hooked on drugs.

Who was she?
Who contaminated her?
What was in her documents?
What happened to them?
Was she murdered?

All the answers aren't in yet. Some may never be.

Chapter 1

Her name was Karen Gay Silkwood, and she was born in Longview, Texas, in 1946. She grew up in Nederland, halfway between Port Arthur and Beaumont, in the petrochemical heart of Texas, where the night sky burns a smoky orange from oil refineries aglow with waste gases.

Her mother, Merle, says Karen had an ordinary childhood, happy, calm, and healthy, except for asthma and hay fever. The honeysuckle creeping over the fence next door used to make her wheeze in late spring and summer. Her friends remember her laughing a lot, with a great appetite for a good time.

She was a baby-sitter at the First Baptist Church Nursery, played the flute from the fourth through the twelfth grade, loved tennis and volleyball. She studied and read a lot, and didn't date as much as her sisters, Rose Mary and Linda, did later.

She had some traits she'd wear like stripes all through her short life. They would get her into trouble and eventually kill her. She wasn't a rebel or a joiner, but she liked to make up her own mind. According to her friends, she was loyal, stood up for what she believed in, and was as tenacious as an abalone on a rock. She was the kind of person who couldn't stand by and ignore something if it was wrong.

She was no ordinary student. Straight A's all through high school, a member of the National Honor Society, and an honor graduate in the class of 1964. Science, especially chemistry, was her best subject, and she was the only girl in the high school chemistry class. When she went to Lamar College in Beaumont to study medical technology, the Business and Professional Women's Club paid her way with a scholarship.

In the summer of 1964, Karen met Bill Meadows. They were both vacationing on their grandparents' farms near Longview. Bill was her age, one year behind her in school. He was from Los Angeles, where his father worked as a division manager for the Mobile Pipeline Company.

Bill dated Karen that summer, and she wrote him a few times during her first year at Lamar. After graduating in 1965, Bill came back to Longview as a machinist-helper for the Mobile Pipeline Company. Karen was waiting for him. They were both nineteen, and within three weeks they eloped.

Karen Silkwood's life with Bill Meadows was predictable. He was the breadwinner; she had babies — three of them — and worked part time, here or there, wherever she could find work between pregnancies. Bill Meadows describes her as a good mother, a sexually faithful wife, and an emotionally and physically healthy woman except for asthma attacks and migraines every few months. She drank moderately, took no drugs, didn't smoke grass, and loved her children.

But it wasn't a happy marriage. Her mother noticed that Karen was getting more and more nervous. She seemed unhappy when she came home for visits, though she didn't complain. She didn't want to burden her parents, Merle says. Once, Karen packed up the children, Beverly, Michael, and Dawn, and came home, saying she wanted a divorce. Her mother, who had never trusted Bill Meadows, encouraged her. But Bill raced to Longview and talked Karen into going back with him to Duncan, Oklahoma, where Mobile had sent him.

Still, their marriage was doomed. Bill raced motor bikes, drank a lot, and spent money faster than they could make it. When he had exhausted all their credit, he tried a credit counselor, but then threw in the sponge. Bill admits Karen had nothing to do with their bankruptcy. It was all his fault; he had expensive tastes.

Besides, there was Kathy. Bill had been seeing her "whenever the opportunity presented itself." When Karen demanded that he give her up, he refused and instead offered an uncontested divorce if Karen would give him the children. She refused, and he continued to see Kathy.

Then, one morning in August 1972, Bill woke up and found Karen gone. After seven years of marriage, she had packed a few clothes and walked out — no note, no phone number, no address. He wasn't surprised she had left, he says, but he couldn't understand why she ran away. Two days later, she called from Oklahoma City, where she was staying with Janet White, a friend from Duncan who worked at the Medical Hospital Center. She had had it, she said. She wasn't coming back. He could have it his way; she wouldn't stop him from taking custody of the children.

Bill filed for a divorce almost immediately. Karen didn't appear in court. All she asked for was her name — Silkwood . . . Karen Gay Silkwood. Bill married Kathy.

It had hurt Karen to give up her three children, the oldest six, the youngest just two. She never talked about them, not even to her mother. Nor did she ever tell her mother why she had walked out.

Over the next two years, Karen saw Beverly, Michael, and Dawn every few months in Duncan or in Oklahoma City, where she kept them overnight. Friends who saw her with the children say they could feel her love and her pain. After her death, she would be accused of abandoning her children and showing little interest in them.

□ □ □

Karen Silkwood found a job clerking in an Oklahoma City hospital, but before she could type the first doctor's bill, she heard that the Kerr-McGee Nuclear Corporation was hiring laboratory analysts at its plutonium plant on the Cimarron River, near Crescent. It looked like a real opportunity to develop her technical talents and make more money. In August 1972, she began work in the Kerr-McGee Metallography Laboratory, which tested plutonium pellets.

Plutonium is a relatively new element, discovered in the forties and first used at Nagasaki in 1945. The metal doesn't exist as such in nature. A by-product of neutron-bombarded uranium, it is now collected from the waste of nuclear reactors. For years it was used only in bombs — thirteen pounds is enough to build an atomic bomb.

Kerr-McGee's plutonium came from the Atlantic Richfield Company in Hanford, Washington, transported in bulletproof vans with four drivers and gun ports. In the beginning, the Atomic Energy Commission (AEC) rode shotgun in special station wagons to make sure terrorists wouldn't hijack the Tri-State semis. Later, Wells Fargo armored cars led the caravans.

The plutonium-nitrate solution rode from Washington to Oklahoma in three-foot-high steel bottles resting comfortably inside a shield made of two drums, welded together, that stood five feet tall. Each bottle held about 4.5 pounds of plutonium. The AEC had licensed Kerr-McGee to keep up to about 700 pounds of plutonium at the Cimarron plant, enough to make more than fifty atomic bombs.

At the Kerr-McGee receiving dock, technicians from the health physics office, wearing respirators that looked like World War II gas masks, checked the van for radiation. The workers, protected against radiation by their forty-pound lead aprons, tore off the barrel tops one at a time. Inside, the bottles sat in a birdcage surrounded by a plastic bag, like a fetus in the amniotic sac. Workers attached a hose and air filter to the bag and sucked the air out so that potential contamination would not escape into the room. The health physics technicians also checked the air filter for radiation.

If the plastic bag was not radioactive, two workers climbed ladders, sheared the birdcage bolts, and lifted the bottles into a glove box that was sealed with negative pressure. If the box or the shoulder-length rubber gloves that reached into it leaked, air would flow in but not out.

With hands inside the glove ports, workers vacuum-hosed the plutonium nitrate into a forty-foot-high tank scale that reached from the basement floor to the roof. They added uranium to the solution until there were seven parts uranium to three parts plutonium.

When the workers on the other end of the assembly line needed more plutonium pellets, Kerr-McGee pumped out the soup, added ammonium hydroxide, and sent the mixture cascading down shelves. The liquid ran off into storage tanks, was treated, and later buried at AEC-approved radioactive dumps. The heavy plutonium and uranium metals got trapped as slimy green mud, which was baked and dried inside a calciner. The plutonium clinkers were then dumped into glove boxes, where a hammermill ground them into a powder that was then tossed in a giant blender until the plutonium and uranium were evenly mixed. The solution-to-powder process took fifty to sixty hours, and the K-M employees worked twelve-hour shifts.

The powder was slugged and pressed into soft green pellets one inch long and one-half inch thick. Workers fired the pellets like clay bullets until they were hard enough to be handled without chipping. If the pellets were too small, the workers rejected them; if they were too big, they were skinned down to size.

Workers visually inspected each pellet to see if it was cracked or chipped and then loaded them all into eight-foot-long, pencil-thin, stainless steel rods. They washed the rod tips with alcohol to remove all radioactive contamination. The rods were fragile — about as heavy as a cup of coffee — and the workers had to be careful that they didn't kink.

Before the rod tips were welded tight, workers stuffed a small cylinder of xylon gas inside each one. Later on, if a rod began to leak, the gas would escape and warn technicians. Workers then wheeled the rods into a huge x-ray room, where electronic eyes read the contents of each rod as a last check. If the rods passed inspection, workers wrapped each one in plastic, bundled 100 rods together, and lowered them into floor wells to await the AEC shotgun riders or Wells Fargo for the ride back to Washington.

The rods, now fully inspected fuel pins, were destined for the AEC's Fast Flux Test Facility (FFTF) at Hanford, Washington, under the management of Westinghouse. The FFTF tested fuel cores for the fast breeder reactor the AEC hoped to build in Clinch River, Tennessee. Fueled by plutonium instead of pure uranium, fast breeders are designed to make more plutonium than they burn. Kerr-McGee had a fixed-price contract to make 12,916 rods for the first test core and 3380 for the second. NUMEC, later purchased by Babcock and Wilcox, the firm that designed and built the Three Mile Island reactor, supplied Westinghouse with the other half of the fuel pins.

Because the AEC expected Kerr-McGee to report and pay for missing plutonium, K-M ran a scrap-recovery program at the Cimarron plant. Workers sorted the waste into "keepers" and "throwers." The throwers were buried after workers logged the amount of plutonium they were tossing out with the item. Keepers would pass through ion exchange, where the plutonium was saved and all other metals discarded. In the beginning, K-M found it was cheaper to bury contaminated items than to clean off the plutonium. But as the AEC raised the price of the metal, K-M found more and more keepers.

In the Metallography Laboratory, Silkwood did quality-control checks. She randomly selected pellets from a lot and then held unexposed x-ray film against them to test for gamma rays. The plutonium was supposed to be evenly distributed throughout the pellet. If it wasn't, the developed film would show "hot spots."

Silkwood also polished randomly selected fuel-rod welds to see if there were any cracks or inclusions. If the pellets or welds flunked, she would run more tests on that pellet lot. If she found a pattern of flaws, she'd reject the whole lot.

Drew Stephens worked across the hall in the General Chemistry Lab. He was Karen's age, bright and handsome, with a reddish-blond beard, a thin mustache, straight hair that hugged his ears, and intense eyes that gave a hint of popping. Drew's father, Donald, had held a top management position at Kerr-McGee, but a difference of opinion sent him packing to the marketing division. Donald had finally quit in a huff, but Drew stayed on.

Drew's marriage was on the rocks, and he began to date Silkwood. "Karen became just as intrigued with the technology necessary to produce nuclear fuels as I was," he wrote later. "Because of the warmth and concern Karen showed for other people, Karen and I became close in the following months."

Karen became the "catalyst," as Drew put it, that forced him to face his marital problems. He moved in with Karen and her friend Janet White. When his wife remarried shortly after their divorce and left their Oklahoma City home, Drew and Karen moved into the house. He shared her interest in music; she shared his in cars.

Drew helped her select a new car — a white Honda Civic. He taught her to slip and slide and straighten out without losing control. He was an autocrosser, a slalom racer on wheels, squealing between pylons against a clock. She joined the Sports Car Club of America, and Drew coached her in parking lots until she could weave and slice as well as he. She entered races for women and won a foot-tall silver trophy. Drew still has it.

They took trips to St. Louis and Kansas to watch autocross gymkhanas. They bought dirt bikes and sped along trails in the woods or down the Cimarron River mud flats.

□   □   □

In November, three months after Karen began polishing plutonium rod welds, she was pacing outside Kerr-McGee's chain-link fence carrying her first ON STRIKE placard. She was no union activist. In fact, she had shown little interest in the Oil, Chemical and Atomic Workers Union (OCAW), which represented the 150 Kerr-McGee rank and filers. She had joined the OCAW, as had a hundred others, because it was their only protection against one of the largest energy conglomerates in the United States. It was her duty to picket, and Karen Silkwood took her turn.

The OCAW International had advised the impotent Crescent local not to strike because it wasn't strong enough, but Local 5–283 took on Goliath anyway, demanding higher wages, better training, and improved health and safety programs. The strike went on for ten weeks, and the union came out barely breathing. The Kerr-McGee Nuclear Corporation — a subsidiary of the Kerr-McGee empire — wasn't hurting in the fall of 1972, so it could hold out against the strike. And because the winter was exceptionally cold, jobs were scarcer than usual. The farms for miles around Crescent were filled with unemployed nineteen- and twenty-year-olds, and Kerr-McGee had no trouble picking strikebreakers to keep the pellets moving. Three dollars an hour seemed like a lot of money to them.

It was an unspectacular strike. No violence; just a war of attrition. As the strike stretched into its second month and Christ-

13

mas loomed ahead, more and more union members scrapped their strike placards and crossed the line, until only a score of diehards were left. Silkwood was one of them.

Two months after the strike began, Karen was back at work under a new, two-year contract written by Kerr-McGee. For the twenty OCAW members left in the battered local, it had been a total defeat, but for Silkwood, it had been an awakening. Taking a stand against Kerr-McGee, walking the line, living off part-time wages as a clerk in a building supply company, watching OCAW members one by one knuckle under to the pressure — all of this had forged her ties with the union. Her relationship with Kerr-McGee would never be the same.

Karen's relationship with Drew Stephens was changing, too. He didn't want to get married, he said. He was feeling stifled, needed some freedom, wanted to date other women. So she moved out in the spring of 1973. It was his idea, and it hit her very hard. She visited a counselor three or four times to try to put her life together again.

One night in September 1973, she called a friend, Connie Edwards, and said she had just tried to kill herself with an overdose of drugs. Edwards rushed to Karen's apartment and found her on the couch in a stupor. She roused Karen and tried to coax her to go to a hospital, but she refused. Edwards then helped Karen vomit and took her home for the night.

Karen moved back in with Drew, but it was a mistake. They worked two different shifts; they became jealous over each other's lovers; they fought. By the end of 1973, they had drifted apart. For the next ten months, they would see each other once or twice a week, sometimes spending the night at her apartment, sometimes at his house.

After the strike and all during 1973, Karen showed little interest in the OCAW. But in the spring of 1974, when the wheat was just poking through the red soil, Kerr-McGee speeded up production. There were twelve-hour shifts, seven-day work-weeks, rotations from day to night shifts, and spills and contaminations. Karen became more and more worried about health and safety, about nineteen-year-old farm boys with tractor grease under their fingernails treating plutonium like fertilizer, and about a manage-

14

ment that used them up and sent them back to plough the fields with plutonium in their bodies, unaware that they were hot. The fire was a perfect example.

Two glove-box operators had filled a plastic bag with plutonium-contaminated waste (a standard procedure), when they spotted smoke coming from a hole in the bag. Plugging the hole, they ran out of the room, but not before they and five other workers had inhaled 400 times the weekly limit of insoluble plutonium permitted by the AEC.

Airborne plutonium contamination is more dangerous than contamination by direct contact of plutonium with the skin. Once lodged in the lungs, the particles are slow to be flushed out, and each new contamination increases the amount of radiation settling on delicate tissues.

The AEC was irked by the way that Kerr-McGee handled the incident. Besides chiding the corporation for not giving prompt attention to the health of the seven workers, it complained that evaluation of the radioactivity levels was hampered because room air monitors had run out of paper the day before the fire.

William J. Shelley, Kerr-McGee director of regulation and control, chafed at the criticism. "We believe," he wrote, "that the attitude that a commercial operation must be equipped to completely research the cause and effect of any such incidents is beyond the requirements of sound business judgment."

Soon after the fire, two maintenance men repairing a pump were splashed by a rain of plutonium particles that settled on their hands, faces, hair, and clothes. The plumbers left at noon to eat in Crescent, unaware of their contamination until they returned to the plant. Kerr-McGee immediately scrubbed them and their car, but didn't bother to check the restaurant for radiation or to inform the proprietor. The AEC heard about the incident from Ilene Younghein, a local environmentalist. By then, there wasn't much it could do for the restaurant patrons who may have eaten plutonium with their lunch.

While all the contaminations were taking place at the plutonium plant, something was gnawing away at Karen Silkwood. Whether it was the Kerr-McGee working conditions, a problem in her personal life, or a combination of the two is not clear. In May

1974, she told her physician, Dr. Clarence Shields, Jr., that she was depressed and couldn't sleep during the day when she worked nights. Dr. Shields's medical records show that he prescribed methaqualone, first under the trade name Parest, then as Quaalude.

In 1974, Quaalude was a popular new sleeping pill that produced drowsiness within ten to twenty minutes. One 300 milligram tablet before retiring was the recommended dosage. If the drug is taken over a long period, the user can develop a Quaalude tolerance, and its effect is lessened. Dr. Shields, like most physicians, no longer prescribes Quaalude as a sleeping pill, because methaqualone turned out to be addictive.

At some point between May and November, Karen began taking Quaaludes as tranquilizers. Drew Stephens told the FBI after her death that she had become "dependent on them as downers to keep her head together; that is, she was using them as a sedative and not as a sleeping pill."

□ □ □

The first bizarre event in the Silkwood story happened on July 31, 1974. She was working the 4:00 P.M. shift in the Emission Spectroscopy Laboratory, where she pulverized plutonium pellets in sealed glove boxes, cut in a chemical carrier, and ran the dust through a spectrograph, which reads the light each metal emits. Although the pellets were pure plutonium and uranium, traces of other metals, such as nickel or chromium, were permitted. If the trace was too high, the pellet lot had to be rejected.

Sometime after Silkwood had finished her work and had gone home to bed, health physics technicians checked the air-sample filter papers used in the lab during the three previous shifts. If the Emission Spec Lab air had been radioactive at any time during that twenty-four-hour period, the filter papers would be radioactive, too.

The technicians found that the filters used before Silkwood began working and after she quit were clean, but those used during her shift were highly contaminated. The technicians were puzzled, for the contamination pattern was not logical. If the laboratory air had been contaminated during Silkwood's shift, as the filters indicated, why wasn't the air still contaminated during the following shift? No one had decontaminated the lab between shifts because no one had reason to believe the room was radioactive.

The health physics technicians surveyed the Emission Spec Lab and found only a little contamination around a flange on one glove box. There was no significant contamination on the floors, exhaust filters, or equipment. The health physics office told Silkwood and two other workers who were in the lab with her during the 4:00 P.M. shift to provide weekly fecal and urine samples that would be tested for radiation by outside consultants. Test results indicated that only Silkwood had been contaminated — but insignificantly, by AEC standards.

The health physics technicians never found the source of the contamination, and health physics director Wayne Norwood concluded that the whole incident was a put-up job. He took a picture of one of the radioactive filters and noted that the contamination was unevenly distributed over the filter's surface. He thought he saw a thumbprint, as if someone wearing a contaminated plastic glove had passed by the filter, which was only five feet off the ground, and dabbed it.

Norwood had been suspicious of contamination frauds ever since the fall of 1972, just before the long strike, when he had discovered five or six contaminated filters. To Norwood, it looked as if someone had taken a cotton swab with plutonium on it and drawn an X across the filter faces. He even thought he saw the number sign, #, on one.

Fraud or not, it seemed clear something was going on at the Cimarron plutonium plant.

Chapter 2

When union elections rolled around the week after the incident in the Emission Spec Lab, Karen Silkwood was ready. She had not been an active member of the Oil, Chemical and Atomic Workers union since the strike two years earlier, but she was concerned about the sloppy conditions at the plant. New contract negotiations with Kerr-McGee would begin in three months.

Silkwood didn't campaign for a spot on the three-person union bargaining committee, but she let it be known that if elected, she wouldn't turn it down. She won — the first female committee member in Kerr-McGee's history. Her assignment was health and safety.

Jack Tice chaired the committee. He was short, mild-mannered, with an easy laugh; a blue-collar gentleman as honest and direct as Oklahoma sunshine. He was balanced and calm where Karen was brash and excitable. He was the pro with seven years of union activism; she was the beginner. But both of them were worried about the worsening health and safety conditions at the plutonium plant.

Just after the election, Jack Tice wrote to Elwood Swisher, the vice-president of OCAW International. Things are bad down here, he said. Respirators designed to filter out plutonium particles in the air aren't working, the turnover of workers is very high, and more and more untrained people are handling the radioactive metals. Tice told Swisher something had to be done.

The OCAW was a feisty union, with 185,000 members in the

United States and Canada. It fought hard for the health and safety
of oil and chemical workers, but it hadn't shown much concern
for the atomic workers. Tice believed this was mostly Elwood
Swisher's fault, but he wasn't sure why. On his kinder days, the
committee chairman thought Swisher just didn't understand how
dangerous uranium and plutonium were. And on his angrier days,
he thought the vice-president was afraid to disturb the complacency of the nuclear industry and of the AEC, which controlled
the former with about as much strictness as a doting father.

So Tice was pleasantly surprised when Swisher responded to
his letter by saying that it was time to complain to the AEC. Take
notes, Swisher said. Document everything you can. Be specific.
The union's legislative office in Washington will set up a meeting
with the AEC in September. Don't tell Kerr-McGee where you're
going or why.

Tice told Silkwood to keep her eyes open in the lab. And she
did. All through August and September, she followed contamination incidents, asked the health physics technicians questions, and
interviewed workers during her breaks and lunch hours. She wrote
her observations neatly in the small spiral notebook she carried in
her purse.

In August, before the three committee members could leave
for Washington, a Kerr-McGee worker launched a drive to decertify the OCAW as the bargaining agent for all workers, union
members or not, in the Cimarron plant. To get the National Labor
Relations Board (NLRB) to supervise a vote that would decide the
issue, 30 percent of the workers had to sign a petition requesting
decertification. More than one-third signed, and, as the current
contract was due to expire December 1, 1974, the NLRB set the
vote for October 16.

The OCAW's chances of winning were indeed slim. Union
membership was down to an all-time low, and many of the current workers were those who had crossed the picket lines during
the 1972 strike. To have any chance of surviving, the local had to
prove that the OCAW made a difference at the plant.

Tice, Silkwood, and Jerry Brewer (the third committee person) left for Washington on September 26. Tice simply told Kerr-
McGee they would be away from work on union business. Silk-

wood was excited. It was her first trip east. She had just been elected seven weeks ago, and here she was, meeting with top union officials and the AEC itself. Something was *finally* going to be done.

When the three walked into the OCAW Washington office across the street from the Washington *Post* and the Soviet Embassy, Anthony Mazzocchi was there to meet them. He was a short, tough-looking, wiry man in his early forties, who had worked his way up from the streets of Brooklyn to a job in a chemical factory and then to the OCAW's legislative office. When he talked through the side of his mouth, he sounded more like a Brooklyn gangster than a union trouble-shooter who tried to keep Washington's regulatory agencies out of the warm beds of industry.

Tony Mazzocchi felt the OCAW atomic workers were in a squeeze between their own union and the industry. He would soon challenge Elwood Swisher for his job as vice-president of the union and win. Now he was up to his elbows in asbestos, fighting to get the fibrous mineral labeled cancer-causing, and pushing for legislation to protect his OCAW asbestos workers.

As Mazzocchi asked questions, Tice, Silkwood, and Brewer told him story after story about accidents at Kerr-McGee, poor training, contaminations, corporate callousness. He had heard it all before, and he was as worried as they. Even more so, because he knew about the link between plutonium and cancer.

When he described the danger, Silkwood became upset. She was bright, had had some college training, and had worked in a scientific laboratory with plutonium for two years; and no one at Kerr-McGee had ever told her the metal was carcinogenic. No one had ever taken an anatomical drawing, pointed to lungs or liver or lymph nodes or bones, said that plutonium can cause cancer there in twenty-five or thirty years. It angered her, for she had been in a contaminated room without a respirator just two months before.

Mazzocchi was surprised that Karen didn't know how dangerous plutonium really was, but he wasn't surprised that Kerr-McGee wanted the workers to believe it wasn't very hazardous. The less they knew, the fewer the problems for the corporation.

But more was going on at the Kerr-McGee plant than workers breathing plutonium or getting it into their blood through a scratch on their fingers. "Another thing that bothers me," Karen said as the meeting was breaking up, "is how the company is tampering with quality control."

She told Mazzocchi that fuel-rod quality-assurance records were being doctored. She wasn't sure what faulty fuel rods might do when they were used at the Westinghouse Fast Flux Test Facility, but tampering was serious. Brewer knew about the cheating, too. Tice, who worked in fuel fabrication, didn't.

Some nuclear scientists say defective rods need not be dangerous. If rods leak, they argue, the radiation inside them will just escape into the sodium solution that cools the reactor. If they continue to leak, the reactor will have to be shut down, radioactive water and steam will have to be drawn off somehow, and the leaking rods will have to be replaced — expensive but not dangerous.

Other nuclear scientists disagree. At best, they say, no one knows what would happen. Therefore, leaking fuel rods should be taken very seriously. At worst, they could cause a meltdown and a nuclear explosion of some kind.

"This is taking on a whole different dimension," Mazzocchi said. "If we can prove that, if we can win the decertification election first, we can get Kerr-McGee against the wall with publicity."

Brewer later told the FBI that Karen had suggested that she and he collect Kerr-McGee records to prove the quality-control fudging, but that he had advised against it. He said he had argued that if they were caught, K-M could prosecute them, and the union would be "subject to criticism." He said that they all decided "to play it straight."

But Mazzocchi had a different idea. He told Silkwood and Brewer not to mention the quality-assurance problem to the AEC the next day. It was a special question, he said, and it should be handled separately.

The next morning, Mazzocchi delegated the agenda for the AEC meeting to his assistant, Steven Wodka, an impatient, hard-working twenty-five-year-old who rarely smiled. Wodka had joined the OCAW legislative staff five years earlier, right after finishing college, and he looked more like a labor relations professor than an oil, chemical, or atomic worker. Unlike Mazzocchi, Wodka had never had his skull cracked by union-busting goons; but like Mazzocchi, he was cynical about corporations and their concern for industrial safety. And he had little respect for the Atomic Energy Commission.

Wodka sifted through the Silkwood, Tice, and Brewer allegations and prepared a four-point agenda for the AEC meeting. Then he drove them to AEC headquarters in Bethesda, Maryland, where John Davis and a half-dozen staff members waited. Wodka explained why they had come, accusing Kerr-McGee of failing to keep levels of exposure to plutonium as low as praticable, provide proper hygienic facilities, educate and train workers adequately, and monitor worker exposure. Silkwood, Tice, and Brewer cited thirty-nine examples to illustrate the allegations:

☐ "On July 23, 1974, an employee became contaminated. As he was led out, contamination was tracked through the facility. The action of the health physics technician and the operator indicated a lack of training for both . . .

☐ "Regular production is conducted in contaminated areas using respirators twelve hours/day up to ten days without cleanup. As in Pellet Manufacturing, room 124, for the last one and a half years. Source of contamination is not identified. Masks required or not depends on the continuous air monitor results. When the continuous air monitor alarms, the only action is to press the reset button . . .

☐ "Workers are not advised of required respiratory protection by posting — as on May 19, 1974, when an operator was not advised to wear respirator for work in a basement area . . .

☐ "There is no routine survey of respirators. Individuals decide when masks need cleaning and place them at a station for health physics to wash. There is no system for routine replacement of filter cartridges. It is believed that cartridges are reused . . .

☐ "A vacuum cleaner was allowed to 'sit' for about fourteen days in a work area. It was decontaminated only when a worker complained. The worker was instructed to clean it and was contaminated because she had not used proper protection and not been told to do so. The vacuum cleaner was sufficiently contaminated to cause the area to be designated as requiring respiratory protection . . .

☐ "There have been misvalving errors. They have resulted in contamination. They could have caused a criticality problem . . .

☐ "Waste jugs are filled in excess of company procedures. After being half filled, instead of absorbent material, more waste is added . . .

☐ "Plutonium samples are stored in desk drawers. Some were stored on a shelf for a period of two years . . .

☐ "On July 7, 1974, when a criticality alarm sounded, workers were told via the public address system that the alarm was being tested and they were not to respond. The workers were not permitted to respond. On other occasions, workers were told to evacuate only upon verbal instruction . . .

☐ "There are only two shower heads for 75 workers per shift. There is no company rule for taking showers and no time is provided for taking showers . . .

☐ "There is no one to repair instruments that become inoperable during non-day shift."

It was not a pretty picture. The AEC recorded every example, and Davis promised that the commission would investigate. (It would report three months after Silkwood's death that twenty of the thirty-nine examples — including those listed above — were at least partially substantiated.) Wodka was skeptical. Congress had created the AEC in 1946, once it became clear the country needed a program to stimulate the production of uranium and to regulate the development of atomic energy. With blind trust, Congress had given the AEC broad powers: ownership of all substances from which atomic energy can be produced; control of all atomic energy source minerals; ownership of all factories making the atomic energy materials; development and production capabilities for atomic power; and research and development facilities for the peaceful uses of atomic energy. What bothered Wodka was that the AEC was both the promoter and the regulator of nuclear energy.

Wodka told Davis to look into the allegations but not to tip off Kerr-McGee that the commission was investigating it or to

spill the information he was given at the meeting. He asked Davis
to assign investigators who were not regular Kerr-McGee inspectors, and not to give advance notice of inspections. It was about
all Wodka could do, and he wasn't very hopeful.

But Local 5–283 had more immediate problems than the
AEC investigation. If it lost the decertification vote the following
month, there would be no one at the Cimarron plant to stand up
to Kerr-McGee on health and safety issues. And if the OCAW
won the decertification drive, the new contract would have to be
negotiated by December 1. Kerr-McGee was in the driver's seat.

Mazzocchi suggested that the Crescent local invite two experts to an open meeting just before the decertification vote to talk
about the hazards of plutonium. If the Kerr-McGee employees
saw how the company had lied to them and how cheaply it re-
garded their health, they just might vote to keep the union as their
bargaining agent. The strategy was sound. Even though the odds
seemed against a union victory, Tice, Silkwood, and Brewer were
ready to give it a try.

Mazzocchi had a second strategy, which he kept from Tice
and Brewer. The union would wait until it won the decertification
vote. Then, just before the contract negotiations were heating up,
it would give the story on quality-control tampering (which it had
not told the AEC) to *The New York Times*, on condition that the
paper would promise not mention the OCAW as the source. The
adverse publicity would apply the kind of pressure on Kerr-
McGee that the weak local couldn't. Mazzocchi had used the
strategy before, and it had worked.

Mazzocchi was impressed with Silkwood. He decided to tell
her his plan and get her help.

"If the charges are substantiated," he told her privately, "it
would be dynamite. But there's no way we can give them to the
Times without substantiated information."

He explained that the *Times* would need specifics: which fuel
rods were unsafe, which weld numbers, on which day, who's
doing what, who ordered it, and why. He built up the *Times* and
his contact there, the reporter David Burnham.

"Facts!" he told her. "No assertions. They won't fly in a
newspaper."

24

"I can get them," she said.

Mazzocchi emphasized that the disclosure of the doctored quality-control documents was the key to the two-point union strategy. He told her that the only way she could succeed was to work quietly and not do anything to draw attention to herself. He warned her not to tell anyone, not even Tice or Brewer, but to report to Steve Wodka in Washington.

Silkwood told him that he shouldn't worry; she understood.

But she really didn't. No one told her it might be dangerous, that her phone could be tapped, her bedroom bugged. That someone could start following her or try to frighten her. That anything could happen. Even in Oklahoma.

So she went back to Crescent. Back to the lab, back to trying to embarrass a corporation with $1.2 billion in assets, back to squeezing health and safety concessions out of number 120 on the *Fortune* list of the 500 biggest corporations in America.

She never thought she'd lose. She was young and smart, and when she bit, she didn't like to let go.

Chapter 3

Silkwood took her undercover assignment seriously. Meeting Mazzocchi, Wodka, and the AEC staff had given her a sense of focus and direction. Her friends noticed she was getting passionately concerned about the workers' health and that the OCAW was becoming more important in her life.

After she finished her shift, she pored over records in the Metallography Laboratory and studied x rays of rod welds. She found that someone was touching up weld negatives to hide defects, that quality-control data were being manipulated so that fuel rods would pass inspection, and that pellet inspectors were cutting corners. "There is no way that they can evaluate these pellets," she wrote in her notebook. "Possibly this is the reason why so many of our pellets are being rejected by Hanford . . . Many employees told by company, 'Do not talk to AEC or Westinghouse reps or else.' "

Karen called Steve Wodka regularly. "They're still passing high welds no matter what the picture looks like," she told him on October 7 in a call that Wodka taped with her consent. "We grind down too far and I've got a weld I would love for you to see, just how far they ground it down till we lost the weld, trying to get rid of the voids and inclusions and cracks."

Wodka changed the topic. What about those kids joking about getting hot, he asked.

"In the laboratory we've got eighteen- and nineteen-year-old boys," she said. "You know, twenty and twenty-one, I mean. And

26

they didn't have schooling, so they don't understand what radiation is. They don't understand, Steve. They don't understand."

What about Jean Jung, Wodka asked.

Karen had found holes in five gloves after the health physics workers had inspected and passed the lab where Wanda Jean Jung had been working.

"Instead of getting the staff out of the room after they found out it was hot," Karen told Wodka, "the supervisor overruled health physics and he said, 'I want these gloves changed and this bag changed so that we can continue with production.'

"They didn't tell them to get a nasal smear. They didn't say *one* thing to them and they were in there from ten till eight . . . in that air. And the samples . . . verified that it was hot in that room."

Karen had suggested that Jean Jung get a nasal smear to see if she had breathed plutonium. Jung was scared, and she paged Karen as soon as she got to the health physics office. Health physics supervisor Don Majors was talking to Jung when Karen walked into the room.

"What are you doing here?" Majors asked Silkwood.

"Jean asked me to come up."

"You don't have any business in here," he barked.

"Yes, I do. If you check the contract, I think I have a valid reason. She wanted someone here, and she looks a little bit scared."

Jung was crying.

"Well, I'm not doing any more talking," Majors said.

"Oh, yes, you are," Jung told him. "You haven't explained to me what's going to happen to me."

Majors told Jung that Kerr-McGee would check her urine to see if she passed any plutonium. More than likely, he said, it all came out in the nasal smear.

Jung stared him down. "But you don't *know* that," she said. "I could have got some of that down into my lungs."

"Well, maybe."

"What about these other boys that were working back there with me?" Jung asked. "They're only eighteen and nineteen."

"What difference does that make?"

"Well, they're younger and their genes are still growing and changing," Jung said. "The younger you are, the worse it is for you. That's what I'm led to understand."

Majors gave Silkwood a dirty look to let her know he knew where Jung was getting her information.

"Yes, that's a fact," he said. "It has *some* effect on genes, but you don't have near that much to worry about."

One of the young men who had been contaminated with Jung was Don Kirk. He had just finished Kerr-McGee training, and no one had told him plutonium was carcinogenic. Kirk was called into the health physics office after Jean Jung's nasal smear showed she had been contaminated. Kirk's smear count was high, too. Silkwood asked him about the count the next day. He didn't seem the least bit concerned.

"A high nasal smear? — Oh, yeah," he told her. "I was on the A.M. shift and I worked in the room with six hot gloves during production and they didn't check them until the next day."

Wodka was shocked at the story.

"This is a nineteen-year-old kid," Silkwood said. "Steve, this shit is going on every day . . . I had to hold Jean's hand. She was shaking like a leaf."

□ □ □

Not everyone appreciated Silkwood's concern. Some fellow union members complained that she rocked the boat. Management considered her a troublemaker. And nonunion members said she wanted to close down the plant. They were afraid of losing their jobs, and there was nothing around Crescent besides Kerr-McGee.

Labor-management tensions began to mount as the decertification vote drew near. Silkwood and Brewer had been separated and transferred within the plant. They saw it as a disciplinary action. Kerr-McGee personnel director Roy King had placed a notice on the bulletin board saying that NLRB rules forbade soliciting of union members on company hours. Tice, Silkwood, and Brewer confronted him and argued that coffee breaks and lunch hours were not "company time." King said they were. Tice grabbed the phone and called the NLRB. King lost.

On October 10, Steve Wodka brought Dean Abrahamson

and Donald Geesaman, both of the University of Minnesota, to talk to the Kerr-McGee workers about the dangers of plutonium. It was part of the OCAW's strategy to stay alive.

Abrahamson and Geesaman were nuclear scientists who didn't work for the AEC and who believed the commission was lying about the little gray metal. Dr. Abrahamson was a physician and a nuclear physicist; Geesaman, a biophysicist who had worked at the AEC's Lawrence Radiation Laboratory at Livermore, California, for thirteen years. Wodka had been impressed by them. "These doctors are going to flip out when they hear these stories that you have to tell," he told Silkwood.

Abrahamson and Geesaman gave two lectures in the Crescent American Legion Hall so that workers from each twelve-hour shift could attend. About fifty people came, and the two professors scared the complacency right out of them. The scientists pounded away at one theme: plutonium causes cancer. No one, they said, knows how much plutonium will cause cancer, or where and when. They said it is difficult to describe just how toxic plutonium really is because it's so different from other poisons. Fiendishly toxic, even in small amounts. Twenty thousand times more deadly than the potassium used in the Auschwitz gas chambers.

They said the AEC "safe" standards were meaningless, set on faith. The standards for the general population had been tightened by a factor of 100 during the past few years, but those for the men and women who worked with plutonium and who breathed it had not been made more stringent.

The plutonium at the Cimarron plant, they told their listeners, gets into the air as a nitrate mist or as a fine oxide dust. You breathe it through your mouth or nose. If you breathe it through your nose, they said, the cilia in your nostrils catch the biggest chunks; the rest enters the body. If you breathe it through your mouth, some plutonium sneaks down the windpipe; some is trapped by mucus and forced into the esophagus. It ends up in your stomach.

The plutonium-oxide dust, they explained, is insoluble. It doesn't pass into the blood, but settles in the lungs and lymph nodes. It does not show up in a urine test, but will show in feces because the cleansing action of the mucus causes some of it to be

29

excreted as solid waste. It may show up in a nasal smear. But nasal smears must be interpreted with caution, they said. If a smear shows no contamination, that does not mean insoluble plutonium-oxide dust was not inhaled.

The plutonium-nitrate mist, on the other hand, is soluble and is dissolved by the blood. It travels mainly to the liver and bones. Some gets into the urine and is eliminated.

Both the soluble and insoluble plutonium in the body, they stressed, emit alpha particles, which can hit the cell right in the nucleus and either kill it or damage it so that it goes berserk. If enough cells start going berserk, you can get cancer in thirty or forty years. Even if the cells are killed rather than damaged by the 3000 alpha particles that go shooting out of plutonium every day, cancer can develop. Cells are knit together like a family. When enough die, the healthy cells get riled up and start going berserk. This can cause cancer, too.

If you get plutonium on your skin, can it cause problems, someone wanted to know.

Abrahamson and Geesaman answered with a story. A trucker was carrying barrels of plutonium solution. When he arrived at the factory, he noticed that a barrel was leaking. He had been exposed, so the company cleaned him up. In a few years, he developed an extremely rare form of cancer in his hand. Part of his arm was cut off in a series of operations. Then the whole arm. In the end, he died, with one suppurating wound where his arm had been. The lawsuit never reached the court; it was quietly settled.

But would he have got cancer if he hadn't had a cut on his hand? the worker asked.

Well, think of it, they said. If you were working on a loading dock, can you imagine a hand without some cut or scratch?

Another worker spoke up: "There was a girl in the General Lab who was sickly and she passed out. It wasn't plutonium or anything. But they went to get something to treat her — a Scott Air Pack — and it wasn't functioning. The ambulance they were going to use had two flat tires."

The discussion turned from safety to the OCAW. "This is the first instance I've had where the union has come forward," one man said. "Where are we going from here? Why, a week from now, should I say let's keep the union in when this is the first evidence that I've had that the union had any interest, and it seems primarily generated by the fact that everybody's talking about getting rid of it?"

Another man asked, "Well, right now with thirty percent membership or so in the union, who's going to support the union? We go out on a walk . . . and the rest of them are going to sit up there and laugh."

They argued. If the union lost certification, wouldn't the workers be better off? Wouldn't Kerr-McGee offer a 10 percent increase across the board to keep the union out? Yes, but wouldn't Kerr-McGee then fire the union members, the trouble-makers like Silkwood, Tice, and Brewer?

No matter how they argued, they always came back to health and safety. And on that issue, the OCAW was the only buffer between Kerr-McGee and the workers.

Silkwood took the floor. "If there is something going on," she said, "if we're going to be susceptible to cancer and we're not going to know about it for years, something's got to be done. And that's the situation we need documented."

After the meeting, Wodka reminded Silkwood of the quality-control documents for *The New York Times*. He told her the OCAW still needed proof — facts, pieces of paper, negatives. The OCAW's credibility with the news media was at stake. He said the union was not about to lead David Burnham on a "wild goose chase." She must come through with documents.

She said she wouldn't let the OCAW down.

The battle for votes was on. The OCAW International wrote to the Kerr-McGee workers, saying that if the union were voted out, there would be no contract after December 1. And no contract meant no guaranteed wages or hours of work. No vacations, holidays, sick leave, retirement. Kerr-McGee could fire or punish without showing just cause. The workers had nothing to gain and everything to lose by voting out the OCAW.

Morgan Moore, manager of the Cimarron facility, also wrote to the workers, urging them to vote. He said the ballots of just a few would decide the future of them and their families. K-M had been working hard to make the Cimarron plant a better place, he

31

wrote, and win or lose, K-M would continue to do so. He told them that workers didn't need a union to get the best Kerr-McGee had to offer, and the only way the OCAW could force K-M to do something it didn't want to was to strike. Strikes cost money. They cost workers money. They cost the community money. The only way workers could be sure of no strike was to vote "no union."

The workers didn't buy it. A week after the Abrahamson-Geesaman talks, they voted eighty to sixty-one to keep the OCAW as their bargaining agent. It was a tremendous victory for the union — and the OCAW was ready for the second part of its strategy.

Contract negotiations were set to begin on November 6. Silkwood was to have her documents ready by November 13. She, Wodka, and the reporter, Burnham, would meet in Dallas. The story would appear in the *Times* before December 1, when the old contract expired.

She continued to recruit people for the union and to record contaminations and management-union problems. She kept sifting through documents. She called a friend and fellow Kerr-McGee worker, James Noel, and said she was concerned that the plutonium workers were not taking her seriously. She told Noel she had found out that forty pounds of plutonium, enough to make almost three atomic bombs, were missing from the plant. Noel made a note of their conversation in his desk log. Karen never told him how she learned about the material unaccounted for (MUF) or whether she had records to prove it. That she even knew about the MUF was significant, for investigators would later speculate that there was a plutonium smuggling ring at the Cimarron plant and that Karen Silkwood was part of it. If she was, why would she tell someone she had learned the plutonium was missing?

As the tension mounted, Karen couldn't sleep. The union, Kerr-McGee, and fears about health and safety were gnawing at her. She began to lose weight, dropping from 115 to 94 pounds. Drew Stephens tried to get her to quit Kerr-McGee. The plant and the union were consuming everything she had, mentally and physically. But she couldn't let go.

Drew had let go. He believed the plant would close itself down in six months, and he had quit his job just before Karen went to Washington. "I felt that things had gotten out of hand there in terms of radiation release and health physics control," he says.

What worried him the most was rotting equipment, like glove-box gaskets that leaked and had plastic taped up underneath to stop the plutonium nitrate from dripping on workers and the floor.

Although Karen wouldn't quit, Drew tried to get her to relax a little, to ease up. She told him how important her union was, how the workers needed her. Karen and Drew squabbled. Eventually, they signed a truce. Kerr-McGee, the union, health and safety were forbidden topics.

During October, something was happening to Karen. No one is sure what, but her dependence on Quaaludes became obvious. Her friend Evelyn Emerich told the FBI after Karen's death that one night in mid-October, after a union meeting at the Hub Cafe, Karen looked "like death." Emerich said Karen was pale, moved very slowly, and spoke with slurred speech.

"What's wrong?" Emerich had asked. Karen had not been looking well for weeks.

Karen said her doctor had given her some medication for depression and she felt sick from it. She said she was afraid she wouldn't be able to drive home to Edmond, about twenty miles away, so Emerich invited Karen to spend the night at her home, which was close to the plutonium plant.

Later in October, Karen called her mother and asked her to send job applications for the oil companies around Nederland, Texas. She was thinking of quitting, as Drew had done, she said, and coming home. But not until her job in Crescent was finished.

At the end of October she called her sister Rose Mary Porter. Karen was scared, very scared. She was crying, and the words wouldn't come. Rose Mary had never heard Karen so hysterical before. Through the crying and gulping, Karen said someone was trying to do something to her. Something was happening to her. She couldn't talk about it over the phone, she said, and begged Rose Mary to come to visit her. She needed to talk, to see her sister. Rose Mary said she had a family to care for and couldn't come.

Karen called Rose Mary once again. This time she told her she was quitting — soon after December 1, when the new contract was to be signed. She said she was scared of the plant and really needed the oil company applications. She asked Rose Mary if she could stay with her until she got settled; Rose Mary told her she was welcome any time.

Rose Mary never found out what was happening to Karen or who was trying to do something to her. Kerr-McGee would later deny it knew Karen was snooping. But lab analyst Leonard R. White saw her and Jerry Brewer examining files in the Metallography Laboratory. "It appeared," he would later tell the AEC, "they were obtaining and recording information for some purpose." Lab analyst Gary Longaker would later testify: "It was sort of a running joke there about watching Karen. We used to kid about that she should have been wearing a double-billed cap and cape from England, because she would come into a room where you were working, and she would write something down on a notebook, and then leave." And James Smith, a former K-M supervisor, would later say in a deposition that supervisors routinely reported to Morgan Moore, the Cimarron facility manager, what union leaders were saying and doing at the plutonium plant.

Chapter 4

Robert Samuel Kerr never let anyone forget he was born in a four-teen-foot-square log cabin when Oklahoma was still Indian Ter-ritory, or that he grew up with Ada prairie dust behind his ears and read borrowed books by the glow of a log fire.

He had set his goals early in life: "A family, to make millions of dollars and to be governor of Oklahoma — in that order." He achieved all three by the time he was forty-six years old.

After graduating from high school, Kerr found a job teaching in a country schoolhouse and saved enough from his meager sal-ary to pay for a two-year correspondence degree from East Cen-tral State College. His father had always told him that "the precise practice of law was the quickest route to public life," so Kerr quit teaching, borrowed $350, and enrolled in the University of Okla-homa to study law. It was 1915, and he was just nineteen years old.

Bob Kerr ran out of money the next year and clerked at a law office while continuing to read law. After a stint in France during World War I, he returned to Ada, married, and opened a whole-sale grocery business. The following year, 1920, his twin daugh-ters died at birth. In 1921, fire destroyed his produce warehouse, and in 1924, his wife died in childbirth. The baby, a boy, also died. Kerr started over again.

He passed the Oklahoma bar exam, hung out a shingle, and married Grayce Breene, a graceful, six-foot blonde from Tulsa. He was $10,000 in debt; her family was in oil. Grayce fulfilled his first ambition — three sons' and a daughter's worth. It took him ten years to make his first oil million.

Bob Kerr and his brother-in-law James Anderson each bought a 20 percent interest in a small contract oil-drilling company. Three years later, in 1929, they borrowed $30,000, bought the business, and called it Anderson-Kerr.

Kerr gave up law and moved to Oklahoma City, where A-K had modest offices. Anderson and Kerr were a good pair. Anderson could smell oil; Kerr, money. Anderson underbid his competitors; Kerr attracted investors.

Oklahoma City was sitting on seas of oil. But "town lot" drilling was dangerous in populated areas, so the city demanded a $200,000 bond for each city well. Bob Kerr raised enough cash, despite the Depression, to sink two, and he made a deal with Continental Oil Company for four more. Continental staked Kerr with $360,000 for a half-interest in all six wells, plus an additional $90,000 to finance the drilling. All six came in.

Kerr's big break came in 1935, when "Boots" Adams, assistant to the president of Phillips Petroleum, asked Kerr for a favor. Oklahoma City had passed an ordinance requiring voter approval of any more drilling inside city limits. Phillips held a lot of leases. Adams asked Kerr to lead a campaign against the ordinance, and Kerr agreed to — in exchange for the chance to drill Phillips' wells.

Kerr won the campaign, the ordinance was killed, and Anderson-Kerr made a small fortune by taking a percentage of each well that came in instead of a flat drilling fee. When James Anderson retired in 1937, Kerr hired Dean Anderson McGee away from Phillips Petroleum as a 12.6 percent partner. McGee was Phillips' chief geologist and one of the best in the country — some say one of the greatest in the history of the oil industry. McGee minded the drills; Kerr went chasing his political dream. He ran for governor in 1942. It was the Sooner State's corniest gubernatorial campaign, with six-foot, four-inch "Smilin' Bob" in suspenders, telling farmers, "I'm just like you, only I struck oil," and "Baptist Bob" singing "Take Me Back to Tulsy, I'm Too Young to Marry," to the guitars of cowboy bands.

Kerr ran as a Roosevelt Democrat, and he squeaked past his anti–New Deal opponent. His support had come from all layers of Oklahoma society — the party machine, old soldiers, blacks,

Baptists, small-town merchants. His personality had been an important factor. Kerr *was* Oklahoma. Frontier vitality, a sense of humor, pride in being a native son, hope at the end of the Dust Bowl years.

Not satisfied with money and the governor's mansion, Kerr turned his restless energy toward national power and politics. Because of his dogged support of the New Deal as governor and his reputation for eloquence, the Democratic Party chose him as its 1944 convention keynote speaker and thrust him into the national limelight. He was never the same again.. He began to lust for the White House; the presidency seemed to be more than just an oil dream.

After six years as governor, Kerr won a hotly contested seat to the United States Senate in 1948. He had starved himself down to 200 pounds so that he would appear fit and trim before the cameras. He was fifty-one years old and no modest freshman. Within a year, he had earned a reputation as one of the hardest-working, smartest, most eloquent, and sharpest-tongued senators in Congress. "The big boom from Oklahoma," *The Saturday Evening Post* called him. He didn't smile; he grinned. He never walked; he strode. Never stood; just loomed up. And he never just talked. He roared or blasted or bellowed. Kerr fed the legend. He loved it.

In 1952, he thought he was ready for the presidency. Kerr was a wildcatter, a gambler. He took risks, and most of them paid off. The only way to get ahead was to be smarter and to risk more than the next man. Win some, lose some, but win more than you lose and win the big ones.

Kerr lost the big one at the 1952 Chicago Democratic Convention. He never stood a chance, yet he was surprised. He got only 65 votes on the first ballot, compared to Kefauver's 340 and Stevenson's 273. Only Oklahoma and Arizona gave him all their votes.

Kerr returned to Washington, head high, but bitter and angry. There was nothing left for him but to grab power — raw power, Senate power, legislative power that brings Presidents to their knees.

When Lyndon Johnson, a close friend of Kerr's, became John

F. Kennedy's vice-president in 1960, he left a vacuum in the Senate leadership. Kerr stepped in. He had carefully built his power base on the Senate Committees of Public Works, Finance, and Aeronautical and Space Sciences. Now he rolled out the old pork barrel. When other senators wanted a chunk of the bacon for pet home-state projects, they had to come to Kerr — and he always remembered his debtors.

Baptist Bob knew money talked, walked, fornicated, and fathered. "A man who doesn't have money can't operate," he told his protégé, Senate gopher Bobby Baker. "Why, if I don't have at least five thousand on me as pocket change, I'm afraid that taxi drivers won't pick me up."

Kerr spread the cash around the Hill. "Loans," he called them, or "campaign contributions" that would buy votes for the oil, gas, public works, or banking bills he pushed. Kerr handled his own dirty-money deals. "A certain kind of business has to be done behind the door," he once told Baker. "If I didn't handle these deals personally, or if I failed to dress them up with pretty little speeches, most of the fellows would be spooked."

Kerr arranged for several friends and relations of a congressman highly placed on the tax-writing House Ways and Means Committee to buy into solid, producing oil wells at bargain prices. He did the same for a powerful figure on the Senate Finance Committee.

Senator Kerr became a Jesse James in Robin Hood togs. While governor of Oklahoma, Kerr had pestered Congress into authorizing the Arkansas River navigation project to the happy tune of $100 million. Oklahoma was land-locked by Texas, Kansas, and Arkansas. Its only access to water was the Arkansas River, which flowed through Tulsa, southeast to Little Rock and the Mississippi — a 400-mile snake. The Arkansas River Project could open Oklahoma to the Gulf and the cities along the Mississippi.

As senator, Kerr nudged through Congress a $1.2 billion Arkansas River Project bill, which, over twenty years, would link Tulsa to the Mississippi while providing hydroelectric dams, irrigation ditches, and flood-control projects. "Bob Kerr's big ditch," they called it back home.

It just so happened that Kerr-McGee had bought 85 million tons of coal reserves in eastern Oklahoma before construction began on the River Barge Canal, one phase of the Arkansas River Project. The canal would enable Kerr-McGee to move cheaply about 1 million tons of coal a year to the steel centers along the Mississippi.

Bob Kerr had also been quietly gobbling up land along the Arkansas River where the Army Corps of Engineers had planned ports. Kerr had babied the corps through the planning, so he knew which river banks were made of gold.

When Kerr-McGee bought into the uranium business in 1952, Uncle Sam was its only customer, and Baptist Bob was there, standing under the flag, to help his country. He arranged through Admiral Lewis Strauss, chairman of the AEC (Kerr had voted for his confirmation as chairman), for Kerr-McGee to sell the AEC all the uranium the company could dig.

By 1959, Kerr-McGee had an AEC contract worth $300 million. Senator Albert Gore of Tennessee, chairman of the Raw Materials Subcommittee, had to approve the contract. Gore just happened to be Kerr's friend and a co-owner with Kerr of the Kermac Ranch in Oklahoma.

Gordon Weller, president of the Uranium Institute, which represented the hard-pressed uranium companies Kerr-McGee was squeezing out of the mines, decided to take on Bob Kerr. "We believe it can be positively proved," he wrote Attorney General Robert Kennedy, calling for an investigation of Kerr-McGee, "that . . . in the production, processing and sale of uranium materials to the AEC . . . those with political influence got the markets. Members of Congress who chose to personally engage in uranium soon became the leading producers. Independents got the short end of the stick."

G. H. Brodie, vice-president of the Golden-Denver Corporation, also complained. He testified that the AEC bought $4 million worth of uranium from Kerr-McGee, shipping it 430 miles and by-passing five other mills along the way. Congress yawned. There was no investigation.

It wasn't just oil and uranium. Kerr-McGee filled its corporate sails with helium, a gas needed for rockets and missiles and moon shots. Columnist Drew Pearson, who liked to nip at the Oklahoman's spurs, exposed Kerr's helium deal.

> Suddenly . . . the Interior Department almost doubled the price to $35 . . . which permitted the Senator's company to charge more [Pearson wrote]. Kerr-McGee opened its plant for business immediately after the price rise, which it seemed to have been expecting . . . Kerr stayed under cover while his friend, Senator Clinton Anderson, New Mexico Democrat, pushed the price-boosting bill through the Senate. It was Anderson who advised Kerr-McGee to go into the helium business in the first place. Anderson held a financial stake in the business for a while, but kept clear of conflict of interest by selling out to Kerr-McGee before the helium bill came up . . .
>
> Kerr is also chairman of the Senate Space Committee which permits him to offer pork in the sky as well as on the ground and in rivers and harbors.

Pearson pointed out that Ohio congressman Mike Kirwan eased the helium bill through the House. Later, Kerr helped Kirwan get a $10 million aquarium for his constituency.

Kerr wasn't embarrassed. His diamond-studded KERR-MCGEE FOUNDER pin sparkled proudly in his lapel. And when a reporter once asked him to comment on his political philosophy, he quipped, "Son, I'm agin any combine I ain't in on."

"How about abstaining from a vote on personal interest matters?" *The Saturday Evening Post* once asked Kerr.

"Now wouldn't that be a hell of a thing," Kerr drawled, "if the Senator from Oklahoma couldn't vote for the things Oklahomans are most interested in?"

When Kerr died suddenly of a heart attack on New Year's Day, 1963, old I. B. Warner, an Oklahoma cracker-barrel political pundit, summarized the senator's career in two sentences: "Smartest politician you'll ever see. Buys organs for churches, pianos for whorehouses."

Baptist preacher and Kerr confidante Dr. Herschel H. Hobbs was much kinder to the memory of Oklahoma's first native governor, who reputedly gave 30 percent of his income to the church. "One loved by his friends, trusted by his colleagues, honored by his church, has gone down," he eulogized, "as when a great cedar goes down and leaves a lonesome place against the sky."

Tucked in the mighty cedar's safe-deposit box in the National Savings and Trust Company a block from the Rose Garden was just under $50,000 — almost entirely in $100 bills. The money was part of a bribe, and just a fraction of the $22 million Kerr fortune.

The story came out long after Kerr's death. Bobby Baker was indicted for influence peddling and tax evasion. To save his skin, he squealed on his former boss. Baker now tells the following story:

In 1962, President Kennedy had sent the Senate a bill that would have taxed savings and loan associations as much as banks. That would have meant a 5 percent tax increase for the associations. According to Baker, the chief lobbyist for a savings and loan group came to him and said, "Bobby, that fucking bill will ruin us. We figure it will cost us a minimum of $43 million *annually.* We just can't live with that. My ass is on the line. Help me."

The lobbyist asked Baker to talk to Kerr. "He's on the conference committee and he can kill or amend the bill."

"Are you crazy?" Baker said. "He's one of the sponsors of that bill."

Baker approached Kerr, but before he could finish the pitch, Kerr cut him off. "I have no sympathy for those bastards," the senator said. "I'm a commercial banker and I feel strongly that they should pay equal taxes. But if you trust the people you're dealing with, tell them it will cost them $400,000 if I'm successful . . . And I'll need my money in cash."

Baker reported back to his lobbyist friend. "Jesus shit," the lobbyist said. "Let me talk to my people and get back to you."

He did. "Tell your man he's got a deal."

Kerr killed the bill, and the lobbyist welched. He could get only $200,000, he told Baker. Half now and half in 1964.

Kerr exploded. "I told you those bastards were no fucking good!" shouted Baptist Bob, who had forbidden his family to cuss in his presence. "Those pricks have always tried to do everything on the cheap and they always will. They'll be sorry for this, those cocksuckers. I'm gonna fuck 'em at every turn from now on. Get me my money."

The cash began to flow in in $100 bills on October 21, 1962 — $51,000 in two envelopes. Baker walked Kerr to the National Savings and Trust, where the senator stashed the loot in his deposit box. Kerr later spent some of it. On October 31, Baker got another $16,200 Kerr payoff.

"Goddammit, Bobby, this has gone far enough,"՝ Kerr said as they went to the lock box. "Tell those pricks to quit dribbling this in."

Baker got $33,300 more on November 9. Kerr died two months later, and Baker joined Jimmy Hoffa at the Allenwood Penitentiary for influence peddling.

□　□　□

All the while Kerr was serving Uncle Sam, Dean A. McGee was building the Kerr-McGee Corporation into an energy conglomerate. In the late thirties, Phillips Petroleum was willing to gamble on its former star oil mole. Phillips underwrote 35 percent of Kerr-McGee's drilling costs for half of the profit. Kerr-McGee soon had a string of wells dotting the Louisiana coast.

If Kerr and Phillips were gamblers, so was McGee. He drilled where few oilmen dared — in the Louisiana tidelands. "It looked better to us than staying on land, where the first-class spots were already leased and drilled," McGee recalls. "Some said it took courage. Others said we were just foolish."

The roll of the dice paid off in 1947. Kerr-McGee brought in the first big tidelands well in eighteen feet of water and out of sight of land. And did it cheaply. Other companies spent $1 million for floating platforms, crew quarters, and storage space. Kerr-McGee spent $250,000.

McGee pioneered Kerr-McGee into another field in 1949. K-M became the first United States company to sign nondomestic oil-drilling contracts for work in the Saudi Arabia–Kuwait Neutral Zone in the Persian Gulf and in Mexico near the Yucatán Peninsula.

McGee kept expanding the company. He bought an oil refinery and a string of 800 filling stations in twenty-three states. The oil loop was complete. Kerr-McGee prospected, drilled, refined, and sold at the gas station. It was time to expand into other energy fields.

McGee bought a uranium-mining company in 1952. He says he moved into atomic energy because he glimpsed the ghost of the oil crisis deep in his wells. McGee's critics say he burrowed for radioactive ore because Bob Kerr saw the chance to corner the market. Whatever the case, McGee considers his venture into the nuclear age one of the most important decisions he made at Kerr-McGee.

A Navajo had discovered uranium in the late 1940s on a cliff face 8500 feet up into the Lukachukai Mountains in Arizona, about forty miles from the spot where Arizona, New Mexico, Utah, and Colorado meet. The high Mesa find was in what geologists call the Colorado Plateau, which holds most of the uranium ore in the United States.

A Colorado banker opened the first mine there on the Navajo Reservation, then sold it to the Dulaney Mining Company. It is not clear whether the Navajo who prospected the lode ever collected a penny.

Kerr-McGee bought the Mesa mine from Dulaney. Within six months, McGee had a dozen geologists and two airplanes prospecting the plateau in Arizona, New Mexico, Utah, and Wyoming. Within twelve months, he had leased 1,275,180 acres of potential uranium deposits. Within twenty-four months, he had opened a $3 million uranium mill on the desert floor in Shiprock, New Mexico, thirty-five miles east of the mines.

By late 1958, the AEC had cut its uranium stockpile in half. McGee sat tight and watched the smaller uranium companies bite the desert dust. When the freeze was over, Kerr-McGee had few serious challengers. By 1970, the company had gobbled up a second mining and milling operation in Ambrosia Lake, near Grants, New Mexico. It opened a $20 million processing plant near Gore, Oklahoma, to convert raw milled uranium into yellow cake uranium hexafluoride for enriching at AEC plants. And it built a processing plant near Crescent to pound the yellow cake into uranium pellets for nuclear reactors.

The uranium loop was now complete. Kerr-McGee prospected, mined, milled, processed, and manufactured. By 1970, it owned one quarter of all known uranium reserves in the United States and was the biggest uranium producer in the country.

The Navajos saw little of that profit. They were hauling uranium on their reservation for $1.50 an hour. But they did have a barbecue. Dean McGee likes to tell the story of how he sealed the deal with the Navajo Tribal Council for the Shiprock mill that would process the Mesa ore.

"Mr. McGee, I have a question to ask before we vote on this measure," a council member had said. "If we approve your building this mill, will you agree to have a barbecue?"

"Yes," McGee had said. "If we can get this through, we'll have the biggest barbecue you've ever had."

The tribal vote was unanimous, McGee signed the AEC contract for the mill, and the Navajos got all the pork and beef they could eat. Everyone was happy. The AEC got uranium; Kerr-McGee, a desert mill. And the Navajos? Money for the land lease, jobs hauling the ore down the mountain and across the desert, jobs in the mill. And cancer.

As far back as 1546, large numbers of miners digging uranium-bearing ore out of the Erz Mountains in Central Europe had died of lung cancer. It was not diagnosed, however, as malignant neoplasia until 1879. Since then, there have been medical journals filled with evidence to support the uranium-cancer connection: 40 percent of the miners of Schneeberg, Germany, who dug uranium between 1875 and 1912 died of cancer; so did 53 percent of those from Joachimsthal, Czechoslovakia, who worked in 1929–1930. It didn't take a Louis Pasteur to see the connection. The miners themselves had noted over the centuries that the discovery of a rich uranium vein was always followed by a lot of coughing, blood spitting, and dying.

The problem is simple enough. As uranium decays, it releases radioactive radon gas, which is trapped in the rocks and cracks. When dynamite blows the rocks apart, the gas and its radioactive daughters cling to the dust particles. The miners, who didn't wear respirators, gulped down the dust. Once in the lungs, the radioactive dirt bombards the tissue for years with alpha particles, eventually causing lung cancer; or silica particles become embedded in the tissue, causing pulmonary fibrosis — a disease that, like black lung, makes breathing very difficult.

Kerr-McGee never told the Navajos that uranium mining and milling cause cancer, even though its leases with the Indian nation said the corporation had the responsibility to "provide for the health and safety of the workmen." Neither did the Bureau of Indian Affairs, which had the authority to take the leases away from Kerr-McGee.

"We're talking about *Navajo* people," BIA official Tom Lynch, himself a Navajo, told Amanda Spake, an investigative reporter, in 1974. "They wouldn't understand radioactivity! Outside the reservation, people may have understood it, but not here, not Indians."

Nor did the AEC warn the Navajos of the danger. The International Commission of Radiological Protection had been recommending mining standards since 1925, tightening them almost every year as they better understood the toxic nature of uranium. Like a federal Pontius Pilate, the AEC washed its hands. Its responsibility began, it argued, once the ore was "removed from its place in nature." The AEC had no jurisdiction over the mines. Each state should regulate its own pits. Or the Department of Interior's Bureau of Mines should crawl into the mines with alpha counters; or the Bureau of Indian Affairs; or Health, Education, and Welfare; or the Department of Labor.

"Back then there were no jobs around here," recalls former Navajo miner Tony Light, "and people needed money. The company came around and said that there were mining jobs . . . but they didn't tell us a thing about the dangers of uranium mining. The labor came cheap back then. The company started paying the men ninety cents an hour, not even the minimum wages."

The work was hard. "It made us sick to go into those mines," said John Lee, another former miner. "The white men sat outside the mines and pushed us Navajos into those dirty mines right after dynamiting." If the miners objected, Lee explained, Kerr-McGee fired them and hired new Navajos.

During the 1960s, the Arizona state inspector checked the Mesa mines, but his office was understaffed and had little clout.

45

When inspectors did come to the mines, they were taken to a section where the dust levels were low. Workers who knew what was going on said they were afraid to speak up for fear of being fired.

By the time federal regulations were set for uranium miners in 1968, after twenty-five years of laissez-faire digging, the Mesa mines were closed. But other Kerr-McGee mines were still spewing uranium dust, and Kerr-McGee was still cheating. The company usually monitored the air near the air intake, where fresh air from the surface was forced into the mines, Margarito Martinez testified during Bureau of Mines hearings. Martinez was one of 700 Kerr-McGee miners and millers at the corporation's Ambrosia Lake operation. Eighty percent of the workers were Chicano, 7 percent Native American, and 2 percent black.

Radiation readings at the air intake, Martinez testified, were much lower than inside the mines, where fresh air hardly penetrated. And even though Kerr-McGee was required by law to tell miners about the radioactive levels in the pit, the company either withheld the information or made it almost impossible to get. When miners reached the maximum exposure levels set by federal regulations, Kerr-McGee didn't bother to get them out into fresh-air jobs, Martinez testified.

Martinez's allegations were later substantiated by federal inspection studies. The Department of Interior told Congress in 1977 that uranium-mining companies, including Kerr-McGee, had reported contamination levels to be 500 percent lower than they actually were.

Because there were no controls and, apparently, no concern, the early uranium miners had been breathing up to 100 times the amount of radioactivity the government later deemed "safe." Kerr-McGee, among others, had fought the imposition of standards, arguing it couldn't afford to improve the mines. By 1980, 16 of the 700 Navajos who worked the mines in the 1950s and 1960s were dead from lung cancer. Their average age was forty-six; the youngest was thirty-one. Public Health Service researchers attribute these deaths to uranium exposure, since Navajos, who generally don't smoke, rarely die of lung cancer. Research indicates that if these 700 Navajos had not worked with uranium, *one* at most would have been expected to die of lung cancer. No one knows how many more former miners will die of the disease.

Kerr-McGee was and is not liable for the cancer deaths of its miners. The Navajos alleged it had left the reservation in a hurry, without burying its refuse dump or tightly sealing the mine opening. Disabled workers and widows didn't get a penny.

□ □ □

Kerr-McGee had opened its uranium plant in Crescent in 1968. It was a warehouse-like building sitting on a knoll overlooking rolling pastures and wheat fields and oil wells. At the bottom of the hill, the Cimarron River, reddish brown like the Oklahoma clay, cut a swath through the fields. There were no Navajos or Chicanos around Crescent. Just poor farmers.

The Crescent uranium plant took yellow cake, vaporized it with steam, mixed it with ammonium hydroxide, and made a paste. Workers dried the paste back into a solid, baked it into a powder, and pressed it into pellets. They stuffed the pellets into fuel rods for light-water nuclear reactors. The sign out front facing Highway 74 carried Kerr-McGee's red, white, and blue logo. SAFETY FIRST, the sign read.

"I never saw anything so filthy in all my life," recalled James V. Smith, a division manager at the plutonium plant. "It was one big pigpen. If you ever walked past the door, you couldn't hardly breathe from the ammonia fumes and the uranium . . . I personally read temperatures at a hundred and thirty-five degrees."

Yellow cake and uranium oxides are more dangerous than uranium ore in the mines because they release more long-lived radioactive decay products. Kerr-McGee workers without respirators breathed the contaminated air into their lungs, and the particles settled in bronchial lymph nodes. Malignant lymphoma could be just around the corner.

Kerr-McGee didn't tell the workers that uranium can cause cancer. It gave them a thirty-minute lecture and turned them loose, says Ken Plowman, a former Kerr-McGee health physics technician. "There was contamination everywhere," he testified after Silkwood's death, "especially in the lunch room. The men were fairly well contaminated on their arms and hands. There was no way to get it off without peeling their hide, so they went home like this every night."

In 1970, Kerr-McGee opened a plutonium plant next door. It was another boxlike warehouse perched above grazing cattle and right smack in a tornado alley. The National Severe Storm Forecast Center ranks Oklahoma third (behind Florida and neighboring Texas) for killer twisters. Between 1958 and 1973, eight tornados actually touched down in Logan County, home of the Kerr-McGee plutonium plant, and fourteen tore corridors into neighboring Kingfisher County.

Kerr-McGee listened to the radio for tornado watches and alerts. During alerts, its security guards scanned the flat horizon for black funnels while workers were free to run for shelter — to the lunch room or their cars.

In the event of a tornado alert, Kerr-McGee, according to AEC regulations, was to tuck the plutonium into a vault so that twisters wouldn't suck it up and sprinkle it across the state, but it was physically impossible to collect all the nasty little metal. For one thing, there wasn't enough room in the vault for the pellets, powder, and liquid. For another, carting plutonium around the plant became too much of a hassle — an expensive one.

In the beginning, James Smith testified, workers managed to get the powder and pellets into the vault. But later on, there were so many tornado alerts, workers simply ignored them.

By 1974, Dun and Bradstreet had pasted a gold star on the Kerr-McGee shareholder report, naming the corporation one of the five best managed in the country. The shareholders were lucky. Dean McGee had built K-M into one of the top twenty companies for return on investments.

Fortune, Business Week, the *Dun's Review* heaped praise on K-M. Dean Anderson McGee, rawhide-tough like his oil drillers, had steered Kerr-McGee through the reefs and shoals of the energy business. It had become one of the most powerful total energy corporations in the United States, with holdings in gas, oil, coal, and uranium. It owned 200,000 acres of timber, and it processed helium. It made phosphates for fertilizers, asphalt and pes-

48

ticides, potash and boron. And it baked and pounded uranium and plutonium into pellets.

But it had a ninety-eight-pound problem with a soft Texas drawl and a stubborn streak. She was no Navajo who couldn't understand the complexities of radiation, nor an Oklahoma farm girl who didn't care. She wasn't afraid to lose her job or to speak out. She thought she understood how the big boys play the game, and she was beginning — just beginning — to fight back.

Chapter 5

On the last day of October, Karen Silkwood swerved off Highway 33 to avoid clipping a cow. Her car spun around, backed off a thirteen-foot embankment, and hit a fencepost. She paid George Martin of the Martin Wrecking Service $13 to tow her back on the highway. Her white Honda Civic had a one-inch dent on the right side of the bumper. The right quarter panel was pushed in, and the right rear tail- and tag lights were smashed. Drew Stephens, who had prepared the $300 damage estimate for Whitfield Volkswagen, noted that there was no damage to the *left* side of the car. The point was important, because after Silkwood's death the Oklahoma Highway Patrol would speculate that the dents in the left bumper and fender were the result of the October 31 accident.

The next day, Silkwood complained to her physician, Dr. Shields, about a sore back and a headache. She said she also felt very lethargic. In addition to the Quaaludes and Mellaril (a tranquilizer) he had given her, she had been taking Tylenol No. 3, which her dentist had prescribed and which contained codeine.

Dr. Shields suspected Karen's lethargy was due to "overlapping" the Tylenol No. 3, Mellaril, and Quaalude. He didn't think she had mixed the drugs on purpose as a downer, however. "Karen, leave it [the Tylenol and Mellaril] with me," he said. "I'll give it back later." She handed him the prescriptions without objecting.

Silkwood spent the next three days with Donald Gummow, a nonunion Kerr-McGee worker about her own age. Gummow was in love with her, and liked her fiery, passionate personality, as he told the FBI after her death. During her three days with Gummow, Silkwood's apartment remained unlocked. Her roommate worked the graveyard shift, midnight to eight, and had a hard time with the lock on the door, so she left the door open.

Silkwood returned to Dr. Shields on November 4. She said she was feeling better but still had a headache. Dr. Shields returned the Tylenol No. 3 and the Mellaril, and told her to take only one Quaalude before retiring. But she continued to use the Quaaludes as tranquilizers during the daytime.

Silkwood reported back to work on Tuesday, November 5, the day before the new contract negotiations were scheduled to begin. John Carver, supervisor of the laboratories, was waiting for her. Silkwood and Carver didn't get along well, and Carver scolded her for taking three days off, allegedly without informing Kerr-McGee. He also handed her a letter of reprimand for taking prescription drugs on the job without notifying him, as K-M rules demanded. A fellow lab worker had seen Silkwood giggling and weaving while on the job. But rather than telling her to go home or reporting her directly to Carver, he had called his wife and told her to phone Carver about Silkwood and the drugs, but not to give her name.

Silkwood felt Carver's reprimand was unjustified, part of Kerr-McGee's pattern of harassment of her and Brewer. She toyed with the notion of filing a grievance, but dismissed the idea, since she had clearly broken AEC and K-M rules. Both Carver and the FBI would speculate after her death that the reprimand drove her to contaminate herself with plutonium to embarrass the company.

At 1:30 P.M., she donned a smock and went to the Metallography Laboratory to catch up on paperwork. Around 2:45, Silkwood and her immediate supervisor took a break. The supervisor monitored herself as she left the work area. Her plastic shoe cover was contaminated. Both women changed shoe covers, called health physics, and waited while the technician checked their feet and the floor with an alpha counter. They were clean and took their break.

Silkwood returned to the Met Lab at 3:30 and finished labeling plutonium samples. Then she slipped into her Kerr-McGee white coveralls, taped lightweight plastic gloves to her wrists, and began to grind plutonium pellets in glove box 6. She handled the pellets with thick rubber gloves that reached from her fingertips to her shoulders. Later, Silkwood switched to boxes 3 and 4 to clean and polish the pellets ultrasonically.

At 5:30, Silkwood removed her coveralls and took another break, checking her hands and lower arms for contamination. She was clean. Fifteen minutes later she was back at glove boxes 3, 4, and 6, grinding, cleaning, and polishing. She was alone in the lab at 6:30, when she slipped out of the huge rubber gloves in box 3 and checked both hands on the monitor mounted on the box. The monitor began to click.

A lab analyst in another room called a health physics technician for her. The right sleeve and shoulder of her coveralls were contaminated forty times more than the level deemed safe by the AEC. Contamination is measured in disintegrations per minute (d/m). The AEC safe limit is 500 d/m; Silkwood, in her clothes, read 20,000 d/m.

The health physics office sent Mary Smith, a safeguards clerk, to help her, because there were no female health physics technicians on duty. Silkwood stripped in the decontamination room across the hall from the Met Lab so that Smith could check to see if her skin was contaminated, too. Smith found contamination on Karen's left hand, right wrist and upper arm, neck, face, and hair. The highest level was on the right wrist — 10,000 d/m, or twenty times over the limit declared safe by the AEC.

The health physics office took nasal smears with two cotton swabs, dried them under a heat lamp, and tested them with a low-level counter. As Dr. Abrahamson had told her three weeks earlier, a positive reading indicates plutonium probably has got into the lungs, but a negative reading doesn't mean it hasn't. Thus, any nasal contamination is potentially serious.

Silkwood's reading was positive, and high for a nasal smear. The right nostril showed 150 d/m; the left, which had been blocked ever since she had broken her nose as a child, showed 9 d/m. Whenever the nasal smear was higher than 50 d/m, K-M routinely used DPTA, a chellating aerosol chemical that acts as a catalyst to combine with the soluble plutonium in the body and is excreted in the urine or feces. To be effective, DPTA has to be used within a few hours after the contamination, before the plutonium settles in bones, tissues, or organs.

No one in the health physics office gave Silkwood DPTA. She had been working with pellets of plutonium oxide — the insoluble kind. Because DPTA combines only with soluble plutonium, whatever she had in her lungs would stay there.

Silkwood showered and washed in a mixture of Clorox and Tide to remove skin contamination. After she had spent thirty minutes under the hair dryer, her skin was surveyed again. No contamination above 500 d/m.

A health physics technician labeled a urine and fecal kit with her name and K-M badge number. The AEC required collection for five days after a contamination with airborne plutonium to see how much had got into the body. Workers picked up the home testing kits in the hallway between the health physics office and the Metallography Lab. At each end of the corridor was an air lock to keep any airborne plutonium inside the plant. The hallway was heavily used. Everyone passed through it going to and from their lockers, the lunch room, and the parking lot.

The health physics office placed the Met Lab on respirator status — everyone who entered had to wear respirators just in case — while they searched for the source of the contamination. The air did not appear to be contaminated, for the air filters were not radioactive, and the monitor buzzer, which is supposed to sound when it detects airborne plutonium, hadn't gone off. Had someone spilled something or tracked in radiation from another room?

Health physics technician Dennis Ford found two contaminated gloves in glove box 3, where Silkwood had been working. He changed the gloves and filled the ones Silkwood had used with water. They didn't leak. How had they got contaminated? No one could tell.

Kerr-McGee never discovered how Karen Silkwood was contaminated on Tuesday, November 5, just as it never learned how she had been contaminated on July 31. John Carver told the FBI after her death: "There are many possible contamination sources in the plutonium plant. Silkwood would not be limited to having

53

become contaminated in the area in which she worked." He said she could have "obtained the source material" from any number of places, implying that she may have contaminated herself.

After she showered, Silkwood returned to the Met Lab, finished her paperwork, and developed quality-control film in the darkroom next door. She stayed away from the glove boxes. Shortly after 1:00 A.M., she stopped working, picked up her specimen kits, and monitored herself before leaving the women's locker room behind the health physics office. She was clean. It had been another twelve-hour day at K-M.

"Well, it happened again," she told Drew Stephens after work. She wasn't terribly upset. Drew had been contaminated several times himself, and he just passed it over.

Chapter 6

The next day was the first day of the contract negotiations, and Silkwood was excited. She checked into the Met Lab at 7:50 A.M., slipped into her smock, and began to wade through quality-assurance papers. Wayne Norwood, the health physics director, called her into his office at eight o'clock. He was a grandfatherly man with silvery hair combed back and slicked down. He had graduated from Oklahoma State University more than twenty-five years before with a B.S. in poultry science, but since then he had taken courses in radiation biology and fundamentals of radiation measurement. The General Electric Company had hired him as a radiation monitor journeyman, and he had been a senior technician at Battelle Northwest Laboratories and at Hanford before Kerr-McGee hired him in 1969.

Wayne Norwood was concerned about health and the increasing incidence of contamination at the K-M plutonium plant. He expressed his concern privately to K-M managers such as James Smith, but he never stood up to Kerr-McGee at the daily staff meetings. Norwood was off when Silkwood had been contaminated the previous evening, but he read health physicist Don Majors' report the first thing that morning. Norwood assumed Silkwood had got contaminated on the shoulder somehow, then touched and scratched her nose.

He was concerned about Silkwood. Was it true, he asked, that her left nasal passage was blocked? Yes, she said, and showed him that she couldn't breathe through it. The point was impor-

55

tant. If the left nasal passage was blocked, she would have breathed less plutonium into her lungs.

Silkwood went back to her paperwork until 8:50. She monitored herself before she left for the nine o'clock negotiation meeting, and the needle jumped. Her right forearm was hot. Soap and water wouldn't remove the contamination.

"I'm going to be late for the union meeting," she told Norwood. "Can't I go out?"

Norwood asked her to wash again. When the contamination wouldn't come off with soap and water, he told her she could go to the meeting, but warned her to come back as soon as it was over. He wasn't worried about her contaminating anyone else. If soap and water couldn't wash off the hot spot, the contamination must have been deeply embedded in the pores of her skin. There was no danger that the minute particles of plutonium would rub off on anyone else.

Silkwood joined Tice and Brewer in the uranium plant conference room in the next building. She had three proposals to push. The OCAW wanted K-M to improve health and safety facilities at the plant. K-M negotiators resisted, arguing it was not necessary, since Kerr-McGee was already bound to comply with state and federal health, safety, and sanitation regulations.

The OCAW wanted K-M to submit annually to the union all available information on sickness and death of present and former employees. The company refused, on the grounds that to do so would open the company to lawsuits.

The OCAW wanted a joint union-management health and safety committee to keep the plant on its toes. Kerr-McGee negotiators said the joint committee wouldn't contribute a thing to accident prevention.

The negotiations broke at noon, and Silkwood lunched offsite with the OCAW negotiating team. At 1:30, the meeting resumed, and continued until 3:15 that afternoon. The OCAW and Kerr-McGee agreed to meet the following Wednesday, November 13, for another session.

Silkwood returned to the health physics office to finish decontamination. She still had 5000 d/m (ten times more than the AEC allowed) on her right forearm, neck, and face. No one knew where it had come from; she had been in the plutonium plant for only one hour that morning and had worked only in the office, the least likely place to be contaminated. Silkwood showered again. But this time she scrubbed with a sandpaper-like potassium-permanganate and sodium-bisulfate paste until her skin was pink and raw. She was getting worried and asked Norwood if she could have another nasal smear. The right nostril read 171 d/m — slightly higher than the day before; the left, 30 d/m — also higher.

"Where do you think it's coming from?" Norwood asked. The 171 d/m reading was very strange, given the short time she had been in the plant that day.

"I think it is coming back out of my lungs," she said. "Back out of my nose."

That seemed far-fetched to Norwood. He asked her to stop by first thing the next morning for another smear. Silkwood asked to have her car and locker surveyed. The results were negative, and she left the plant at 7:00 P.M. after an eleven-hour day.

She called Drew that night. "Gee, I don't know where it's coming from," she said. "I don't understand it."

Later she called Dr. Abrahamson. Karen was in tears; she said she was afraid of dying of cancer, and asked him for legal and medical advice. He told her he did not practice medicine by phone.

There was no doubt in Abrahamson's mind that Silkwood wasn't contaminating herself, even though she had asked him in October whether it would be dangerous to swallow a plutonium pellet. He had told her it would probably not be harmful, because the pellet would pass through the body without seriously contaminating it. "I wouldn't ever have thought, nor do I now think, that she had [contaminated herself] purposely," he would later tell ABC Television. "The nature of her concern, the kind of questions she was asking, her concern with her health . . . just doesn't fit with that idea."

Drew Stephens spent the night with Karen. He came in at ten and woke up Sherri Ellis, Silkwood's current roommate. She was a blonde with stringy hair, innocent baby-blue eyes, and a love of horses and prairies. Her parents owned a rundown farm off Highway 74, one mile from the Kerr-McGee plant, with an oil well pumping day and night behind a gray barn that barely held its own against the wind. Ellis had moved in with Karen two months before. Like Karen, she was a K-M lab technician. Unlike Karen, Sherri did not belong to the union. Ellis and Silkwood hardly saw each other and were not close friends.

While Ellis was getting ready for the midnight shift, Drew and Karen went out and bought pizza and Dr. Peppers. All three ate together. Sherri made a ham sandwich for lunch and left for work at 11:30. Drew and Karen listened to records and went to bed around one or two.

□ □ □

Silkwood reported for work on Thursday morning at 7:50, carrying a Wednesday night fecal sample and four urine samples. Wayne Norwood took nasal smears again. The results were a frightening 45,015 d/m in the right nostril and 44,988 in the left, even though it was blocked. There were 40,000 d/m on and around her nose, and 1000 to 4000 on her hands, arms, chest, neck, and right ear. She was hot; it was very serious. Even after a health physicist irrigated her nose, she still showed contamination, indicating she might be exhaling contaminated air from her lungs. A special counter placed against the fecal kit measured 30,000–40,000 d/m. Wayne Norwood was worried. It was now clear that she wasn't getting contaminated at the plant.

Silkwood went back to scrubbing with the potassium and sodium paste. The health physics people were careful to trap Silkwood's shower water and soap in tanks for analysis before discharging them into a sanitary lagoon, which was supposed to be sampled regularly for contamination. Technicians checked the women's locker room, the employee lunch room, and Silkwood's car. All were negative, except for a few hundred d/m on a car switch. There was only one place left to check.

When Sherri Ellis walked through the apartment door, the phone was ringing. She had quit work at 8:00 A.M. and then stopped off to feed her horse. It was Karen on the phone. She told Ellis that K-M inspectors would be over later to check the apartment. "Stay out of the bathroom and kitchen," she said. Sherri mumbled, "Okay," and went to bed.

Silkwood, Wayne Norwood, and two K-M health physicists — Judy Ward and William Rogers — got to the apartment at 1:30 that afternoon. Ellis was sleeping. Norwood and Rogers walked their alpha counters around the apartment; Ward woke up Ellis and checked her. Ellis' hands and buttocks read 2000 d/m.

Norwood surveyed the kitchen. He had his alpha counter set on the 1000 d/m scale, where it could read up to 20,000 d/m. The needle flew off the scale. Norwood reset the counter. The stove read 25,000 d/m and the refrigerator door 20,000. But inside the refrigerator, the wrapper covering some bologna and cheese read 400,000 d/m.

Norwood was shocked. "Where in the world could all this come from?" he asked Silkwood, who had followed him into the kitchen.

Karen didn't reply directly. "I spilled urine in the bathroom," she said.

"Did anyone . . . eat any of the cheese and bologna?" Norwood asked.

Silkwood looked at the package and could tell by the way it was sitting in the refrigerator that no one had touched it since she had. "No," she said.

Rogers surveyed the bathroom. The black toilet-seat cover read 100,000 d/m, the matching black floor mat 40,000, and the floor 20,000. There were between 500 and 2000 d/m on Silkwood's pillowcases and bed sheets.

Silkwood later told the AEC that on Thursday morning, November 7, she had spilled some of her fresh urine sample from the kit onto the floor, wiped the container and the floor with paper towels, and flushed them down the toilet. She took some bologna from the refrigerator and put it on the closed commode top to remind her to make a sandwich for lunch. Then she remembered that she had left some food at work and decided not to bring a sandwich. She put the bologna and cheese back in the refrigerator. Health physics technicians found that her lunch still at the plant — Doritos, a container of fruit salad, and a cheese on rye — had not been contaminated.

The health physics office also learned that the plutonium in her November 7 urine sample was insoluble. Since insoluble plutonium does not dissolve in blood and pass through the kidneys, there is no way it could be eliminated in urine. Therefore, someone must have spiked her urine kit either at work or at her apartment. But one thing was certain — if the person who had salted the kit wanted to make it look as if Silkwood passed hot urine, he or she didn't understand the difference between soluble and insoluble plutonium.

Norwood assumed Silkwood had spiked her own samples, that she didn't realize how much plutonium she was playing with, and had got it on her hands when she spilled urine in the bathroom. That's what he later told the FBI. It would be easy to get her face and nose contaminated once her hands were hot. A scratch here, a touch there. A deep breath, and it's in the lungs.

Norwood, who was suspicious of Silkwood, based his assumption on a couple of little things. When she had called Ellis from the plant, hadn't she told her to stay out of the kitchen and bathroom? Why would she have said that unless she *knew* they were contaminated? And at the apartment, when he had asked her where she thought the contamination had come from, hadn't she said she'd spilled urine in the bathroom? It was obvious she *knew* the source of the contamination, and therefore did the spiking herself to embarrass Kerr-McGee.

The fact is, security at K-M was so lax, anyone could steal plutonium at any time and never get caught. Up until April 1974 — six months before the contamination — the AEC did not require K-M to have a security organization of uniformed, armed guards. In May, K-M started deploying two guards at the main entrance inside the gate. They were trained at K-M headquarters in Oklahoma City by the Police Training Academy. One of the first armed guards was Leonardo Garnell Crusher, who turned out to be Garland Buford, a convicted bank robber. When he was discovered through a routine credit-bureau check, K-M argued that the AEC did not pay for a personnel security check, and that, although the AEC forbade hiring ex-cons as guards, the corporation could not request name and fingerprint checks from the FBI and city police.

Before 1974, Kerr-McGee also had no way of telling whether workers were stealing uranium and plutonium pellets. "In your pocket, if you want," former K-M worker Ron Hammock put it; "all you could carry." One worker brought a uranium pellet home to his son, who took it to school to show his classmates. Although only mildly radioactive, the K-M uranium pellets were still dangerous, because radioactive particles from the pellets could enter the body through the nose and mouth or through cuts in the skin.

In another instance, two boys around six or seven years old showed some pellets to a neighbor working in her garden. They got them from their father, they said. Her husband told the boys to get rid of the pellets, so they threw them into John's Creek. AEC's Gerald Phillip investigated the incident much later. "There being no confirmation of the theft and no concern for health and safety by the public," he wrote in his report, "no further action is necessary."

K-M eventually installed an exit monitor at the plant, but the device could not detect plutonium in amounts less than half a gram, enough to give 767,500 nuclear workers the maximum exposure permitted by the AEC for a fifty-year period if they all breathed air contaminated with it. Since, even with the monitor, it would be easy to smuggle plutonium out of the plant, K-M guards were supposed to inspect all packages. In fact, they frequently didn't. And even if they did peek into packages, it would still have been easy to steal a few grams or ounces or pounds. Wrap some in the lead tape available at the plant and walk out. Put some pellets into a plastic bag, toss them out the window, and pick them up later. Hide some pellets with the waste, cart the waste outside, and sort the pellets later. Monitors are shut off before waste passes through.

Even John Carver admitted that Kerr-McGee security was poor. In an attempt to make it sound as if Silkwood had contaminated herself, Carver told the FBI, "She could have obtained the necessary material at numerous locations in the plant and thereafter smuggled the material from the plant with little or no difficulty in view of security procedures which were then existent." Carver added he had heard rumors that Drew Stephens might still have some of the stolen plutonium, but he offered the FBI no evidence.

Chapter 7

Wayne Norwood and his health physics team returned to Silk-wood's apartment at 5:40 P.M., wearing full-face respirators, special galoshes, and gloves taped to their white coveralls. They began to pack Silkwood's and Ellis' hot things into fifty-five-gallon drums lined with plastic, for burial in a radioactive-waste dump . . . one pound of frozen strawberries, a two-pound bag of stew vegetables, one box of Raisin Bran, a three-speed J. C. Higgins bicycle, a jar of Avon Bath Oil, a box of Q-tips, one pair of panties with 1000 d/m on them, sixteen Crayola crayons, a bottle of Wind Song cologne, *I Never Promised You a Rose Garden.*

Norwood found in Silkwood's bedroom a lot of small items she had apparently pilfered from the lab. Glassware, mainly — graduated vials and cylinders, flasks, and sample bottles. Several items were stashed inside K-M fecal kits. The discovery only confirmed Norwood's belief that she must have stolen some plutonium from the plant, perhaps in her fecal kit.

Kerr-McGee didn't send Karen a doctor, even though an afternoon nose smear still showed 6000 d/m — a very serious contamination level. But the company did send an attorney. While Norwood's team decontaminated her apartment, the lawyer had Silkwood cornered in a company car outside. Karen was crying. She was afraid of dying, she kept saying. The lawyer kept trying to squeeze a statement out of her, Karen later told Wodka. The Kerr-McGee attorney succeeded. In a two-page document that began "I, Karen Gay Silkwood, being of legal age, do say," she told her story. "I have no knowledge of what happened," she concluded. "But I feel the contamination is coming out from my body."

During the afternoon, Karen called Drew Stephens at work. She told him she was getting shaky, and she also asked him to come to the apartment because the health physics people wanted to check him. He should think about where he had been since he left the apartment that morning, in case he had been contaminated.

Drew finished work at six and went home to change. He made it a point not to shower so that if he was contaminated, he wouldn't wash the radioactivity into the Oklahoma City sewers. When he got to Karen's apartment around seven, Norwood and his team were still there, but Karen was gone, no one knew where. She had just got into her car and driven off. Drew watched the men point their alpha counters at Karen's belongings and listen through earplugs to the radioactive clicks like rapid-fire static electricity. Then he went into Sherri Ellis' room and stripped. Will Rogers monitored him. He was negative. Rogers went home with Drew to check his house. It was clean, too, so Rogers went back to Karen's apartment and the packing drums. Drew worried and waited for Karen to call. He knew she would, and there was nothing else for him to do.

In a phone booth not far from her apartment, Karen called Steve Wodka in Washington. She was so frightened and was crying so hard, Wodka could barely understand her. She told him she couldn't get into her apartment. She mumbled phrases and contamination numbers — d/m, nasal smears, DPTA, bologna and cheese. She said she believed someone had walked into the unlocked apartment and contaminated the food in her refrigerator. She told Wodka that Kerr-McGee was harassing her, "forcing" a statement from her.

Wodka was angry at Kerr-McGee. He told Silkwood to stay away from the company until he got there and could act as her representative. He warned her not to talk to the AEC, either.

About nine o'clock, Drew Stephens' phone rang. He grabbed it.

"This is Steve Wodka."

Stephens and Wodka had never met. Drew had left Kerr-McGee by the time Wodka became involved with the local and Karen. She had told Drew she liked Wodka and had slept with him a few times.

"What's going on down there?" Wodka wanted to know. He said Karen had called and was scared half to death. She was hysterical; didn't make much sense.

While Drew was filling Wodka in, Drew's neighbor, Alan Davidson, came to the door. Karen was on his phone and wanted to talk to Drew. "Your phone was busy so she called me," Davidson said.

Drew hung up and ran next door. Karen was still in the phone booth, crying, but she wouldn't say where she was.

"Tell me where you are," Stephens said. "I'll come and get you."

"Stay there." She hung up.

Then Karen called her mother, Merle. She was still crying. She said she had been contaminated. She was dying. There was no treatment for it. She was coming home.

Should she come up there and get her, Merle asked.

No, Karen said. Drew was going to take care of her.

Wayne Norwood was the last to leave Silkwood's apartment. It was about eight o'clock, and all the really hot items had been packed and sealed in drums. It would take him another month to clear the apartment completely. Silkwood's bedroom was still a mystery. She had refused to allow him to enter the room after the preliminary check earlier in the day. Norwood assumed she wouldn't let him in because she was hiding narcotics or contraband of some kind.

Norwood backed through the living room to the front door, monitoring his steps as he went. He left the $600 alpha counter just inside the door so that he could check himself in the morning. Once outside the apartment, he tried both the back and front doors to make sure they were locked. When he returned Friday morning, he later told the FBI, the alpha counter was gone. He figured Silkwood or Stephens took it when they came back for the hidden drugs. It would be useful to her in covering her radiation tracks, he suggested to the FBI.

After he left the apartment, Norwood called Dr. Charles Sternhagen, a former full-time Kerr-McGee physician and a current K-M consultant working at the Lovelace Clinic in Albuquerque. They decided to put Silkwood on intravenous DPTA to clean out of her body as much of the plutonium as they could. By that time, Norwood had no idea whether Silkwood had been contaminated with soluble as well as insoluble plutonium. If she had soluble plutonium in her body, the DPTA could still clean out some of it. Norwood tried unsuccessfully to reach Silkwood to tell her Dr. Sternhagen would be in town the next day.

When Silkwood returned to her apartment, Norwood and the decontamination team were gone. It is not clear whether she had been watching her apartment and waiting for them to go. She told Drew Stephens later that night that the back door was unlocked and that she went into the bedroom to get something. She never told anyone what it was, but her landlord said later there was a trap door in the ceiling of her closet.

By the time Karen reached Drew Stephens' house, she was shaking like an aspen, hysterical, and incoherent.

"I'm going to die," she sobbed. "I'm going to die."

Stephens tried to calm her. "No, you're not," he said, giving her two Quaaludes. She settled down. The whole thing had finally hit her, Stephens thought. The nasal smears, the apartment, the lungs.

Stephens called Wodka back after he had calmed Silkwood down. There are going to be some pretty heavy sessions down here, he told Steve. They'll grill everyone until they find out what happened at the apartment.

Wodka said he'd come the next day. He had already talked to his boss, Mazzocchi, who was in Nevada on business, and Mazzocchi had given him the nod to go to Oklahoma. They were both worried. Wodka had asked AEC's John Davis to make arrangements for immediate medical care. But the OCAW leaders were also worried about their plan to leak documents to *The New York Times*. The contamination had put the spotlight on Silkwood, and that is just what she and they had agreed to avoid. Her cover was blown, and that gave K-M the excuse to search and seal her apartment. Neither knew what the next move would be.

□ □ □

Drew took Karen to work with him Friday morning. The AEC had called his house, looking for her. They wanted to interview her, but Karen refused to talk to them, as Wodka had instructed. Stephens knew Wodka was coming that evening, so he told the AEC that after work the three of them would come to the Northwest Holiday Inn, where AEC people always stayed.

While Drew pounded fenders at Whitfield VW, Karen sat on a bench and watched. She was pensive and quiet and looked scared. Stephens found it difficult to concentrate on his work. Midafternoon, they drove to the airport to pick up Wodka.

Silkwood gave Steve a big hug.

Was she all right, he asked. She told him she thought so.

They went to the Holiday Inn to meet AEC investigators Jerry Phillip and William Fisher, who checked Silkwood and Stephens for contamination. They were both clean. Then they all settled back to listen to Silkwood describe her Tuesday, Wednesday, and Thursday contaminations. It took the better part of four hours. She was emotional but not hysterical. She was scared she was going to die, she said. Later, she told Wodka, "It hurt to cry because the tears burned the skin on my face from the salt." The AEC reported that she appeared "honest and candid," and they had "no reason to doubt her account of the situation."

Wodka asked her later that night if she'd have her quality-control evidence ready for *The New York Times* on November 13. She said she would. Wodka impressed on her that the *Times* was no newspaper to fool around with, that the quality-control allegations were serious, and that she had to produce. She told him not to worry. Wodka did not ask to see her documents, and she didn't volunteer to show them.

On Saturday afternoon, with Wodka present, the AEC's consulting physician, Dr. Neil Wald, and K-M's consulting physician, Dr. Sternhagen, interviewed Silkwood, Ellis, and Stephens. The two physicians thought it would be good for Silkwood to go to Los Alamos, New Mexico, for a full-body count. Since plutonium emits alpha rays for only a short distance, internal contamination can be read only by highly sensitive equipment. Los Alamos, where the atomic bomb was developed in the 1940s, had some of the best radiation-detection equipment in the country. Dr. Wald felt that if the biosamples taken so far were accurate, the Silkwood case had great medical significance.

Karen didn't want to go to Los Alamos alone, so Wodka suggested Drew go along.

"What about me?" Ellis asked.

It was settled that all three would go. K-M, which was not represented at the meeting, agreed to foot the bill.

Silkwood, Ellis, and Stephens went to Baptist Hospital that night. Dr. Sternhagen supervised the examination. Karen was given a blood test and took a quick-acting laxative so that her feces could be checked for radiation. The test showed that she was still contaminated internally, but the reading had dropped by 90 percent. The cheese and bologna themselves had been tested by that time, not just the wrapper. They had up to 2500 d/m, five times the level deemed safe by the AEC. From her feces, it was clear she had eaten some plutonium, but no one knew when, where, or how.

Dr. Wald, who had defended the AEC in many of the personal-injury lawsuits filed against the commission, told the FBI, when they interviewed him one year after her death, that Silkwood had been very uncooperative with him and Dr. Sternhagen. She was more than an hour late for her appointment at Baptist Hospital and offered no explanation. She refused to take the DPTA treatment, saying that Kerr-McGee had suggested it and that all the physicians she had seen so far were K-M consultants. She said she didn't trust them, including Dr. Wald.

Wald told the FBI that Silkwood said she was having her period and was concerned that her sanitary napkins might contaminate someone else. He said he told her to save them. Wald also told the FBI that some of the weekly urine samples Silkwood had been giving Kerr-McGee since her July 31 contamination were hot. He said the October 25 specimen in particular, taken fourteen days before Kerr-McGee found out her apartment had been contaminated, showed a "high significant reading."

It was clear that the deliberate contamination of Karen Silkwood had begun sometime in October.

□ □ □

Silkwood, Stephens, and Ellis flew to Albuquerque on Sunday morning and rented a car for the hundred-mile drive northwest, past Santa Fe, to the Los Alamos Scientific Laboratory, owned by the AEC and staffed by the University of California. Karen drove, Sherri sat up front, and Drew slept in the back seat.

Karen and Sherri were drinking wine. They missed the turnoff to Los Alamos.

"God, what's happening?" Drew said when he woke up.

They were bumping along on the smallest and ruttiest road he had ever seen, in the middle of the Pecos Wilderness Park, some sixty to eighty miles out of their way. A forest ranger was waiting to greet them at the other end of the park.

"Did you all drive all the way through there?" he asked.

"Yes," they said.

"That road is usually closed at this time of the year, and it's amazing that you got through."

When they finally pulled into the tiny town at the edge of the forty-square-mile Los Alamos Scientific Laboratory (LASL), it was dark. Stephens called Dr. George Voelz to tell him they had arrived and to apologize for being late. Dr. Voelz had a lab team waiting, but because it was so late, he canceled the full-body and chest counts until the first thing Monday morning. The Oklahoma trio got a bite and turned in early. Silkwood was apprehensive and tired, but mostly just plain tired.

The Los Alamos Scientific Laboratory is perched on the Parjarito Plateau, 7150 feet above sea level, and almost touches the Jemez Mountains. The snow-capped Sangre de Cristo Mountains poke up from the desert floor in the distance. Other than the Santa Fe National Forest, the Rio Grande running between the mesas, and the Pecos Wilderness, there is nothing but canyon after canyon all the way to Santa Fe, thirty-five miles southeast.

In 1974, there were almost 6000 AEC employees at LASL, 85 percent of whom worked on nuclear projects. LASL was routinely doing full-body and chest counts on about 600 of its own plutonium workers. From time to time, it tested outsiders like Silkwood at the request of the AEC.

Dr. Voelz joined Silkwood, Stephens, and Ellis for coffee on Monday morning. He explained what the tests were and how long they would take. He wanted to put them at ease, and he was surprised to see how calm Silkwood was. Dr. Sternhagen had told him to expect the worst.

Voelz was forty-eight years old and had been an AEC physician for twenty-two years — all of his professional life. He had 500 people working for him. His primary responsibility was the health and safety of LASL's employees and anyone outside LASL who could be hurt by fire, radiation leaks, or LASL wastes. He also had a good bedside manner. He was soft-spoken and patient, almost fatherly. He knew Silkwood was scared, and he wanted to keep her calm until he knew exactly how badly she had been contaminated.

LASL technicians did a nose and mouth swipe on Silkwood when she arrived at the measurements lab in the sub-basement of the gray, boxlike Health Research Laboratory building. She was clean. Then she showered to get rid of any skin contamination and donned a pair of white paper pajamas with an elastic waistband and buttons down the front, and brown paper slippers. Technicians surveyed her carefully before they allowed her to step into the nine-foot-square cell made of seven-inch-thick pre–World War II battleship steel. The room looked like a giant bank vault painted green on the outside and covered on the inside walls, ceiling, and floor with disposable brown paper. There was a bed covered with a cotton blanket and a chair that looked like a dentist's chair.

The three plutonium scanners inside the vault were extremely sensitive to radiation. Because the New Mexico mountains were high in uranium and thorium, even the bricks and cement blocks made from New Mexico rock and soil had slight radioactive traces in them. That was why the lab insisted on showers and paper clothes, and why the room was built with steel that could not have been contaminated by radioactive fallout from the bombing of Nagasaki and Hiroshima.

Silkwood slid under the blanket, and a technician placed a

sodium-iodide detector over each breast. The detectors were attached to swinging steel arms and looked like two giant floodlights. They could measure the gamma rays coming from the plutonium and the americium in her lungs. Americium is a daughter of decaying plutonium. Since it is more energetic, it penetrates farther and is easier to read.

The technician closed the vault door and flipped on the intercom in case Silkwood wanted to talk or got claustrophobic. She lay there for thirty minutes while readings clicked into the computer outside, programed just for her. Before she had entered the vault, the technician had scanned her ultrasonically. The twelve measurements taken, principally the thickness of tissue layers, would help the computer calculate the strength of the gamma rays coming from her body.

By noon, Dr. Voelz had a preliminary reading on Silkwood. She was as calm as anyone waiting for an important test result could be. Dr. Voelz told her the first results looked "very good." She had been exposed, he said, but the levels seemed "quite low." She had less than one half of the body burden the AEC allows for plutonium workers, he said. He wanted to run her through the tests again to be sure.

After lunch, Karen had a full-body count. She sat in the chair in the vault, and the technician aimed the nine-inch-diameter, five-inch-thick sodium-iodide counter housed in a spotlight casing. Karen, in her white paper pajamas, sat there for another thirty minutes.

Dr. Voelz met with her at 4:30. The test was completely negative, he said. The technician would run her through it again on Tuesday. She was relieved, in the mood to celebrate.

There was a scientific conference at LASL the next day, so Silkwood, Stephens, and Ellis had to move out of their motel. They drove to a Santa Fe Ramada Inn and ate at a little Mexican place. The Ramada Inn had a band; Karen loved music. The three sat around for a few drinks. There was something on Drew's mind, a question he had held back. He blurted it out.

"Did you eat a pellet?" he asked.

She stared at him with wide eyes and started to cry. "No, I didn't," she said. Drew felt terrible about asking.

It had been a long day and was getting late, nearly eleven. Drew was tired; he wanted to go to bed. Sherri and Karen wanted to party. Sherri had already found someone and was on the dance floor. Drew thought it would be better if Karen got a good night's sleep. They had words, and, in the end, Drew gave her the motel room key and went to bed.

The next morning they were back at LASL by nine. The lab did a radiation count on Karen's abdomen because technicians had found some level of radioactivity in the preliminary analysis of her feces. Then the lab recounted her chest and ran a second count on each lung separately to see if there was any difference between the right and left lungs. By one o'clock, Dr. Voelz was ready for the last conference. He had no final results on the urine and fecal samples. Just roughs. But he had all the chest, lung, abdomen, and full-body count results. He met Silkwood, Stephens, and Ellis together.

He told Stephens and Ellis their tests showed some positive contamination, but the results were "statistically insignificant." In layman's language, that meant they were clean for all practical purposes, he said.

Silkwood wasn't so fortunate. The counters could not directly read any plutonium in her lungs, Voelz told Silkwood, but they did detect .35 nanocuries of americium on the first test, and .33 on the second. He said that since the two tests were essentially identical, the reading was probably accurate. The .33 to .35 nanocuries of americium indicated that she had about 8 nanocuries of plutonium in her lungs. He explained that the amount of americium in the body tells the amount of plutonium there, as well. He told her that the AEC maximum permissible lung burden for plutonium workers was 16 nanocuries of plutonium. Her contamination would therefore amount to about half the permissible lung burden.

Voelz reassured Karen. He had seen many, many LASL workers contaminated with higher levels, and they didn't die of radiation or develop cancer. Besides, he said, the final results would not be completed for another ten to fourteen days. He was sure the analysis would show less than 8 nanocuries and that she would not have any significant health problems — although he admitted that her actual lung count could be three times higher (24 nano-curies) or three times lower (2.7 nanocuries) than the readings. Twenty-four nanocuries is one and a half times the AEC's permis-sible lung burden.

Karen asked to talk to Dr. Voelz alone. She was concerned about what the plutonium would do to her genes. The Abraham-son-Geesaman talks about genetics had frightened her. Could she still have babies, she asked. Would they be born deformed? Voelz assured her she could have normal children. Would he run tests on her menstrual pads, she asked. Just to make sure her ova and womb weren't contaminated? He said that tests would be useless; she was fine.

Dr. Voelz believed in his nanocuries, and he believed the AEC's safe limits were indeed safe. But what he failed to tell Karen Silkwood was that there was a group of scientists — most of whom did not work for the AEC — who thought the standards were ridiculous. Dr. John Gofman, who helped develop the plu-tonium for the A-bomb, thought the limits should be reduced 8 times. Others wanted them lowered by 100 to 1000 times because they believed that the AEC safe limits were based on incomplete, unreliable data from small samples.

Furthermore, scientists such as Dr. Gofman point out that the 16 nanocurie lifetime lung burden is the AEC safe limit for an *average* person; that plutonium is more dangerous for women, persons with respiratory problems, the young, and skinny people; that heavy smokers are ten times more susceptible to lung cancer from plutonium than nonsmokers; that the sodium-iodide coun-ters don't distinguish between soluble and insoluble plutonium; and that insoluble plutonium is at least twenty times more toxic.

Silkwood was female, twenty-eight, asthmatic, ninety-four pounds, a heavy smoker, and contaminated with insoluble pluto-nium. Dr. Voelz did not take her personal history or adapt the numbers to fit her.

Karen accepted Dr. Voelz's judgment to a point. She felt bet-ter, more confident that at least she wasn't going to die right away. But she continued to keep her used tissues and sanitary nap-kins in a plastic bag so that she wouldn't contaminate anyone else. And she wouldn't kiss anyone. Not even Drew.

Chapter 8

They drank complimentary Almadén Grenache Rosé on the plane back to Oklahoma City, and when they landed at 10:00 P.M. on Tuesday, they picked up Sherri's friend Debbie at a club and went to Drew's house. He had a rosé nightcap and went to bed. Karen, Sherri, and Debbie sat up for a while — no one is sure how long — and drank Bloody Marys — no one is sure how many.

Karen called Steve Wodka before she went to bed. She told him about her Los Alamos test results. She said she was ready for the meeting with David Burnham on November 13, the next night.

"If you can't put it together," Wodka said, "I won't bring him down."

"Let's do it," she said. It was her last conversation with the union leader.

Karen was late for the Kerr-McGee–OCAW bargaining session the next morning, Wednesday, November 13. She was supposed to meet Jack Tice and Frank Murch, who had replaced Jerry Brewer on the union bargaining committee, at the Hub Cafe in Crescent for coffee and strategy. Instead, she went straight to the bargaining session in the uranium plant conference room. It was the second bargaining session, and K-M didn't budge on health and safety.

Later that day, the AEC questioned her again for two hours about her contamination, and by five o'clock she seemed de-

74

pressed. The OCAW had scheduled a 5:30 meeting at the Hub Cafe to tell the members how the bargaining had gone and to listen to their suggestions. Silkwood drove Murch the six miles from K-M to the Hub. He noted that she looked anxious, nervous, and tired but not drugged or "downed out." She had no problems driving and didn't nod at the wheel.

The Hub Cafe is right on Highway 74 in the middle of Crescent's two-block business strip. It is a typical small-town greasy spoon with a lunch counter and stools along one wall and booths along the other. In between are tables with sugar shakers and plastic-covered menus with fingerprints. The meeting was held in the back room, a narrow hall with about ten tables. It served as a dining room when the Hub was busy.

About ten of the local's twenty members attended the debriefing. Jack Tice chaired the meeting and told them that Kerr-McGee wasn't budging. It claimed it couldn't afford to increase salaries and upgrade health and safety measures. What it really wanted, he said, was to rip up the OCAW's contract and write its own.

For the Kerr-McGee Corporation, 1974 was a whopping good year, with a net income of $116.5 million. But Kerr-McGee claimed that its Nuclear Corporation contributed little to that profit. It told the AEC it was losing money on the $7.2 million fixed-fee fuel-rod contract, and the AEC tacked on another half-million to the contract to keep the rods rolling from Crescent to Hanford.

Silkwood was quiet during the meeting. She made a brief presentation and sipped iced tea. Jean Jung sat eight feet away. "During the meeting, she was leafing through papers in a folder," Jung said in a sworn affidavit signed less than three months later. Steve Wodka had asked Jung for the statement, and the union had paid her way to the OCAW legislative office in Washington, where the affidavit was drafted and signed. "I watched her doing this, and noticed that some of the papers were quite heavy — almost like cardboard — and smaller than typewriter paper. These looked to me like they might be photographs."

Silkwood stepped out of the meeting at six o'clock to call Drew Stephens on the public phone hanging on the wall next to the back room just inside the main café. She told Drew not to forget to pick up Wodka and Burnham at the airport. Wodka had set up the meeting with the *Times* reporter for eight o'clock at the Northwest Holiday Inn in Oklahoma City, about thirty miles south of Crescent. The contamination incident and the subsequent Los Alamos trip had killed plans for a Dallas rendezvous.

"She was very excited and looking forward to the meeting," Stephens later told the FBI. "In good spirits."

After the union meeting, both Brewer and Murch offered to drive Karen home. They knew she had been contaminated and had gone to Los Alamos. Everybody did. It was a small plant. Why each of them made the offer turned out to be significant. Murch later said under oath: "She was depressed. And I . . . was just trying to be a good guy. If she needed a ride home, I'd be more than glad to give her a ride home." He did not say then and denies ever saying that she looked sleepy.

Brewer had a similar reaction. "She was somewhat nervous," he later swore under oath. "She had been through quite a bit. We assumed she was tired after all she had been through the past two weeks or so." Brewer did not say then and denied ever saying that she looked sleepy. On the contrary, he testified there was nothing unusual about her behavior.

Silkwood thanked them both and said she was all right. She joined Jean Jung and Alma Hall at the back of the meeting room. Jung worked in the fuel-fabrication room, where she sometimes ran the plutonium pellet grinder or pellet presser. Karen had helped her get decontaminated two months before.

Silkwood trusted Jung and had told her, against Wodka's orders, that she was collecting documents showing how K-M was doctoring quality-assurance records. She was trying to get the plant straightened up, Karen had told Jung. She was working in the lab after hours. Try to keep it quiet, she had said. Don't tell anyone.

Jung said in her affidavit: "She told me one day she had photographs of defective welds on sample fuel pin claddings taken from lots which were passed by quality control. She once told me about a particularly bad batch of rods. I believe it was lot 233 — which she said should never have been allowed to leave the plant."

76

Alma Hall was not Silkwood's friend. Hall thought Karen was too pushy, a troublemaker, but she recalled that Silkwood looked scared and nervous that evening. Karen was clutching a thick folder and said something about quality control, Hall says.

Jung was eager to get home. It was seven o'clock and dark. "See you at work in the morning," she told Silkwood and Hall.

Karen followed Jung into the main part of the café. "I want to talk to you," she said. She started to cry quietly. She told Jung she was frightened; that she had been so badly contaminated, she would eventually get cancer and die from the plutonium in her lungs. She said someone had deliberately contaminated her.

"I can't believe who would do such a thing like that," Karen said. "It has got to be somebody that works for Kerr-McGee that can get it out."

Her voice was normal. She wasn't trying to whisper as if to hide something or share a secret. Jung felt sympathetic. She had attended one of Abrahamson and Geesaman's talks, and they had "scared the fire" out of her.

Karen pointed to her documents. "She then said there was one thing she was glad about," Jung swore in her affidavit. "That she had all the proof concerning falsification of records. As she said this, she clenched her hand more firmly on the folder and notebook she was holding. She told me she was on her way to meet Steve Wodka and a *New York Times* reporter . . . to give them this material."

The folder was dark brown, legal size or larger, made of heavy material like cardboard. The spiral notebook was 8½ by 11 inches and reddish brown.

Silkwood and Jung walked toward the door. "I've got three kids," Jung said. "I have to go on home."

"Well, I've got to go, too," Karen said.

It never crossed Jung's mind that Silkwood might not be capable of driving safely. To Wanda Jean Jung she seemed determined to deliver to the *Times* whatever she had in her folder and notebook.

It was cold outside, partly cloudy, with a light wind blowing across the fields. The rough asphalt of Highway 74 was dry, and the road rolled gently in an almost straight line south toward

Oklahoma City. Silkwood was wearing a brown leather coat and leather half-boots, with a black blouse and red, white, and black plaid slacks.

It was about ten past seven. She would be late for the eight o'clock meeting at the Holiday Inn on the northwest side of the city, but not by much. She hopped into her 1973 white Honda Civic with new trick tires to give the car more traction, front wheel drive to prevent drifting, and a large white racing mirror.

The Oklahoma Highway Patrol said later that she drove between fifty and fifty-five miles an hour down the road she traveled almost every day. Five miles south of Crescent, she crossed the Cimarron River. Kerr-McGee sat on a knoll half a mile ahead of her, on the left. The plant's lights commanded the horizon. She passed the gate. SAFETY PAYS ON OR OFF THE JOB, the sign read. She crossed over Highway 33. She was surrounded by pastures. Logan County was farm country, caught somewhere between rolling hills and flat fields of red dirt.

Exactly 7.3 miles from the Hub Cafe, her Honda crossed from the right side of the highway to the left and traveled 240 feet along the grass shoulder. Then it flew for twenty-four feet over one wingwall of the concrete culvert running under the road and not visible from the highway, and smashed into the other wall at a speed of forty-five miles an hour. The car flipped onto its left side in the culvert entrance.

Inside her purse was the little notebook she always carried around. The last entry read: "Try to reach a contract by November 26. Meeting again Thursday, November 21."

Around 7:30, James Mullins backed up his red rig to the ditch. His boss, John Trindle, raced in his pickup truck to a gas station pay phone a mile and a quarter away and called the Oklahoma Highway Patrol (OHP).

Soon after Trindle left, Kerr-McGee employees Fred Sullivan and Law Godwin stopped to see what was going on. Godwin later told the FBI that his wife had had a flat tire on Highway 74 about twelve or fifteen miles south of Crescent. Shortly after seven, he said, he picked up his friend Sullivan, fixed the tire, and was driving back toward Crescent when he saw people standing at the side of the road.

Neither Godwin nor Sullivan belonged to the union. Godwin was a supervisor in the quality-control division of the plutonium plant; Sullivan was a document-control manager. Godwin told the FBI that neither he nor Sullivan was sent to the accident site by Kerr-McGee. "They came by chance," the FBI agent who interviewed them reported.

Godwin sidestepped down the bank into the ditch, as Dalton Ervin, Mullins' brother-in-law, had done before, while Mullins, Sullivan, and Ervin watched from the road. Peering into the car with a flashlight, Godwin saw what Ervin had seen — a woman in the driver's seat. It looked as if the steering wheel had pushed against her and almost pinned her to the Honda's roof. She was motionless, and the blood on her face had partly dried. Like Ervin, Godwin could not detect any sign of breathing.

Godwin thought the woman looked like Karen Silkwood. Not knowing that Trindle had just driven off to call the police, Godwin shouted to Sullivan to get help. Sullivan sped off to call the Guthrie police from a nearby farmhouse while Godwin stayed with the wreck.

By the time Trindle got back to the car, Oklahoma Highway Patrol officer Rick Fagen was there with a Guthrie Fire Department ambulance. It was about 8:15 — an hour after Karen had left the Hub Cafe. Trindle, Fagen, and the two ambulance drivers lifted the 1638-pound Honda to its wheels. Trindle cut the door open with his Porta-Power tool. Mullins and Ervin stood above on the shoulder and watched. Mullins noticed some papers on the ground and Silkwood's purse resting against the retaining wall two feet in front of the Honda.

The two ambulance drivers lifted Silkwood onto the stretcher and raced her to Logan County Hospital. She was pronounced dead on arrival. The nurse noted that her pockets contained a $5.00 bill, two singles, $1.69 in change, a blank check, Dr. Sternhagen's card, a Mickey Mouse pocket watch, two used Kleenex, some pills, and her Kerr-McGee I.D. badge and electronic security key.

□ □ □

Tow-truck driver George Martin just happened to be at the Guthrie Police Station when Sullivan's call came in. He was dispatched immediately to the wreck. For Martin, it was just another routine assignment. Over the past four years he and his wrecker had chased police cars to more than 1200 accidents.

When Martin was about five miles from the accident scene, the Guthrie police radioed him to head for home. He turned back. "It just didn't make sense to turn me around . . . when there was a report that someone was pinned in a car, particularly when I was running right with the ambulance," Martin later told a writer for *New Times* magazine. "You just don't pull someone off a Code Two alert."

About 8:30, just as the drivers were lifting the stretcher into the ambulance, Ted Sebring pulled up with his wrecker. Sebring owned the Ted Sebring Ford dealership in Crescent, up the street from the Hub Cafe on Highway 74, and ran a part-time wrecking service, as well. After closing the office and garage at five that evening, he had gone to Kenneth Vallequette's house for a drink. Vallequette had just bought a new car from Sebring. Kenneth Hart was there, too.

At 8:15, Vallequette's phone rang. It was the OHP District One dispatcher. There was a wreck on 74 just 7.3 miles south of Crescent on the left, or east, side of the highway, the dispatcher said. Could Ted Sebring haul the car from the ditch? Sebring asked Vallequette to drive him back to the garage for the wrecker. Hart came along.

Sebring climbed into coveralls, jumped into the wrecker, and took off from town, over the Cimarron, past Kerr-McGee, and across Highway 33 to the accident. Hart and Vallequette went back to Vallequette's house for some flashlights.

Before Sebring arrived at the accident, Rick Fagen had looked inside Silkwood's car. He was a rookie, on the job only five months. He had investigated fewer than fifty accidents. He noted a hat, raincoat, spare tire, jack, and wrenches in the rear of the Honda. In the back seat, a plastic bag filled with used sanitary napkins, two half-inch-high stacks of K-M–OCAW bargaining papers, and a plastic flask. In the glove compartment, he found the

Honda's title and an estimate for car damages. Fagen picked up the purse near the car and handed the wallet to Guthrie police officer William Clay. Clay, who had once worked for K-M, said he knew Silkwood, and told Fagen to call K-M personnel director Roy King for the name of her next of kin.

Fagen also noted, inside the purse, two cigarettes that appeared to be marijuana, a pill, and half of a tablet. He would later send them to the Oklahoma State Criminal Laboratory for tests.

Fagen picked up the papers lying around the car and tossed them into the front seat. He did not recall exactly how many papers there were or anything that was written on them.

Sebring pulled up on the west side of the highway, walked across the road to the ditch, and waited for Trooper Fagen to give orders. When the ambulance and six or seven bystanders had left, Fagen authorized Sebring to haul out the Honda. Sebring asked Fagen to patch a call through to Harold Smith, his sales manager. Smith usually helped his boss on tow jobs, and it wouldn't be easy to get the Honda out without damaging it.

When Smith arrived, he hooked the wrecker's log and chain to a ring welded to the Honda's frame. Sebring watched as Smith made the tow taut. The car's rear end swung around, and the driver's side hit the concrete wingwall. As Smith pulled, the rear end scraped against the soft dirt. Sebring was certain no damage was done to the left rear bumper or the left fender. This turned out to be an important observation. Harold Smith would contradict Sebring, but from where he operated the wrecker, Smith could only hear the grind and rasp of metal on cement and stone and dirt. Fagen would also contradict Sebring.

Sebring got back to his garage in Crescent about 9:30. Leaving the smashed car on the wrecker hook, he went home.

Rick Fagen stayed at the accident site. He followed the Honda's tracks back through the grass for 155 feet, to the point where it had left the road. He walked up the highway for another 100 feet and didn't see any glass or debris indicating a hit-and-run. To Fagen, it looked as if Karen Silkwood had just fallen asleep at the wheel. But it was too dark to see much, and Fagen had other things to do. He would come back the next day and have a closer look. He and Officer Clay went to the Guthrie Police Station and called Roy King at Kerr-McGee for the name of Silkwood's next of kin.

Fagen went home at midnight. About fifteen minutes later, Irene Henning, the Crescent Police Department dispatcher, called him to say the AEC wanted to check Silkwood's car and needed his authorization. Fagen got dressed and went to Sebring's garage, where he found Crescent police officer Joseph McDonald and three or four men in moon suits waiting. Fagen later said under oath that they had identified themselves as AEC inspectors, claiming they had to check the car for radiation.

Sebring had let them into the garage. He, too, had got a call from Irene Henning, saying that Kerr-McGee (not the AEC) wanted to see the Honda. When he said he wouldn't let them in without the okay of the highway patrol, she told him a trooper was on his way.

Wayne Norwood appeared to be in charge. According to Fagen's sworn testimony, the "inspectors" ran the alpha counter over the Honda and read Silkwood's papers. Norwood found some red liquid inside a white plastic flask in the back seat. Thinking it might have alcohol in it, he took a sample.

Fagen opened Silkwood's purse for Norwood. The two rolled cigarettes and the medication fell from the coin side to the floor, and Norwood waved the alpha counter past the purse. Everything was clean, Norwood told Fagen. Sebring locked the garage, and they all left.

□ □ □

Steve Wodka had landed at Will Rogers World Airport at seven that evening, before David Burnham. The OCAW official had been doing some trouble-shooting in Texas and had flown in from Dallas. He and Drew Stephens waited at the airport about forty-five minutes for the *Times* reporter.

Wodka was secretive. All he would tell Drew was that Burnham was interested in hearing what Karen had to say about health and safety conditions at the plant. Wodka said nothing about quality-control documents, and Stephens had no idea what Karen was going to bring or say. They chatted about the Los Alamos trip.

Stephens, Wodka, and Burnham couldn't fit into Drew's sports car, so Burnham rented a car. At 8:30, they got to the Northwest Holiday Inn, where they were to meet Karen. Wodka recognized two AEC inspectors in the lobby.

When he checked at the desk, Wodka found that someone had canceled his reservation. He thought that was strange; he had been traveling thousands of miles a year and staying mostly at Holiday Inns and had never lost a reservation before. There was a vacancy, though, so he registered.

The three men had dinner in Wodka's room and waited for Karen to ring from the lobby or rap at the door. They hadn't really expected her to be there at eight, but it was getting close to ten o'clock and they were getting nervous. Wodka tried to dial the Hub Cafe to see if the union meeting was over. His room phone was dead.

Wodka rushed to the front desk. The clerk made a quick check. There's no problem from down here, he said. The phone should be working. When Wodka returned to the room, it was. He reached Jack Tice at home. Where was Karen? Had Tice heard anything?

Indeed Tice had. Denny Smith, a Kerr-McGee worker, had told him there had been a car accident on Highway 74 near the plant and that it looked like Karen's car. "They towed it over to Ted's," Smith said.

Tice had gone to Sebring's. It *was* Karen's car. He called the Crescent police. "How is she?" he asked. "What hospital is she in?" Tice was worried not only about Karen. He knew she had been contaminated badly, and he was concerned she might have contaminated the whole hospital. But dispatcher Irene Henning said it was against the rules to talk about the accident, and hung up. She called Tice back a few minutes later, saying she had got to thinking about contaminating a whole hospital. She told him Karen had been taken to Logan County — dead on arrival.

Tice told Wodka what he knew. He said that the Oklahoma Highway Patrol had found her car off the left side of the road south of Crescent. Wodka hung up.

"She's dead," he said. Drew, who was sitting next to him on

the bed, began to cry. He got up and went into the bathroom. Wodka just looked at Burnham.

Drew's first reaction was that someone had murdered Karen. She was a good driver. She couldn't just have crashed. Stephens, Wodka, and Burnham were stunned. They decided the best thing they could do — the only useful thing — was to go to the accident site and look for clues, for something that could explain what had happened. Broken glass, skid marks, tire treads . . . anything. Before they left, Wodka went to the AEC inspectors' room, but Jerry Phillip and Bill Fisher had left for Crescent.

By the time Stephens, Wodka, and Burnham reached the concrete wingwall, the Honda had been towed away. It was about eleven o'clock, and no one was there. They pointed Burnham's headlight into the ditch and poked around. Like Trooper Fagen, they followed the Honda tracks along the grass parallel with Highway 74. They tried to figure out where she had left the road, what her trajectory might have been. They looked for skid marks on the asphalt, documents in the weeds. All they found were a piece of tire, a paperback novel, and Karen's paycheck stomped into the mud.

Suspecting she had been forced off the road, they drove into Crescent to see if the Honda could tell them anything. The lights were on in Sebring's garage, and they peered through the small windows on the huge automatic lift doors. The white Honda still hung on the back of the wrecker. Its left front side was squashed in, and the license plate read OKLAHOMA IS OK.

NOVEMBER 14 AND 15, 1974

Trooper Rick Fagen was back at the accident site at nine the next morning. Once again, he traced the Honda tracks in the grass, but he didn't take any pictures. To Fagen, it appeared that the Honda had hit the left shoulder at a 45 degree angle, then suddenly straightened out and headed straight for the wingwall. Therefore, Fagen assumed, she had also cut across Highway 74 from the right lane to the left shoulder at a 45 degree angle. He drew a 45 degree line on his green accident-report form.

Fagen checked again for skid marks. There were plenty at the corner where Highway 33 crosses 74. There were rubber scuffs between the corner and the culvert, too, near the oil road that led to Sherri Ellis' father's farm. But none of them matched his 45 degree angle theory, so he didn't bother to check them closely. He drove his patrol car slowly up the road for 500 feet, past the point where the Honda hit the left shoulder, and saw nothing that interested him. He had no idea of exactly where Silkwood had left the west side of the road and crossed to the east side.

Fagen then went back to Ted Sebring's. Harold Smith showed him where the Honda was parked on the back lot. Fagen said he wanted to check for any valuables he might have overlooked. Smith watched him. Fagen found a cigarette roller and some letters, one to Karen from T. K., bearing Canadian postage and postmarked Ontario. Smith later told the FBI that Fagen read the letter, quoting parts of it aloud to him. The letter explained how to use the roller. Karen was supposed to have the art mastered by the time T. K. arrived. The letter contained references to grass. When Fagen finished, Smith and Sebring packed Karen's personal effects into an oil case and sealed it.

Sebring checked the Honda title and noted a November 5, 1974, repair ticket issued by Eskridge Oldsmobile-Honda. The ticket specified body work on the right rear quarter panel, and from force of habit Sebring ran his hand over the Honda's rear. There was no damage, he later told the FBI, as further proof that he did not dent or nick the car when he pulled it from the ditch the night before.

A Kerr-McGee worker came to Ted Sebring's later and asked for another sample of the red liquid in the plastic flask with a silver cap. Smith broke the case seal and gave him some. The liquid turned out to be spoiled tomato juice.

When Fagen got to his Oklahoma Highway Patrol office in Guthrie, Wodka, Stephens, and Burnham were waiting. Wodka talked to the trooper alone. What happened, the union official wanted to know.

Although the investigation wasn't finished yet, all evidence indicated that Silkwood had fallen asleep at the wheel, Fagen said. He added that unless he found evidence to the contrary, that's exactly how he'd report it. Wodka told him he suspected foul play. Fagen said there was no evidence of that. Wodka, Burnham, and Stephens then went to Sebring's to get the Honda. They wanted to make sure no one tampered with their only evidence until they could figure out what to do next. Someone had to do something. The OHP didn't seem very concerned or interested. Burnham waited in their car; Stephens and Wodka went into the office.

Sebring wouldn't release the car or Silkwood's belongings without authorization, so Drew called Karen's father, Bill Silkwood, on Sebring's phone.

"There may be something important in there for Steve Wodka," Drew said.

Bill Silkwood hesitated. "Well, okay," he said.

Stephens handed the phone to Ted Sebring. Bill Silkwood told him to give Karen's car and belongings to Drew. If need be, Bill said, he would grant Drew power of attorney. Sebring said that wouldn't be necessary.

Bill and Drew talked some more. Drew suggested that an AEC doctor assist in the autopsy, and Bill agreed. Drew told him Karen had wanted to be cremated. Bill said he and Merle would decide that.

Drew carried the oil case to Burnham's car, where they ripped it open. A picture of Karen, old OCAW papers and 1974 contract proposals, some articles on nuclear hazards and industrial ecology, and a 4-by-5-inch notebook. There was no thick, dark brown file folder and no 8½-by-11-inch reddish-brown spiral notebook. There was no letter from T. K. in Ontario, or from anyone else.

Drew's friend Jack Bennett towed the Honda to the home garage of another friend, Bob Ivins, in Oklahoma City. They all agreed to keep the hideout a secret. Drew took pictures of the car with Karen's Instamatic, which she had bought on the Los Alamos trip because she loved the New Mexico scenery so much. It was among the belongings Sebring had packed in the oil case.

Stephens was suspicious. He wanted the accident story in the press, but he didn't want anyone to be on his tail or to find the

Honda. He called the *Daily Oklahoman* from a public phone booth and told them he had pictures and that they would find the film taped underneath the phone booth shelf.

□ □ □

While Drew was negotiating with Ted Sebring for Karen's car and belongings, Rick Fagen was at the Cimarron plant. Roy King had offered the previous night to make Hub Cafe witnesses available if Fagen wished. The trooper said he'd be there by ten that morning. He was about an hour late.

King was waiting with Jerry Brewer and Frank Murch, but not with Jean Jung or Alma Hall. Brewer and Murch told Fagen that they had offered Karen a ride after the union meeting. Fagen later told the FBI that Brewer and Murch admitted Karen Silkwood had been in an "extreme emotional condition."

The next day, November 15, Fagen filed his accident report. He put an X in the box labeled "drinking — ability impaired." Next to the box labeled "sleepy," he typed "drugs." Under the section labeled "unsafe, unlawful," he typed "under the influence of drugs." On the bottom of the report, he wrote, "Witnesses interviewed stated that they had advised the driver was in no physical condition to operate a vehicle." And his diagram showed the Honda crossing the highway from right to left at a 45 degree angle, straightening out, and homing for the wingwall.

Fagen's official accident report was the only one written by the OHP, and it was based on hearsay and assumptions. Fagen hadn't yet learned that the red liquid in Karen's car was spoiled tomato juice. The autopsy wasn't completed. And the Oklahoma State Bureau of Investigation hadn't yet reported that the two cigarettes in Karen's purse were marijuana, the pill was methaqualone, and the half-tablet was too small to analyze.

What, then, was Fagen's evidence? That there were no skid marks in line with his 45 degree trajectory theory; therefore, she had fallen asleep and drifted left. That someone had told him she had driven all the way to and from Los Alamos and had pulled in the night before her accident; therefore, she must have been sleepy. (Actually, she had flown back.) That someone had told him Karen had been drinking at the Hub Cafe (she'd had iced tea). That she was taking sleeping pills and that she was in an "extreme emotional condition"; therefore, her driving ability had been impaired by alcohol and she was under the influence of drugs — asleep behind the wheel.

Who told Fagen that Silkwood had driven back from Los Alamos, that she was drinking at the Hub, and that she was taking sleeping pills? Later, Fagen, by then a lieutenant, would swear under oath that he couldn't recall. His superiors in the Oklahoma Highway Patrol never challenged his report.

The Investigations

Chapter 9

To Tony Mazzocchi back in Washington, it wasn't so simple. The more he listened to Wodka, the more he felt the November 13 crash might not have been an accident. Mazzocchi discussed Silkwood with the OCAW's president, A. F. Grospiron, who was in Washington on other business, and suggested the union hire an accident investigator. Grospiron gave the go-ahead.

Wodka called Fred Baron, a Dallas attorney who had represented, in a third-party suit, former OCAW workers exposed to cancer-causing asbestos. "The State Highway Patrol aren't going to help us," Wodka told Baron.

Baron suggested that the OCAW hire A. O. Pipkin, the sole owner and employee of Accident Reconstruction Laboratories, in Dallas. Pipkin had been investigating accidents for twenty years. He had poked around at the scenes of more than 2000 crashes and testified in more than 300 court trials. Head-on collisions and truck accidents were his specialties. His most famous case was the accident, between Biloxi and New Orleans, that killed actress Jayne Mansfield in 1967. Baron considered Pipkin an expert.

Drew Stephens met Pipkin at the Oklahoma City airport on November 16, three days after Silkwood's death. Pipkin was a potbellied, round-faced man, and he carried all his equipment in a thick gray case filled with foam rubber into which he had cut holes to hold his tools. There was a Pentax camera with lenses; a pocket-sized transit compass for angles and road grades; a tape measure, folding tripod, and stopwatch; a camera clamp that bites onto trees or fenders or bridges.

Stephens and Pipkin drove to Bob Ivins' garage. Pipkin climbed into the Day-Glo orange jumpsuit he kept folded neatly in his tool bag. He jacked up the Honda and went over every inch. "He was under it, in it, and all over it," Stephens remembers.

It was unusual for Adolphus Pipkin to be called three days after an accident; he generally enters a case weeks or even months after crack-ups. He doesn't interview eyewitnesses even if they are available. The vehicles, the skid and tread marks, the accident site — they all leave clues. Pipkin just finds the clues and applies the laws of physics and common sense.

He doesn't put much faith in police reports, either. He had joined the motorcycle division of the Albuquerque Police Department in 1950, after completing two years of engineering at the University of New Mexico. The department had just introduced mobile radios, but the old-timers didn't like being bothered by the dispatcher, so they kicked in the squawk boxes. That left Pipkin as the only motorcycle cop the dispatcher could send to accidents.

Pipkin began carrying a camera and selling pictures on the side to newspapers and lawyers. The department sent him to a six-week course in accident reconstruction, and, in 1955, he struck out on his own. No one had to lecture Pipkin about the flaws in police accident reporting.

Pipkin didn't say much while he poked around Silkwood's Honda Civic. When he finished, he took pictures and asked Stephens to drive him to Highway 74 and the concrete culvert. Pipkin preferred to study accident scenes after the ambulance and wreckers had left. "You can find more evidence a few days after the accident than at the scene," he claims. "There's too many people around; litter and fuel spilled all over the highway. Wait till all that gets cleared out and that's when you find the stuff everyone else misses."

Pipkin took out his Role-tape, measured and studied the Honda tracks on the left shoulder and grass. He measured the skid marks on the highway, studied the culvert wingwalls, and took more pictures. Then he asked Stephens to take him back to Silkwood's Honda.

After examining the car, he told Stephens there were fresh dents on the left rear fender and on the bumper just above it. He pointed out that the dents were concave, not the kind made by a wrecker pulling a car from a ditch.

Stephens hadn't noticed the dents when he'd checked the car inside Ivins' tiny garage. The only way he could have seen them would have been to jack up the car, as Pipkin had done. And as far as Stephens knew, only he and Ivins had seen the Honda since Jack Bennett had towed it from Sebring's two days before. There was no evidence that anyone had broken into the hideout.

Pipkin also told Stephens that the sides of Silkwood's steering wheel were bent forward. When an unconscious body falls against a steering wheel on impact, it bends the *top* and *bottom* forward, not the sides. That meant Silkwood was conscious and holding the wheel, Pipkin said. She had locked her elbows against the crash.

Pipkin checked the Honda once again, then flew back to Dallas. There was enough evidence, he told Mazzocchi, to suspect Silkwood was forced off the road.

Mazzocchi wired Attorney General William Saxbe the next day, November 18, and told him the OCAW had hired Pipkin:

> I spoke to him this morning. He has told me there is evidence to suggest that Ms. Silkwood's car was hit from behind by another vehicle which caused her to leave the road and hit the concrete culvert . . . I recognize the full gravity of my suspicions and urge your immediate attention to this matter.

Unfortunately for Pipkin, the OCAW gave a copy of the wire to David Burnham and the *Times*. The Oklahoma papers picked up the story on the nineteenth, and the press began to hound the investigator. Kerr-McGee called. The Oklahoma Highway Patrol called. And Pipkin hadn't even written his report. "Bad timing," Pipkin recalls.

The next day, November 20, someone from Dallas phoned, asking him if he'd like a job. What were Pipkin's rates?

The investigator was interested. What company wanted to hire him, he asked. What was the company's phone number? What was the caller's name? The caller was evasive at first, then gave Pipkin the information.

Pipkin waited a few minutes, then dialed the number. The receptionist answered the phone with a different company name. Whoever the caller was, Pipkin thought, he has to be pretty damn dumb. He hung up and phoned a friend in the state licensing bureau in Austin. Did the friend know the caller? Yes. Strange Pipkin should ask. The caller was a Pinkerton detective who had just checked to see if Pipkin had a private investigator's license.

Pipkin called Drew Stephens. Be careful, he said. Somebody has turned the Pinkertons loose.

The Pinkertons were well known to unions. They had helped management bust the Molly Maguires, a violent secret society of Pennsylvania miners who, from about 1856 to 1877, organized against cave-ins and unsafe working and living conditions. The Pinkertons worked for management during the bloody Homestead strike of 1892. In fact, Pinkerton was synonymous with labor spying and strikebreaking until 1936, when Congress investigated management's use of private police to deprive workers of their right to speak freely, organize, and bargain collectively. Congress passed a resolution saying: "The so-called industrial spy system breeds fear, suspicion and animosity, tends to cause strikes and industrial warfare and is contrary to sound public policy." After the hearings, Pinkerton announced it would stay out of labor disputes. Since 1936, Pinkerton detectives have worked as uniformed security guards and plainclothes gumshoes. They had guarded the Kerr-McGee Cimarron facility during 1973 and the first half of 1974, until K-M trained its own guards.

In a report dated November 21, 1974, the Pinkertons said they checked out Pipkin through phone books, criminal, court, and public records, newspaper files, and private-eye firms. They said Pipkin had no criminal record in Dallas, but that he had sued his wife for divorce. The suit was "highly amicable," they said. They pointed out that Pipkin was not a licensed private eye, that his business was not a licensed private-security agency in Texas, and that Pipkin was not a registered Texas engineer.

The Pinkertons reported that they had searched "clip files, art

94

files and microfiche records" from 1925 to the present and had found no record of Pipkin or Accident Reconstruction Laboratories. Of the twenty-six private-eye firms contacted, only one knew Pipkin, the Pinkertons reported. He charged $300 a day, $40 an hour, and his specialty was heavy trucks.

The top of the Pinkerton report read:

Office of Origin:	Oklahoma City
Reporting Office:	Dallas
Character of the Case:	Investigation
Status:	Final
Client:	Kerr-McGee

The Pinkertons advised Kerr-McGee that a Rice University professor, Dr. Tonn (first name not given), was the best regional authority on accidents and could testify in court from photographs. The detective agency also advised K-M to check out Pipkin in Albuquerque, where he had been a cop and a private accident investigator.

James Reading, Kerr-McGee's security director, later denied under oath that the corporation had hired the Pinkertons. He said someone in the Pinkerton company — he wasn't sure who — had just volunteered the information. "The call was initiated by him," Reading swore. "He informed me on the telephone of his findings. He asked me if I would like to have a copy of it, and I said I sure would. And he sent me a copy. I was never billed for it. It was never requested. I had information — I had information from another source."

That source was a Captain Wimberly of the New Mexico State Police intelligence unit in Santa Fe. Reading had contacted him the day the Pipkin story broke in Oklahoma. Wimberly promised Reading he'd sniff around.

He called Reading the next day. Pipkin had flunked his New Mexico private-detective exam, Wimberly reported. Pipkin also had had a scrape with the Internal Revenue Service twenty years ago and moved to Dallas. Wimberly told Reading that a former New Mexico State Policeman (he gave name and phone number) knew more about Pipkin's "character and business operations." Wimberly said the source had appeared against Pipkin in many civil suits involving accidents. "In the event of a lawsuit in this area," Wimberly reported, "he [the source] would probably make a good rebuttal witness."

W. Spot Gentry also checked on Pipkin. Gentry, a one-time FBI agent and former Kerr-McGee security director, did free-lance sleuthing for K-M. Reading had hired him to protect K-M against Silkwood-related lawsuits, Reading said in a sworn deposition in October 1977.

Pipkin told Gentry that when he agreed to investigate the Silkwood accident for the OCAW, he had warned Steve Wodka there would be "no hanky-panky . . . He'd report the facts as he found them." Pipkin promised Gentry he would give K-M a copy of his report "at the same time he submitted it to the OCAW."

The Kerr-McGee–Pinkerton reports on Pipkin were filled with errors and false assumptions. Texas did not require an accident investigator to have a private-detective license. After the Silkwood case began to heat up, Pipkin applied for one to keep Kerr-McGee off his back. He passed the test the first time around.

Nor did Texas require accident investigators to be engineers. No attorney had ever tried to impugn Pipkin's court testimony because he wasn't an engineer. If an engineering degree were a must, OHP trooper Rick Fagen would have been in trouble. Pipkin, at least, had taken two years of engineering in New Mexico as well as engineering courses in Dallas. That was two years more than Fagen had.

It was true, however, that most of Pipkin's accident work in 1974 involved truck crashes. That's where the money was. But it is not true that a truck-accident investigator can't handle a car crash. "One is just bigger," Pipkin says.

One K-M report made it sound as if Pipkin had packed up his tool kit one night and slipped into Dallas to escape the IRS. In fact, he moved to Dallas because there were more accidents in Texas than in New Mexico. And he never flunked the New Mexico private-detective exam. The state didn't have one.

□ □ □

The *New York Times* story on Pipkin stirred up the media hornets, and they buzzed the chief medical examiner's office. "The phones were ringing off the damn walls," Dr. A. Jay Chapman recalls. He had to feed the media something, even though his autopsy report, like Pipkin's accident report, hadn't been written yet. In a November 21 news release, the medical examiner said:

☐ Cause of Silkwood's death was multiple fractures, contusions, lacerations, and abrasions.

☐ No trace of radioactivity was found at the autopsy and the accident was not related to radiation exposure.

☐ Silkwood "was under the influence of a sedative-hypnotic drug, methaqualone, associated with a trace of ethyl alcohol." She had .35 mg in her blood. A therapeutic dose is .25 mg; a toxic dose, .50 mg.

☐ There were .50 mg of methaqualone in Silkwood's stomach, still in the process of being absorbed.

☐ The rear-bumper dents were made when the car was towed by the wrecker.

☐ Silkwood's death was, therefore, an accident.

The chief medical examiner's office has never wavered on those conclusions. But Dr. Chapman later said under oath that whether Silkwood had been "under the influence" of methaqualone or not was strictly a judgment based on the assessment of his toxicologist, Richard Prouty.

What about the "trace" of alcohol in her blood?

The amount was so minute, Dr. Chapman said under oath, that its effect would have been "infinitesimally small."

Well, then, how did Dr. Chapman know it was a one-car accident?

The OHP said it was, Dr. Chapman explained. "We take their report at its value. That does not mean it is correct, but we assume that it is as nearly correct as possible."

Why didn't Dr. Chapman investigate the accident on his own? Look for contradictory clues?

"Quincy we ain't," he replied.

DECEMBER 8 TO DECEMBER 15, 1974

Dr. Chapman's autopsy report put the heat back on Pipkin, and the investigator was not about to take any chances. He returned to Oklahoma on December 8 with Dr. B. J. Harris in tow. Harris was a Dallas consulting engineer, a former University of Oklahoma professor, and an accident analysis expert. Pipkin had hired Harris to challenge his own deductions.

Harris studied the Honda in Ivins' garage. He and Pipkin noted marks on the bumper and fender that the FBI had left when they took paint scrapings a few days earlier. Pipkin took pictures of the FBI tool prints. Then he and Harris went to the accident site to recheck treads and trajectories.

Pipkin finished his report on December 15, a little over a month after Silkwood was killed. The first thing he noted was that the Honda went off the left-hand side of the road. "In most one-vehicle accidents where the driver has gone to sleep, or because of impaired abilities," Pipkin noted, "the vehicle has always gone off to the right because of the contour of the road, namely the crown."

The second thing Pipkin reported was that the Honda tracks in the grass showed the car did not *drift*, but was actually out of control before it left the highway. "The only way that this car could have been put in that attitude," he wrote, "was either an impact by an unknown vehicle or a combination of an impact by an unknown vehicle and driver over-reaction and subsequent loss of control."

The third thing Pipkin reported was that the rear bumper was covered with a film of road dust except for a two-inch dent made by an object moving from the rear to the front. "Careful study," he reported, "ruled out any possibility of this dent being made by stones or the concrete on the wingwalls."

The Pipkin analysis was followed by Harris'. The engineer had studied the dents with a special magnifier. He concluded: "The direction of the scratches in the dent definitely ruled out their being made by the bumper hitting the concrete wingwalls or something on the ground during the impact since any such blow would have made scratches in the opposite direction to those noted. The possibility of banging the vehicle during extraction from the ditch was also considered carefully. There does not appear to be any way the direction of the scratches in the dent could have been caused by handling during removal."

Harris said the dents were made by something moving from the rear to the front. The blow was a glancing one that would not have changed the direction or speed of Silkwood's car. But it could have caused her to overreact and lose control, he concluded.

The next move was up to the Oklahoma Highway Patrol. Once the Pipkin story hit the papers, the OHP yanked Fagen from the Silkwood case and gave it to Lieutenant Larry Owen, the OHP accident expert and Fagen's former teacher.

The OHP announced in a press conference chaired by Public Safety Commissioner Roger Webb that the department had found nothing to "substantiate" Pipkin's findings. Webb said the OHP spent six days investigating, reconstructing the accident, and testing Pipkin's theories. The new studies, Webb said, convinced the OHP that its own initial theory was correct and that Pipkin's was "without foundation."

The OHP still argued that the dents were made when the Honda was towed from the ditch. There was white Honda paint on the wingwall, and the dents had rough sandpaper-like scratches, Webb pointed out. If Silkwood had been hit by another car, the dents would have been smooth. Furthermore, the dents were only thirteen to fifteen inches from the ground. The OHP couldn't think of any car whose bumper rode so close to the road.

The OHP argued that if Silkwood had been hit from behind and had overcompensated, there would have been skid marks. There were none. And if Silkwood had been fighting to get back on the road once she hit the grass, the tires would have chewed up the turf. They hadn't.

Finally, Silkwood's car had drifted left instead of right while she dozed, the OHP argued, because Silkwood had been in a minor accident two weeks before her death. "The Silkwood car was likely not in proper alignment," the OHP said. In sum, Pipkin had "no real evidence to support the theory of foul play."

But the OHP was clutching at straws. The dents *were* concave, and the scratches in them were *not* sandpaper-like. It was true there were no skid marks at the accident scene, but burned rubber on asphalt was not the only accident clue. Pipkin had investigated many car and truck crashes where the driver was awake and there were no skid marks. And Silkwood's tires *did* chew up the turf. Hadn't his pictures shown that?

Furthermore, how did the OHP know the Honda was not properly aligned? In the earlier accident, Silkwood had only backed into a fencepost, pushed in a quarter panel, and smashed a tail- and tag light. As for the bumper and fender dents being only thirteen to fifteen inches from the ground, there were plenty of cars with bumpers hanging that low. All the OHP had to do was visit a parking lot with a yardstick and measure a dozen cars.

Pipkin told Drew Stephens to bring him Silkwood's bumper and fender, and explained how to pack them to avoid new scratches. Then Pipkin cut and sent the evidence to Dr. Gerald U. Greene, a Harvard metallurgist and a former professor at the New Mexico School of Mines and Metallurgy.

Greene did a megascopic study of the fender and bumper dents. "It is my opinion that the [bumper] dent was not made by the concrete wingwall during removal of the car by the wrecker," he reported to Pipkin. "The force lines were from rear to front of the automobile. If concrete had made this dent, there would have been particles embedded in the force lines or scratches."

Greene concluded the same about the fender dent. However, he reported, there appeared to be black smudges in some of the scratches. He suggested a chemical analysis to find out what the smudges were.

Pipkin next sent the fender piece to Dr. Ernest L. Martin, formerly a professor of chemistry at the University of New Mexico. Martin did a chemical test and confirmed there was no sand or other residue in the dent. He reported that the black marks across the top of and inside the dent were soluble in organic solvents but not in water. The smudges were rubber, Martin said.

100

Chapter 10

On December 1, three weeks after Karen Silkwood's death, Kerr-McGee got the union contract it wanted. Union demands for better training and health and safety precautions were buried like plutonium waste. But Kerr-McGee was not smiling. AEC inspectors were all over the plant, interviewing management and checking quality-control records.

As if to retaliate against the already demoralized union, Kerr-McGee closed the Cimarron facility for two weeks at Christmastime. Company officials told workers the halt was necessary to discover who had "contrived" the latest incidents at the plant. The OCAW called the action "intimidation."

The day before the shutdown, two Kerr-McGee employees had found twenty-five "low-enriched" uranium pellets scattered outside the plant but inside the main security fence. The FBI was called in to investigate. It concluded that someone stood either on the shipping-loading dock or near the uranium lab and tossed out the pellets, but it couldn't learn who the culprit was. Kerr-McGee claimed the workers scattered the pellets to embarrass the company. The workers believed K-M did it to discredit the OCAW.

The day after the pellet incident, a safeguards clerk picked up a package of contaminated waste. She weighed the bundle according to AEC regulations, then placed it on the table near the scale while she removed her gloves to make a log entry. Her left hand brushed the package; she felt moisture on her skin. A quick monitor showed 1 million d/m.

The same day, four workers were at the glove boxes when liquid containing plutonium leaked through a thermocouple. All four were exposed to airborne plutonium. The Oklahoma City papers reported the pellet-scattering incident and the contaminations.

Not only did Kerr-McGee close the plant; it asked every worker to take a polygraph test. "The Company has an obligation to make sure the facility is operated safely and reassure employees and the public of its safe operations," Kerr-McGee Corporation president James J. Kelly announced. He later claimed the test was "voluntary."

Among Kerr-McGee's official list of polygraph questions were: "Have you ever had any contact with Steve Wodka? Do you know anyone in the process of filing a civil suit against Kerr-McGee? Have you had contact with Drew Stephens since November 4? [Silkwood discovered she was contaminated on November 5.] Have you participated in anti-nuclear activities? Since November 4, have you furnished any outside personnel information about the Cimarron plant? Have you ever committed an act you considered to be detrimental to the Kerr-McGee Corporation?"

The test also asked each subject if he or she had ever used narcotics or drugs since being employed at Kerr-McGee, if he or she knew anyone who used them, if he or she had stolen anything from the plant or knew anyone who had. Kerr-McGee had a legitimate concern about drugs. A small group of workers was smoking marijuana on the job, and Kerr-McGee managers could smell the grass in the dining room, locker rooms, and hallways.

The OCAW alleged that Kerr-McGee also asked workers questions not on its typed list, for example: Did they belong to the union, or have an affair with anyone at Kerr-McGee, or talk to Karen Silkwood? "The message thus conveyed to the employee was obvious and unmistakable," the OCAW emphasized. "Being a union member, or talking to such union representatives as Steve Wodka and Karen Silkwood, were dangerous and undesirable activities . . . The polygraph interrogation of employees about their union membership activities is inherently coercive and unrelated to any possible company objective."

The OCAW advised the members of the Kerr-McGee local to

refuse the test as an invasion of privacy. "The results . . . are notoriously unreliable which is probably the reason that very few labor arbitrators and almost no courts allow evidence of such tests to be introduced or even alluded to in cases before them," President Grospiron wrote.

JANUARY 1975

The plutonium plant reopened in January 1975, with a Kerr-McGee caveat. "The Cimarron facility is a private enterprise having contracts with other privately owned companies," Kerr-McGee said in a directive to all employees. "Information concerning its operation, practices, and personnel is proprietary or 'company confidential' and is to be released to the public only when authorized through the procedure . . . which provides for company review, verification of accuracy, and approval of information to be released."

Tice, Brewer, and five other union members who refused to take the lie detector test were transferred to an isolated warehouse in which no nuclear materials were stored or handled. They were under constant supervision. They couldn't chat with the other workers during breaks or lunchtime.

At the end of January, Brewer was fired for an alleged time-card irregularity, going to his car during a break, and not having the "proper" attitude. The same day, Tice was transferred to the "wet end ceramic area" of the uranium factory, one of the dirtiest places in the Cimarron plant. Kerr-McGee assigned him a supervisor to make sure he wouldn't even urinate alone. To Tice that was harassment and a humiliation.

Kerr-McGee leaned hard on Tice, appealing to his sense of loyalty. It told him the workers respected him and that if he submitted to the test, others would, too. Tice had nothing to fear, the company said. It knew he didn't smoke grass or steal. Besides, Tice had a duty to improve the working conditions at the plant. Wouldn't he answer a few questions for the record? K-M showed

103

him the basic list, promised that the test would be administered by an independent polygraph testing company, and told him that his answers would be kept confidential. Kerr-McGee did not tell him the test was "voluntary" or that the examiners would ask questions not on the official list. Tice refused to take the test.

James Reading, the K-M security director, later told FBI special agent Lawrence Olson, Sr., that 243 workers had taken the test and 13 had refused, including Tice and Brewer. "Several instances of drug abuse and narcotics usage were detected during the course of these interviews," Reading said. All those who had admitted smoking grass or taking drugs on the job were fired.

Mazzocchi was angry about the way Kerr-McGee had treated Tice and Brewer, so he fired off complaints to the Nuclear Regulatory Commission (which had assumed the AEC's regulatory duties when that commission was disbanded) and the National Labor Relations Board. He also submitted the firing of Brewer to arbitration under the terms of the Kerr-McGee–OCAW contract.

Arbitrator Paul C. Dugan ruled that Brewer had been fired unjustly, and awarded him reinstatement with back pay. The Nuclear Regulatory Commission sent Mazzocchi's complaint to the FBI, since interfering with the rights of whistle-blowers in nuclear matters is a federal crime. The FBI passed the OCAW complaint to the Justice Department for a decision as to whether the Bureau should investigate the alleged crime.

The National Labor Relations Board was the only agency that had investigated Kerr-McGee. The NLRB told the FBI that from its "extensive investigation," it concluded that Kerr-McGee had committed a civil violation of the National Labor Relations Act. Correspondence between the NLRB and the FBI suggests that the NLRB also found evidence that Kerr-McGee had criminally deprived union members of their rights. The NLRB turned the case over to the FBI, gave the Bureau a copy of its report, and did not press civil charges against Kerr-McGee.

□ □ □

While the OCAW was fighting Kerr-McGee over its treatment of Tice and Brewer, the AEC released three reports — one on Silkwood's contamination; another about the OCAW's allegations of falsification of K-M quality-control records, made after Silkwood's death; and the third about health and safety complaints made to the AEC by Silkwood, Tice, and Brewer.

The AEC's seventy-page report on Silkwood's November 5, 6, and 7 contaminations was cautious. It concluded that Karen had been contaminated outside the Cimarron plant, and that she'd been contaminated deliberately, because there was no radiation trail from the plant to her apartment. Thus, the .03 mg of plutonium (about $5.00 worth) that contaminated her apartment on November 7 did not blow in through the window, nor was it accidentally carried in on Karen's clothes. But the AEC said it didn't know who had contaminated Silkwood or her apartment, and there was no indication from the commission's report that it had tried to find out.

From Kerr-McGee, AEC, autopsy, and urine- and fecal-sample reports, the following inclusive set of contamination facts emerged:

□ Someone had spiked two of Silkwood's urine samples with plutonium pellet fragments. An isotopic study of the plutonium in the spiked samples and the plutonium that contaminated her refrigerator, kitchen, and bathroom on November 7 revealed that plutonium from two different sources at the Cimarron plant was used. No one could find out who had spiked her samples or whether they were spiked at the plant or in her apartment. And no one could determine how the bologna and cheese in her refrigerator had been contaminated and by whom.

□ Silkwood's contamination and / or urine-spiking began sometime in mid to late October. No one could determine exactly how many times she had been contaminated during October and November.

□ Silkwood had eaten and inhaled plutonium. No one could find out how, when, or where she ate it.

□ An isotopic study of the plutonium that contaminated her refrigerator, kitchen, and bathroom on November 7 showed that the metal came from Kerr-McGee pellet lot 29. The plutonium pellets from lot 29 had been shipped to Hanford, Washington, in August 1974, three months before Silkwood's apartment was contaminated.

□ Silkwood did not work directly with pellet lot 29, but she could have stolen some plutonium from it in August and hidden it for three months. After August, a sample from pellet lot 29 was kept in the Kerr-McGee vault. Silkwood did not have access to the vault.

□ No one conducted a thorough investigation to determine who may have had access to pellet lot 29 in August or to the vault. And no one has ever reported whether or not .3 mg or more of the pellet lot 29 sample in the vault was missing.

□ A urine sample, which Karen Silkwood had placed in her locker at the Cimarron plant and which was discovered after her death, was not spiked or highly contaminated. This fact suggests that Silkwood's contaminated samples may have been spiked after she handed them over to Kerr-McGee.

Though there were no answers, the Silkwood contamination at least served one useful purpose. Larry Olson, Sr., wrote in a report: "It is noted that the Silkwood contamination incident is of major scientific importance in that it is the first such time that a human being involved in a contamination incident has been studied in the fashion Silkwood has. She received a full body count at LASL . . . She died on November 13, 1974, an autopsy [being] performed almost immediately thereafter. This had allowed data to be obtained that has never previously been possible inasmuch as previously science had been limited to obtaining this type of information only through the study of animals . . . It was fortuitous [sic] that she did not get DPTA."

□ □ □

In its second report — on allegations that K-M quality-control records had been doctored — the AEC said that a lab analyst, Scott Dotter, had touched up forty quality-control negatives with a felt-tipped pen. Lab analysts worked with sixty-rod lots. Once the ends of a lot were welded shut, two of the rods went to the Metallurgy Lab for tests. If the two rods passed, the lab analyst took a four-by-five-inch black and white exposure of the sample welds. Prints were submitted to Westinghouse as part of the documentation proving the fuel rods met specifications. If either of the two sample rods was defective, further tests had to be made on the remaining fifty-eight rods in the lot. By touching up the negatives, Dotter could provide a flawless documentary print for Westinghouse.

Dotter admitted he had touched up the negatives. He signed a statement saying he did it without the knowledge and consent of Kerr-McGee management, and that he only doctored flaws in the negatives themselves. At no time, he said, did he try to hide cracks in the rods. A later AEC analysis of the negatives supported Dotter's explanation. "None of the markings was used to obscure weld defects themselves," the AEC reported.

The AEC also found evidence that a K-M lab analyst was improperly using analytical data, and that Kerr-McGee workers were inspecting only half of each sample pellet for cracks and chips. But the AEC played down the seriousness of its findings, arguing that K-M's contract called only for the visual inspection of half of each pellet, and that the plutonium rods were inspected once again at Hanford. "Although the investigation showed some violation of quality assurance procedures," the AEC concluded, "the Hanford inspections have revealed no evidence to indicate to date that the quality of the fuel pins has been compromised."

Westinghouse supported the AEC conclusions. It told reporters that initially it had found 3.5 percent of the Kerr-McGee plutonium rods unsatisfactory, mainly because the Westinghouse standards were so high, but that ultimately, because Kerr-McGee had gradually improved the quality, only 0.4 percent of the rods were sent back.

The AEC made a serious blunder, however, which raised questions about the commission's objectivity and honesty. Its report stated that an AEC investigator had interviewed Jean Jung, who had also complained about quality control, and that Jung had denied there was a quality-control problem at the plant. Jung later signed a sworn affidavit that she never met the AEC inspector on the date the AEC claimed, that she had never met the AEC inspector on any other date, and that she had never discussed quality control with any other AEC inspector. Jung again stated under oath that Kerr-McGee quality control was indeed faulty. Mazzocchi asked the AEC to explain the discrepancy to him. It didn't.

☐ ☐ ☐

The AEC's third report — on Kerr-McGee's health and safety measures — confirmed that conditions at the plutonium plant were poor. Investigators substantiated in whole or in part twenty of the thirty-nine "allegations" that Silkwood, Tice, and Brewer had made in Washington. The AEC hastened to add, however, that only three of these were violations of AEC rules:

☐ Kerr-McGee failed to notify the AEC when it had shut down for forty-eight hours to decontaminate the plant.

☐ Kerr-McGee placed more plutonium in a work area than was safe, given the high criticality of the element.

☐ Kerr-McGee was using plutonium in a purer form than the AEC allowed.

The AEC further argued that these violations did not threaten the health and safety of the workers.

Mazzocchi and Wodka had mixed feelings about the three AEC documents. They were satisfied that the AEC confirmed Silkwood's contamination as not being an accident, but they were disturbed that the commission apparently hadn't tried to find out who had contaminated her. They believed the felt-tip tampering with the negatives had proved Silkwood's charge that Kerr-McGee was cheating on quality control, but they did not believe the AEC had dug deeply enough, and they were suspicious about why the commission had not. And they were furious with the AEC's health and safety report.

The commission investigators had relied mostly on Kerr-McGee management for its health and safety information and had approached the investigation with preconceptions. Silkwood, Tice, and Brewer had never made thirty-nine "allegations" against Kerr-McGee. They had presented thirty-nine verbal examples to support four basic *charges*. The AEC investigated the examples instead of the charges, and did a poor job on them. For example, Silkwood, Tice, and Brewer complained that Kerr-McGee had a 60 percent turnover of workers. The AEC labeled the statement "not substantiated," claiming that there was only a 35 percent turnover.

But the AEC had distorted the figure. It counted Kerr-McGee *management* in its tally, whereas Silkwood, Tice, and Brewer had complained about the high turnover of untrained workers. Had the AEC investigated the actual supporting example, it would have learned that approximately 66 percent of the hourly workers had left K-M between January 1, 1974, and October 31, 1974. Furthermore, the AEC did not address the *issue* that Silkwood, Tice, and Brewer raised: a high turnover of untrained, inexperienced workers is a health and safety hazard.

Deep down, Mazzocchi and Wodka were not surprised at the less than thorough and patently biased AEC reporting. For years, they had distrusted the AEC's honesty and integrity. And with good reason.

Just three days after Silkwood's death — after poring over hundreds of memos and letters written by AEC officials and the nuclear industry — *The New York Times* revealed a massive AEC cover-up of the hazards of nuclear power. For a decade, the AEC had been suppressing studies, some written by its own scientists, that suggested that plutonium and nuclear reactors were far more dangerous than the commission was willing to admit publicly.

The *Times*'s documents proved that the AEC had ignored recommendations from its own scientists to investigate further the safe use of plutonium and nuclear reactors. They also showed that on at least two important matters, the AEC had consulted with the very industry it was supposed to regulate before deciding to bury a study critical of nuclear safety.

Citing instance after instance, the memoranda, which *The New York Times* got from AEC whistle-blowers and through Freedom of Information requests, indicated that the AEC was more concerned about public relations than about public health or safety. In September 1971, for example, Steven H. Hanauer, a top AEC official, buried a staff paper that questioned AEC estimates about nuclear reactor safety. "The present goal," Hanauer wrote, "should be a paper that can be published without hurting the AEC and without inciting a cause célèbre for squelching a paper . . ."

The AEC also suppressed in 1964 a $120,000 study, conducted by the commission's own Brookhaven National Laboratory, on how dangerous a major nuclear accident would be. AEC official Stanley A. Szawlewicz pointed out in a November 13,

1964, AEC memo, "The results of the hypothetical Brookhaven National Laboratory accident are more severe than those equivalent to a good-sized weapon and the correlation can readily be made by experts if the BNL results are published . . ."

Szawlewicz had good reason to be worried. Brookhaven reported that a big nuclear reactor accident would cover an area "the size of the state of Pennsylvania."

The AEC met twice with the Atomic Energy Forum — the industry's lobby — about the Brookhaven report. The forum "strongly urged" the AEC not to publish the report "in any form at the present time."

The AEC bought it. In its press releases, it claimed that that particular Brookhaven report was never completed. When parts of the document were made public eight years later under the threat of a Freedom of Information lawsuit, the AEC called the released portions "the final draft." It was a simple lie.

And while the AEC was deceiving the public, it was trying to dupe Congress. The General Accounting Office found that the AEC withheld from Congress information showing that the Fast Flux Test Facility at Hanford had a cost overrun of more than $800 million and that the project was taking five years longer than planned.

Even more important to the union, the relationship between Kerr-McGee and the AEC was a cozy one. Both had something to gain by keeping the fuel rods rolling out of Crescent. The AEC's fast breeder dream would become a reality sooner; Kerr-McGee could hope to make a tidy profit under its fixed-fee contract. The OCAW stood between the dream and the profit.

The AEC got around the OCAW's complaints about health and safety conditions by keeping its "unannounced" inspections an open secret. James V. Smith, who took the minutes during the company's morning management meetings, testified under oath that Wayne Norwood would notify all management personnel at least one full week before any AEC "unannounced" inspection. Kerr-McGee cleaned up the plant as best it could before the "surprise" visit, Smith said, and workers were told not to talk to the AEC inspectors.

Chapter 11

Kerr-McGee workers were tense and frightened in the weeks following Karen Silkwood's death, and relations between them and Kerr-McGee were more strained than usual. Drew Stephens carried a pistol in his car and slept with a loaded shotgun next to his bed. Jack Tice thought he was being followed and began to ride around in his pickup with a shotgun at his side. The day after Silkwood's death, Jean Jung decided to carry a gun in the glove compartment of her car because she was afraid to drive to and from work alone at night, but a friend talked her out of it. Someone also called Jung and threatened to kill her boyfriend, who lived in New Jersey. Very few people even knew she had a boyfriend on the East Coast. Even Dean A. McGee, chairman of the board of Kerr-McGee, received threats in the mail.

But no one seemed more scared than Silkwood's roommate, Sherri Ellis, who eventually moved into her father's abandoned farmhouse on the oil road about a mile from the spot where Silkwood was killed. The house was broken into two or three times after she moved there, and once, when she heard someone trying to get in, she ran up to the attic and sat on the floor with a gun pointed at the steps. No one came up.

Ellis also attacked the Kerr-McGee plant with a .22 rifle. She appeared at the front gate at ten o'clock in the morning and sat in her car for fifteen minutes while a K-M security guard watched her through binoculars. When he turned away, she scaled the seven-foot-high fence and walked up the sidewalk toward the plant. Using the rifle as a cane, muzzle down, she shouted, "I want to be killed! I want to be killed!" Morgan Moore, the facility manager, opened the door for her and then tried to grab the gun. She hung on to it. Four other K-M employees joined Moore, and they struggled with Ellis for two minutes before they could get the weapon. Moore took it into the guard station and pulled back the single bolt. The gun was empty.

The Logan County sheriff searched Ellis' car, found some grass, and hauled her off to jail, but the Justice Department declined to press charges. Ellis said later she hadn't been high on pot but was worn out, and that Kerr-McGee's lies about how good and safe the plutonium plant was had just angered her. She eventually traded her car in for a pickup truck in case someone tried to bump her off the road or she had to cut through fields and brush to escape.

□   □   □

In mid-December, Barbara Newman and Peter Stockton, reporters for National Public Radio, came to Crescent to sift through the Silkwood debris for clues. Newman was an ace investigative reporter, and Stockton was a congressional investigator on leave from Congress with a partial grant from the Fund for Investigative Journalism. Newman and Stockton found little love for Karen Silkwood in the tiny Oklahoma town. Workers blamed her for what they called Kerr-McGee harassment and for the December shutdown. And many called her a "bitch" who was going to cost them their jobs.

Even some union members were angry with Silkwood. The only person who threatened Stockton during the two months he spent interviewing workers was a union member who told him, from high up in the cab of a truck, "Get the fuck out of town before you get hurt."

Tice and Brewer were uptight about the union, too. From a *New York Times* syndicated story on November 20, a week after Karen's death, they learned that Silkwood had been collecting quality-control evidence. The OCAW committeemen were miffed that they hadn't been told about Silkwood's undercover assignment, and they felt Wodka had used the local. It was an open secret in the small-town atmosphere of the plant that he had been sleeping with Karen. Tice thought Wodka's conduct had been unprofessional and had hurt the local's credibility.

Late in December, after Kerr-McGee had closed the plant temporarily, Barbara Newman broadcast that K-M couldn't account for between forty-four and sixty-six pounds of plutonium, enough to make several bombs. The next day, Burnham reported in the *Times* that thousands of pounds of nuclear materials were unaccounted for in the fifteen plants throughout the United States that process nuclear materials. K-M was one of those plants.

"Although officials say there are no unresolved cases of theft," Burnham reported, "the amounts of highly enriched uranium and plutonium that cannot be accounted for means that the AEC is unable to give positive assurance that the missing materials have not fallen into the hands of a terrorist group or hostile governments."

An unnamed federal official told Burnham that there had been two known instances of employees smuggling enough special nuclear material to make a nuclear bomb. Another unnamed official told Burnham that one plant could not account for 9000 pounds of highly enriched uranium. Forty pounds is enough to make a bomb capable of killing thousands.

The AEC argued that the missing plutonium and highly enriched uranium had been lost in the crude accounting system the commission designed to keep tabs on special nuclear materials. The AEC had already authorized the production of 20,000 pounds of plutonium alone and had a projected need for 100,000 pounds by 1990.

Barbara Newman was also the first investigative reporter to challenge the official line on Silkwood's contamination and death. Several pharmaceutical experts, she reported, disagreed with the Oklahoma state medical examiner, who believed that the .35 mg of methaqualone in Silkwood's blood had placed her "under the influence of the drug." Newman's sources said that .35 mg is less than a therapeutic level and, by itself, would not impair driving. If Silkwood had been taking Quaaludes for a long time, they said, she could well have built up a tolerance.

No one knows for sure just how many Quaaludes Karen Silkwood swallowed each day. Dr. Shields had prescribed one a day, but in August and again in October he wrote two prescriptions for thirty capsules. He recalls writing the second August prescription when Silkwood told him she had lost the first thirty capsules. Dr. Shields said that, since patients frequently misplace pills, he had no reason to suspect she may have been lying. He does not recall why he wrote the second October prescription. Records of the Gilliam Prescription Shop, downstairs from Dr. Shields's office, indicate that between August 13 and her death on November 13 Silkwood had purchased 180 Quaaludes. If she had swallowed all of them by the time she died, she would have been taking an average of two a day.

There have been allegations that a second doctor also prescribed Quaaludes for Silkwood, but no one has produced any evidence so far to support that rumor. Whatever the case, Drew Stephens, who had seen Karen under the influence of Quaaludes many times, denied that the drug would have put her to sleep behind the wheel. He told the FBI that "they made her perhaps a little drowsy but still very functional." Special Agent Larry Olson noted in his report on the Stephens interview: "Stephens does not believe the pill would have impaired her driving ability."

Newman saved her sharpest attack for Oklahoma Highway Patrolman Larry Owen, who had replaced Rick Fagen after Pipkin had challenged Fagen's accident report. Why hadn't Owen analyzed the bumper and fender as Pipkin had done, Newman wanted to know.

"We did not have availability of the car after it was released from the authority of her father," Owen said.

"But you could have asked for it," Newman pressed. "You never did."

"We checked to try to find where it was. We could not locate it in either Crescent, Edmond, or Oklahoma City."

"Are you saying that the investigative work of the Oklahoma State Highway Patrol couldn't find this car that I, a reporter from Washington, D.C., *could* find?"

"We should have kept the car in our custody," Owen said. "But that's strictly in hindsight."

Newman also interviewed AEC investigator Jerry Phillip,

who came close to saying that Silkwood had contaminated herself. Phillip told Newman that Silkwood's apartment had been contaminated deliberately between 11:30 P.M., November 6, and 8:00 A.M., November 7 — a fact not in the AEC report. He apparently had reached that conclusion because Sherri Ellis had said she'd made a ham sandwich before she left for work about 11:30 P.M. Ellis' urine and fecal samples showed no internal contamination, even though she kept her ham in the refrigerator found to be highly contaminated the next day.

"Who was at the apartment from eleven-thirty at night to the following day, to your knowledge?" Newman asked Phillip.

"To my knowledge, only Karen Silkwood and Drew Stephens."

"So then, what you think is that one of them contaminated the apartment?" Newman queried.

Phillip chose his words. "It was indicated at one point along the way that at some time the apartment was left unlocked," he said. "And therefore someone could have entered during the absence of those people, and we can't rule out that possibility."

MARCH TO MAY 1975

The Silkwood case intrigued ABC Television enough for it to send "Reasoner Report" correspondent David Shoumacher to Oklahoma. Shoumacher picked on Larry Owen once again.

"It appears that Miss Silkwood braced her arms as the impact occurred," Shoumacher said. "Is that consistent with someone who is practically unconscious from drugs?"

"Well, probably not," Owen told a national audience. "In other words, you're normally in a very relaxed state, being under a depressant-type drug. I was unaware of the fact that . . . I mean where do you have the information, or how do you presume that she braced just prior to impact?"

"Only because of the way the steering wheel was bent,"

Shoumacher explained. "That is, it's bent on both sides, and bent straight forward."

"Okay, this could have possibly been done by her abdomen and chest, depending on —"

Shoumacher cut Owen off. "Well, she can't hit both sides of the post at the same time and bend the wheel forward," he said.

"Well, she could if she pressed it against her chest and her stomach at the same time," Owen argued.

"But then why didn't, for instance, the top or the bottom of the wheel bend in?" the reporter asked.

"Well, depending on what alignment — have you seen the car?"

"Right."

"All right, is there —"

"Have *you* seen the car?" Shoumacher cut in.

"No, I haven't. Is the alignment of the wheel indicating that she bent it with her hands, or bent it with her chest and stomach?"

"My guess, from looking at the wheel, would be that it was pressed forward on both sides," the reporter answered.

"Okay, that would indicate that probably she had hold of the wheel, which is interesting."

"And that would not be consistent with someone who was unconscious because of the use of drugs," Shoumacher said.

"No, probably not . . ."

"But why did his [Pipkin's] investigators find no trace of concrete in the dent?" the reporter asked.

"I don't know," the OHP accident expert said. "I really — that's a very good question."

ABC ran tests on the stretch of the highway where Silkwood was killed, using the same kind of Honda Silkwood drove. Both the ABC experiment and the Pipkin analysis led to the same conclusion: Karen Silkwood was awake behind the wheel. ABC's tests confirmed that if a sleepy driver had drifted left across Highway 74, the car would have come to a safe halt in a pasture yards from the concrete wingwall. Only by forcibly holding the Honda on course along the grass shoulder could the test driver stay in line with, and finally hit, the wingwall.

Contrary to Pipkin's analysis, the ABC test car drifted left — not right — fourteen out of fifteen times. However, what ABC forgot to mention in "The Reasoner Report" is that the road had been resurfaced, making it difficult to tell which way a sleepy driver would have drifted on November 13.

Ms. magazine, *New Times,* and *Rolling Stone* all challenged Kerr-McGee, the AEC, and the Oklahoma Highway Patrol. Most of the articles pointed out inconsistencies and contradictions, and raised questions; but *Rolling Stone* came right out and said what many were thinking: Karen Silkwood was murdered because she had documents somebody didn't want her to have, and she was going to give them to a newspaper. *Rolling Stone* reporter Howard Kohn speculated that there was a plutonium smuggling ring at K-M, that Silkwood knew it, and that she paid for her knowledge. Kohn caused a good bit of unease with his articles, but he couldn't find a smoking gun. Neither could the other reporters, so they let the Silkwood case die.

So did the FBI. On November 21, 1974, two days after Mazzocchi had wired the attorney general about Pipkin's findings, the Justice Department had assigned the Bureau to the case. Mazzocchi didn't know — and no one said — exactly what the Bureau was investigating. Karen's death? Her contamination? The theft of plutonium? Harassment of a whistle-blower? Interfering with the rights of union members?

The only thing Mazzocchi learned about the FBI investigation came from a story in *The New York Times:* on May 1, 1975, five and a half months after the FBI had begun its investigation, the Justice Department announced that the Silkwood file was shut. The *Times* simply quoted an unnamed Justice Department source as saying that Silkwood's accident "did not appear to be murder" and that the case was closed.

Mazzocchi's mind was filled with questions. Whom had the FBI interviewed? How thorough had its investigation been? Had it found any of Silkwood's missing documents? What new facts had it uncovered? Why had the FBI appeared to dismiss Pipkin's evidence? What about the dents?

Although the Justice Department had closed the Silkwood case, Mazzocchi was not convinced that Karen's car crash was an accident. By that time, the killing of Karen Silkwood had taken a personal twist for the union leader. In February 1975, three months after Silkwood's death, Mazzocchi was driving back to Washington from the Airlie House, fifty miles west of the city, where he had attended a church-sponsored conference on worker safety.

The Airlie House is a conference center hidden on 1700 acres of rolling farmland in the foothills of Virginia's Blue Ridge Mountains. Almost every government department has used the center, including the State Department, the Department of Defense, and the White House. So have hundreds of groups like the House Republican leaders, the International Association of Chiefs of Police, and the NAACP Legal Defense Fund.

The Airlie House was closely tied to the Department of the Treasury, which, since 1963, had secretly leased 1733 square feet of space there. Airlie was the department's relocation site and classified communication post in case of a nuclear attack.

Mazzocchi gave a lecture at the conference, had two martinis between five and seven o'clock, then left for Washington. Twenty minutes from the Airlie House, Mazzocchi blacked out at the wheel. His car drifted right, shot over an embankment, missed a group of trees, landed on its roof, and rolled over on its side. Mazzocchi was tossed from the car; by the time the ambulance arrived, he was conscious but was hallucinating about World War II.

Mazzocchi suffered some facial contusions and brain lacerations, which eventually healed. He wasn't suspicious enough at the time to have the hospital check his blood, but later he became curious. Two martinis couldn't black *him* out. Or could they? He asked a couple of buddies to help him take a test. He'd chug-a-lug Beefeaters; they'd count and let him know how many he'd drunk before he passed out. After five, he was still standing. A little tipsy, but on his feet.

Mazzocchi never told reporters that he suspected someone at the Airlie House conference had drugged him and that he had almost ended up dead in a ditch like Silkwood. Mazzocchi had no proof, and he was afraid that the media would consider him paranoid.

Now that the Justice Department had closed the case, he felt he'd never get the answers to his three basic questions — who contaminated Karen Silkwood, what happened to her documents, and how did she die — unless someone could find a way to reopen the investigation.

Chapter 12

Kitty Tucker, the legislative coordinator for the Washington, D.C., chapter of the National Organization for Women (NOW), had been following the Silkwood story with a passion. She, too, was angered when the Justice Department closed the case, and she made up her mind to do something about it.

Tucker was a blonde with a frizzy Afro too big for her small, childlike face. She had a soft voice, and when she smiled she seemed shy, almost timid. But she wasn't. She had a bold plan to press Congress into reopening the Silkwood investigation. To do that, she needed the help of Sara Nelson, the national director of the NOW Labor Task Force.

Nelson was drinking coffee in her kitchen in Adams-Morgan, the Greenwich Village of Washington, when Tucker dropped in, carrying her baby, Amber, and an armful of papers. It was mid-May 1975, just weeks after the Justice Department had slammed its file drawer on the Silkwood case.

Nelson was tall and trim, with a tanned face and short-cropped blond hair. She was the kind of woman who felt lost without a cause, a thirty-two-year-old romantic who still believed in people, justice, and freedom. She had an easy, almost flirtatious smile, but when she talked about the plight of working women and the big corporations that use them, her hazel eyes grew hard and angry.

Nelson and Tucker had met several times during conferences around town, but other than their feminism, they had little in

120

common. Tucker was an antinuke environmentalist and lobbyist. Nelson was a full-time, unpaid NOW labor activist, pushing full employment for women.

"I've been working on this problem," Tucker told Nelson as she nursed Amber. "It's a serious problem. It's about Karen Silkwood. I went to the national office and I asked them if there wasn't some way NOW could get involved in this. They told me I should get in touch with you because this was a labor issue. So here I am."

Tucker outlined the Silkwood case — how someone had fed Silkwood plutonium, stolen her documents, and pushed her off the road. She explained what plutonium was and spoke quietly but eloquently about the dangers of nuclear power.

"They're saying Silkwood contaminated herself to discredit the company," Tucker continued. "I know that Karen Silkwood knew enough about plutonium to make that absolutely ridiculous." Tucker said she wanted to kick up a Silkwood storm but didn't have enough people. With 55,000 members and eighty chapters, NOW had great grassroots power. Would Nelson's Labor Task Force adopt Silkwood?

Tucker left Nelson a stack of articles, and at 2:30 the next morning, Nelson settled back to read them. Nelson always worked late; the night was the best time to reach people, and long-distance calls were cheaper. Only after the phones were quiet was there time to read and reflect and enjoy the peace. She dug into Howard Kohn's *Rolling Stone* stories, B. J. Phillips' *Ms.* articles, and David Burnham's *New York Times* pieces. "The sweat began to roll down my back," she recalls. "I felt as if Karen were in the room . . . I had this very strange feeling that if NOW didn't do something about this, *nobody* was going to. That meant that if *I* didn't do something, nobody would."

Nelson's mind began to race over her NOW priorities. She had to organize a lobby for the Equal Opportunity and Full Employment Act. She wanted to sort through the 1976 congressional candidates and signal the NOW membership who was feminist and prolabor and who wasn't. She had a plan for Women's Employment Councils to create jobs for women and lobby for full employment. When would she find time for Karen Silkwood?

And what about the FBI? If NOW adopted Silkwood, the Bureau would have another excuse to snoop on the women's organization. Nelson wasn't sure how NOW's leaders, already split by an ideological rift, would react to that or to dragging NOW into a nasty fight with the Department of Justice. Besides, Nelson had been in Washington for only three months. What did she know about Congress, the Nuclear Regulatory Commission, the Justice Department, the FBI?

Nelson slept fitfully, with Silkwood under her pillow. The next morning she was up at 7:30. She woke up Jerry Stoll, a neighbor who worked with her on the Labor Task Force.

"I'm dropping a bomb on your desk," she said and left.

Stoll was as angry as Nelson when he finished reading the articles. "Forget the Task Force priorities," he told her later that day. "You can't just make plans, then ignore history."

Sara Nelson called Kitty Tucker. "The Labor Task Force will adopt Silkwood," Nelson said. "When do we begin?"

Overnight, a coalition was born. Nelson was a doer with a network; Tucker, a strategist with a plan. Their goal? To use NOW to spark a congressional investigation into the contamination and death of Karen Gay Silkwood.

□ □ □

Tucker was working closely with Bob Alvarez, an environmentalist on the staff of South Dakota's junior senator, James Abourezk. Alvarez and Tucker thought Senators Abraham Ribicoff and Lee Metcalf, both liberal Democrats, were the best choices to lead the congressional investigation. Ribicoff of Connecticut was chairman of the Government Operations Subcommittee on Reports, Accounting and Management, which had already asked the General Accounting Office (GAO) to review the FBI's, NRC's, and AEC's Silkwood probes. So far, the GAO had received no cooperation from the Justice Department or the FBI.

Before Tucker and Nelson launched NOW at Ribicoff and Metcalf, they wanted the approval of Mazzocchi, who, together with Steve Wodka, had provided most of the initiative in getting the Justice Department, the FBI, the Nuclear Regulatory Commission, and the National Labor Relations Board moving on the Silkwood case. Tucker and Nelson needed to know how Mazzocchi would feel if NOW began leaning on the Hill.

Mazzocchi told them he was as disturbed as they about the unanswered questions in the case, but that his hands were tied. There was a struggle in the OCAW over the nuclear issue, he explained. Members wanted a safe and healthful working environment, but they also wanted jobs. Besides, he said, Karen Silkwood is just *one* dead union member. OCAW workers die on the job every month. The union has only so much money, and he and Wodka only so much time. Unfortunately, the OCAW had already spent all the money and time it could on Silkwood. "But," Mazzocchi said, "I want you to know it would be terrific if the women's movement does something about this."

That was all Tucker and Nelson needed to hear. By July 20, 1975, Tucker had the strategy worked out. NOW would call a chapter action alert for August 26, the anniversary of women's suffrage. The action's focus would be "Stop Violence Against Women NOW," and Karen Silkwood would be billed as a typical case of violence against a woman. "NOW can generate a wave that will insure that the life and work of Karen Silkwood will have not been in vain," the alert notice said.

The alert package sent to NOW's eighty local chapters urged members to write to Ribicoff, Metcalf, and their own congresspersons, calling for a congressional investigation. The package included sample letters as well as a petition for a congressional investigation. The chapters were urged to circulate the petition in stores and factories and at meetings; to alert the local press about NOW's call for an investigation; to try to get on local radio and television; to visit congresspersons while they were home during the August recess; and to prepare for an action on November 13 — Karen Silkwood Memorial Day. The high point of the alert would be a NOW delegation to the Justice Department to demand why the FBI investigation had been closed.

□ □ □

The alert caught on. On August 26, local chapters held rallies and candlelight parades in New York, Newark, Chicago, St. Petersburg, and Cleveland, with members carrying WHO KILLED KAREN

123

SILKWOOD? signs — in all, nineteen actions around the country.

In Washington, Tucker, Nelson, NOW president Karen DeCrow, and Cathy Irwin, NOW's vice-president for public relations, met with the Justice Department. Nelson had called Attorney General Edward Levi in July, requesting the August 26 conference. She did a soft-sell job, telling Levi's secretary that NOW simply wanted to meet with the attorney general to get an update on the Silkwood investigation. The secretary called Nelson back to say that the attorney general would be out of town on the twenty-sixth but that Joseph J. Tafe of the Criminal Division would meet with the group.

NOW had sent out a release, calling the Justice Department's closing of the Silkwood case a "cover-up," and announcing the meeting with the Justice Department officials and a press conference afterward.

When NOW walked into the huge Justice Department office on the first floor, Tafe and three other attorneys were waiting. The men sat in chairs in a semicircle; the women, on the couch. The meeting lasted ninety minutes, and both sides left angry.

NOW demanded to know where the investigation into Karen Silkwood's contamination and death stood. The Justice Department attorneys told them the Silkwood case was closed. NOW demanded an explanation. The Justice Department attorneys said they could not discuss the details of the case. They implied that maybe Silkwood never *had* documents, that maybe she poisoned herself with plutonium to discredit Kerr-McGee, and that maybe the women were watching too much TV — the FBI doesn't solve all its cases.

"They had no answers," Nelson recalls. "On each one of the points, no answers. They were giving us just a little pat and trying to tell us 'Don't worry your little heads about this.' It was condescending, patronizing, and insulting."

The women walked out of the Justice Department into the oppressive Washington heat to face the nearly 100 reporters and TV crew members who had been waiting in the corridor and on the stone steps. "It looks to me as if there's a cover-up," DeCrow said into the network cameras. She was angry and disgusted. "And there is every reason to believe this *is* a cover-up."

The cover-up theme caught, and for two or three days the NOW phones rang off the hook. What evidence did NOW have that the FBI had covered up the facts in the Silkwood case, reporters asked. Did NOW know who had contaminated and killed Karen? The reporters wanted facts and leads, but, like the Department of Justice, NOW gave them theories. Before long, they stopped calling.

Reporter Patricia Welch covered the NOW–Justice Department story for the Nashville *Tennesseean*. Like other reporters, she was looking for facts and leads, so she called her friend Jacque Srouji, a part-time copy editor at the *Tennesseean*. Srouji had been in Washington in May and they had eaten dinner at Mama Ayeesha's Calvert Café, a Mid-Eastern restaurant where many Arabs ate. Srouji had ordered and chatted with the waiter in Arabic.

Welch and Srouji had known each other for ten years, but they were not close friends. Welch always found Jacque (pronounced Jackie) somewhat vague and mysterious. At dinner, she told Welch she was in Washington to do research on a nuclear power book. Welch was interested in the Silkwood case and asked Srouji if she was going to touch on it.

Srouji said she had researched Silkwood thoroughly, and had talked to people who had never been interviewed before; she had original stuff. She had been to Silkwood's home town, Nederland, Texas, and had collected a lot about sex and drugs. Welch was surprised. Srouji, a copy editor and a free-lance feature writer, had never done any real investigative reporting, as far as Welch knew.

During dinner, Srouji had chatted about Larry Olson, Sr., the FBI agent, whom she and Welch had known in Nashville. Olson had once tried to recruit Welch. His approach had been very subtle. "You'll get further ahead if you were to be trusted," he had told her. At that time, Welch had been covering the women's movement for the *Tennesseean*. "There are things you need to know from the FBI. It's important to know what's going on in Nashville." Welch didn't take the bait.

Olson was now working in the Oklahoma FBI office, Srouji said, and had let his crewcut grow out. He had been very helpful to her on Silkwood. He had shown her his FBI documents. She pointed to her bulging briefcase next to the table.

That had been several months ago, and Welch hadn't thought much about Jacque Srouji until the NOW news conference on the Justice Department steps. Maybe Srouji might have a lead or two about Silkwood. Welch called Jacque in Nashville. They talked about the NOW cover-up charge.

Silkwood was murdered, Srouji said, and when her book came out, she'd name the murderers.

Did Srouji have any leads, Welch asked. Any facts? Any clues? Any documents?

Srouji was as vague and mysterious as usual.

□ □ □

Three thousand NOW members gathered in Philadelphia at the end of October for the NOW annual convention. Tucker and Nelson were ready for them with WHO KILLED KAREN? buttons and posters, and a resolution demanding "a vigorous and public Congressional investigation of the case of Karen Silkwood."

The resolution never got to the floor. The NOW convention was torn between the Majority Caucus and the Chicago Machine. The politics were bitter, and Karen Silkwood was caught in the middle. NOW president DeCrow headed the Majority Caucus under the banner OUT OF THE MAINSTREAM, INTO THE REVOLUTION. The slogan was blazoned in gold on red T-shirts. Mary Jean Collins-Robson, a member of the NOW National Board, ran the Chicago Machine. Her supporters wore large bandages on their arms with FACTIONALISM HURTS printed on them.

DeCrow had been elected president of NOW in 1974, but she had had trouble all during 1975 trying to squeak her policies past the National Board, the majority of whom did not share her "Out of the Mainstream" views. The importance of Karen Silkwood was one of those views.

DeCrow came to the Philadelphia convention with a slate of candidates and a platform. Her Majority Caucus wanted NOW to endorse for public office political candidates with a feminist perspective. They refused to limit themselves to such mainstream token gains as affirmative action. They favored street demonstrations, court action, civil disobedience. And they viewed the ideological division in NOW as a sign of health and growth.

The Chicago Machine wanted NOW to stay out of national politics and to support traditional but slower legislative-legal maneuvers. They wanted NOW to present a united front to women who would not join the organization if it appeared too radical.

The convention vote was important to the Silkwood case. Tucker, Nelson, and DeCrow had already felt heat from some National Board members over the August 26 action alert. If the Chicago Machine or any other united-front candidate won, Tucker and Nelson would lose their Labor Task Force launching pad, and Silkwood would be out.

DeCrow was re-elected by a slight margin, but her Majority Caucus carried two thirds of the National Board seats. Silkwood was in.

Tucker, Nelson, and DeCrow had planned to present Karen Silkwood an honorary membership in NOW at the convention. It would be the organization's first posthumous award, and Bill and Merle Silkwood were flown to Philadelphia, at Tucker's suggestion.

Constitution Hall's bleachers were filled with 3000 women and a few dozen men. Nelson led the Silkwoods to the stage steps. Bill and Merle climbed up to the stage, where DeCrow was waiting.

"NOW wants to give the membership to Karen Silkwood to honor her courageous fight to bring safe working conditions to the nuclear industry," DeCrow said. She handed a membership scroll to Merle. Constitution Hall was quiet, as if someone had called for a moment of silent prayer.

"Thank you," Merle said. Her voice broke. "I just want you all to know that if Karen were here, she would be really comfortable with all of you."

Whatever the women thought about Karen Silkwood as a NOW issue, they were still soul sisters, and they were on their feet, cheering, applauding, hugging each other. Some stood silently weeping.

The national press failed to mention the posthumous membership, as if it were a piece of convention theatrics. But in retrospect, the award and Merle's tears were significant. They made Karen Silkwood alive, a sister, a feminist. NOW leaders would carry the spirit of Constitution Hall back to their chapters and breathe life into the coming November 13 Silkwood memorial rallies.

While the crowd cheered and wept, Kitty Tucker was at work. She had organized a team to pass around gallon milk cartons with the tops cut off. Tucker collected almost $1000 for the fledgling organization she and Nelson had founded, Supporters of Silkwood, which worked out of a huge white community house in Washington. SOS badly needed money for paper, stamps, and phone calls to keep the pressure on Ribicoff and Metcalf.

After the NOW convention, letters calling for a congressional investigation poured into Ribicoff's and Metcalf's offices. It was time for NOW to move. Nelson called Ribicoff to ask if he would meet with a coalition of feminists and labor organizers who were calling for a congressional investigation into the death of Karen Silkwood.

Chapter 13

On November 19, a year after Silkwood's death, Nelson and Tucker met with Senator Ribicoff. They brought with them 7000 petitions, requesting a congressional investigation, and a delegation made up of Eleanor Cutri Smeal, chairone of NOW; Pat Ganzi, president of the D.C. Coalition of Labor Union Women; James Cubie of Ralph Nader's Congress Watch; and Bill and Merle Silkwood.

It was an aggressive and angry group, but Ribicoff handled them like a master. With him were Senator Lee Metcalf; E. Winslow Turner, the chief counsel to Metcalf's Government Operations Subcommittee; and Peter Stockton, the veteran congressional investigator who had worked on the Silkwood story in Oklahoma with Barbara Newman of National Public Radio.

Ribicoff had already decided Metcalf would conduct an investigation, but he had not told Nelson or Tucker. In fact, two days before Nelson and Tucker met with Ribicoff, Turner and Stockton had asked the Justice Department for its files on the Silkwood investigation. The department balked, and Turner and Stockton suspected they would have to fight for every Justice and FBI file. "We are getting the old 'hunker down like a jackass in a hailstorm' treatment," Stockton jotted in a memo.

But Tucker and Nelson didn't know the ball was already rolling, and they pitched all they had to Ribicoff. Nelson was nervous. It was her first visit to a senator's office. She explained her

129

concern over the unanswered Silkwood questions — Karen's contamination, her documents, the dents in her car. Others in the delegation chimed in.

When they finished, Ribicoff assured them there would be an investigation and that it would receive his full backing. He promised to cooperate in trying to get FBI and Justice Department files. There would be more staff, if necessary, he said, and subpoena power, too.

But Ribicoff also made it very clear that the investigation would *not* try to find out directly who had contaminated Silkwood, what had happened to her documents, and who might have pushed her off the road. The investigation would focus on how effectively government agencies like the AEC/NRC had done their jobs of protecting workers and the public from nuclear hazards, and how adequately these agencies had investigated the charges of health and safety violations at the Kerr-McGee plant.

Ribicoff then turned toward Metcalf, who sat to the side and just behind him. "Lee, I want *you* to handle this," he said.

Metcalf was a former Montana Supreme Court judge, who came to the House of Representatives in 1953 with the backing of big labor. In 1960, he moved to the Senate. In many ways, Metcalf was ideal for the Silkwood investigation. He believed in conservation, wanted a rational energy policy, tried to regulate the power industries, and fought to increase citizen participation in government. He was concerned about the clout of giant corporations like Kerr-McGee.

But Metcalf was more judge than prosecutor. He was going on sixty-five. And he was sick.

Metcalf pulled out the statement Win Turner had written for him. "This issue has been the subject of increasingly serious concern among various groups representing the rights of women, labor, and the safety and health of individuals involved in the production and transportation of nuclear materials," Metcalf said. "In my opinion, it deserves a thorough investigation."

It was a good show. Nelson and Tucker were elated that their pressure tactics had worked. Merle and Bill Silkwood were hopeful that, at last, they might find out who had contaminated and killed their daughter. Turner and Stockton were chafing at the congressional bit for a go at the FBI. And Ribicoff had just handled a potentially nasty meeting with grace and charm.

□ □ □

Metcalf was a good friend of Michigan representative John Dingell, chairman of the House Subcommittee on Energy and Power. They had worked closely during Metcalf's House days, and both were concerned about the broader nuclear issues surrounding the Silkwood case. In fact, Dingell had called for a congressional investigation almost a year earlier, but nothing had happened. Metcalf and Dingell had decided unofficially to make the investigation a joint one. Dingell's subcommittee would focus on missing plutonium; Metcalf's, on health and safety. Dingell loaned Metcalf his investigator, Peter Stockton, on a part-time basis.

Win Turner, chief counsel of Metcalf's subcommittee, led the investigation team. He was an experienced attorney who knew the federal bureaucracy well. A tall, hulking, cautious man, Turner was used to trying to wheedle documents from government agencies. And the Silkwood documents, tucked in Justice Department and FBI file drawers, were essential if he and Stockton were to find out how well the Justice Department had done its job.

At first, the Justice Department promised to prepare for Turner summaries of the facts about Silkwood's death and contamination. It was a start, and Turner was a patient man, but when he read the summary of facts on Silkwood's death, he was shocked. It was full of errors and inconsistencies.

When he asked for the summary on Silkwood's contamination, Deputy Assistant Attorney General John Keeney told him he couldn't give it to him because the contamination investigation was still open.

Turner was angry. Justice had announced that the investigation was closed; now it was saying the case was still open. The distinction was not academic. According to a twenty-year-old Justice Department policy, when an FBI investigation is still open, all files are closed. So as long as Justice kept the contamination investigation open, there was no way Turner could see so much as a thirdhand summary unless the subcommittee subpoenaed the records. Subcommittees are reluctant to do that, for if a government agency refuses to honor the subpoena, the full House of Representatives must review the reasons for the subpoena, and the full House has rarely voted to uphold its subcommittees' subpoenas.

"In my opinion, they are stalling," Turner wrote in an internal memo. "Possibly because in preparing the summaries, they have learned the inadequacy of their investigation."

Turner wasn't about to give up. He asked Metcalf to lean on Attorney General Levi himself, if necessary. The work of the subcommittee, he told Metcalf, cannot move without a thorough review of the Justice Department's Silkwood files.

It worked. Justice finally invited Turner and Stockton to review the department's files, although not the raw FBI files, which amounted to more than a thousand pages of interviews and summaries. It was only a foot in the file cabinet, but, for the time being, Turner was satisfied. He was a patient man.

The Justice Department, however, had a little surprise in store for the men from the Hill. They could *read* the files, Justice told them, but not take notes. And if that wasn't bad enough, the file drawers were almost empty. Just a few thirdhand reports with no real hard facts, and a lot of references to reports stashed down the street at FBI headquarters in the J. Edgar Hoover Building. The only important document Turner and Stockton uncovered was the Silkwood Death Fact Memorandum.

In important cases like Silkwood, the Justice Department orders the FBI to investigate, and outlines to the Bureau the scope of the investigation. The FBI usually assigns a special agent to direct the sleuthing. Once the special agent gets all the reports from the field agents working for him, he may or may not summarize the investigation. Then he sends all reports and summaries to FBI headquarters in Washington, which, in turn, forwards them to the appropriate United States attorney's office or to the Justice Department for a decision to prosecute or not.

If the FBI reports go to Justice (as they did in the Silkwood case), the department assigns an attorney to study the documents. The attorney then writes a Fact Memorandum, summarizing the case and recommending to his superiors that it be prosecuted or dropped.

132

Turner and Stockton had been eager to get the Silkwood Death Fact Memorandum because it would outline the Justice Department's key facts and logic. To their delight, they also stumbled on an intriguing ten pages of handwritten notes on yellow, legal-size paper. The jottings were penned by the Justice Department attorney who wrote the Death Fact Memorandum, and they were an important clue to the Justice Department's thinking. Turner and Stockton itched to take the notes back to the Hill, where they could study them carefully. So, while Turner kept Justice Department lawyers busy, Stockton crouched over a pocket recorder and dictated eight of the ten pages before his tape ran out.

After Turner and Stockton read everything they could find and interviewed some of the Justice Department attorneys responsible for the Silkwood investigations, they went back to the Hill to sort through what they had learned.

Thomas H. Henderson, they found out, was the first Justice Department attorney assigned to direct the FBI on the Silkwood case. Sometime between November 21, 1974 — Silkwood was killed November 13 — and the end of the year, Henderson had drastically limited the Silkwood investigation by telling the FBI to conduct a "preliminary" probe and not to examine "possible suspects or motives." Early in 1975, Henderson passed the Silkwood case over to Philip Wilens, chief of the department's management and labor division. But Wilens was busy with other cases, so in March he gave the Silkwood death case to Thomas Goldstein, a low-ranking attorney who was tied up with another Justice Department case in New York.

Goldstein read the FBI documents on the Silkwood case during March and April 1975, but he did not interview Kerr-McGee, the FBI, or Larry Olson, Sr., who had written most of the FBI reports. He apparently researched and wrote the Death Fact Memorandum in two weeks while he was packing for Florida and a job with the Miami Transit Authority. Deputy Assistant Attorney General Kevin Maroney's action in officially closing the investigation into Silkwood's death was based on Goldstein's conclusion that her car crash was an accident.

Turner and Stockton were troubled. Why did the Justice Department assign such a relatively inexperienced attorney to review the Silkwood case? Didn't the department think Silkwood was important enough to give Goldstein more time? And Goldstein — why didn't he talk to Larry Olson, Sr.?

Turner and Stockton began to dissect the Death Fact Memorandum itself. It was obvious that the five-page document was hastily written. It referred to "The Reasoner Report" as "60 Minutes," and said Silkwood had had a previous car accident on October 30, 1974, instead of October 31. And it claimed "it was extremely windy" on the evening Silkwood was killed, suggesting that even if a sleepy driver tended to drift right, a strong crosswind would have pulled the car left. In fact, weather bureau reports indicate that only a light wind was blowing that night.

The Death Fact Memorandum never raised the issue of Silkwood's allegedly missing documents. It never challenged the medical examiner's argument that .35 mg of methaqualone would be an overdose in Silkwood's case. And almost two thirds of Goldstein's analysis was an attack on A. O. Pipkin.

"Mr. Pipkin's reasoning seems unsupported by the facts," Goldstein wrote.

What facts?

"The Honda Civic Hatchback tends to drift left and that explains why a sleepy Silkwood would end up on the left shoulder rather than the right," Goldstein reasoned.

Furthermore, "the Honda's front-wheel drive affected steering, especially at low speeds. During acceleration, the car tended to lunge and pull to the left," Goldstein wrote, quoting *Consumer Reports*. "With each blow from our bumper basher, the Honda bounded away more violently than any other car we've tested."

If Silkwood's car had been hit from the rear, Goldstein theorized, and the Honda brakes were as good as *Consumer Reports* said, then Silkwood would have left skid marks on the blacktop. She didn't. Furthermore, according to the Oklahoma Highway Patrol there was no evidence that she ever fought for control of the car once she hit the grass along the road.

Goldstein implied that Drew Stephens might have made the dents in Silkwood's Honda. No one ever saw the dents before Stephens towed it to the hideout, Goldstein reasoned. Pipkin examined the car two days after Silkwood's death. Therefore, the dents must have been made after the accident, while Stephens had the Honda.

Goldstein concluded: "On the basis of the facts produced by the above investigation, it was determined that there was no significant indication of a violation of federal criminal law in the death of Ms. Silkwood."

Stockton couldn't believe how easily Goldstein had dismissed Pipkin's findings. He flipped to the notes Goldstein had made from raw FBI files. The jottings began by raising serious doubts about Karen Silkwood's character and then portrayed her as a kind of mentally disturbed junkie. Next, Goldstein's notes described the union as desperate for a bargaining issue to squeeze concessions from Kerr-McGee and save face. Finally, after destroying the credibility of Silkwood and the union, Goldstein seemed to have dismissed Silkwood's charges about health, safety, and quality-control violations as irrelevant.

Stockton concluded that Goldstein had a "peculiar mind cast," and "appears to be one of the least curious people you could find." It was clear to both Stockton and Turner that the Justice Department had done an embarrassingly poor job of analyzing the death of Karen Silkwood and, therefore, had been reluctant to show the subcommittee its files. They hoped the Contamination Fact Memorandum would be better than the Death Fact Memorandum. But judging from what they had seen so far, they doubted it.

They were wrong and they were right. The Justice Department did not prepare a Contamination Fact Memorandum because the FBI had just completed its investigation. Instead of a Fact Memorandum, the Justice Department gave Turner and Stockton the twenty-one-page summary of the contamination incident prepared by Larry Olson. The FBI summary was worse than the Justice Department's Death Fact Memorandum.

> It is the damnedest document one could imagine for basing a decision to close an investigation [Stockton wrote to his boss, John Dingell]. It is a crude compilation of contradictory and sometimes irrelevant interviews with various witnesses. There is no analysis. There is no conclusion as to who was telling the truth. There is no evidence that the FBI tried to break down anyone's story. There is

135

a good deal of malicious gossip about Silkwood. Based on the document, it is inconceivable that Justice could have closed the contamination investigation. It appears clear that Justice took no interest in attempting to get to the bottom of this incident — whether the contamination was accidental, sabotaging Silkwood, or deliberately contrived by Silkwood.

Stockton pointed out to Dingell:

☐ Olson wrote that "no attempt had been made to include information concerning the automobile accident within" the contamination summary. Olson then went on to say that Silkwood "was under the influence of a hypnotic drug, methaqualone, at the time of her death." And he quoted a statement made by Sherri Ellis about the night before Silkwood died: "They drank Bloody Marys mixed with 190 proof grain alcohol, not measuring drinks but eyeballing them (about four fingers each)." Neither statement had anything to do with Silkwood's contamination and everything to do with her death.

☐ A Kerr-McGee document had used the same language about the alleged drinking the night before Silkwood was killed — Bloody Marys, 190 proof, eyeballing, four fingers. Did Olson lift the information from a K-M report and pass it off as his own. Furthermore, Ellis had told Stockton that her interviews with Kerr-McGee had been taken under great duress and were aimed at getting her to finger Silkwood.

☐ Olson quoted Drew Stephens as saying Silkwood drank "only a little wine" the night before she was killed. Yet Olson apparently had made no attempt to find out which of the two statements was correct or to subpoena Ellis, a key witness, who had refused a voluntary interview.

☐ Silkwood's lover Don Gummow told Olson that "he feels that Silkwood was incapable of contriving a contamination incident acting on her own but that he had considered the possibility that someone had used Silkwood." It was strange that Olson did not say who Gummow thought might have "used" her.

☐ Olson said Kerr-McGee health physics director Wayne Norwood told him that someone stole an alpha counter from Silkwood's apartment after he had locked the place on November 7. Olson apparently never verified Norwood's story, or, if he did, he never tried to find the thief.

□ Olson told the subcommittee that he was suspicious of Steve Wodka and complained that because he did not have the chance personally to "break down Wodka's story," his investigation had a hole. It was puzzling that a Washington Bureau agent, Carlton Broden, had interviewed Wodka instead of Olson, who knew the most about the case. Broden talked to Wodka three months after Silkwood's death and had not asked many probing questions.

□ Olson wrote: "James Reading, Security Officer, KMC, made available background information concerning Silkwood, her roommate Sherri Ellis, and Silkwood's boyfriend, Drew Stephens . . . and furnished background information concerning the results of the KMC investigation to date." The K-M background information the subcommittee had seen so far was "extremely derogatory and effectively destroyed Silkwood's credibility." Apparently that was the reason the FBI did not take the Silkwood issues seriously.

□ Olson did not conclude whether Silkwood had been contaminated accidentally or deliberately, even though the AEC had concluded the contamination was deliberate. If it was deliberate, then a federal crime had been committed — the theft of plutonium. The Justice Department apparently had not instructed Olson to go beyond the preliminary investigation (was a crime committed?) to the next stage (who did it and why?).

The FBI contamination summary was based on eleven reports Olson had written between December 13, 1974 (a month after Silkwood's death), and September 22, 1975. It took Olson four months (September to December) to write the twenty-one-page summary. It was clear that when the Justice Department announced that the Silkwood contamination case was closed, it had not had either an FBI summary or a Contamination Fact Memorandum. And it appeared that when Turner and Stockton demanded documents, Justice reopened the case and told Olson to write a quick summary so that the department would have at least something to show the subcommittee.

To Turner and Stockton it was also clear that the Justice Department did a poor job of analyzing the facts surrounding both the contamination and the death of Karen Silkwood. The question was: Did the Justice Department do a poor job because the FBI had conducted a superficial investigation? The only way to answer the question was to get copies of the Bureau reports themselves.

Win Turner went to work on the FBI. He had contacts there and at Justice, but he kept bumping into Hoover's stone wall. Did the FBI have something to hide? Was the Bureau trying to block the congressional investigation? If so, why? Turner was beginning to lose his patience.

Chapter 14

It was around six in the evening of December 6, 1975. The subcommittee staff had gone for the day, and Turner was working on a speech, unrelated to Silkwood, in his cramped office in the Old Senate Office Building. Jacque Srouji walked in with her eight-year-old son.

She looked like a housewife from down the street, a pleasantly round mother who'd love to lose just seven more pounds but never seemed to have the time, what with children and housework and hamburgers. She was thirty-one years old. Dark eyes, short, wavy, reddish-blond hair, and a smile that put everyone at ease.

She was a journalist from Nashville, she told Turner. She had written a book on nuclear power and had read a wire service story, "Silkwood Probe Reopened." She wondered how the investigation was going and hoped she could learn something that would help her with her book, which was now in galley form.

Turner was not surprised that Srouji had just walked in off the street. People drop by subcommittee offices all the time. As they chatted about the Silkwood case, Turner let Srouji's boy play with the Xerox machine. Jacque seemed pleasant and friendly, strong but not pushy. But soon she began to ask hard, searching questions. It became obvious to Turner that she was trying to find out where the investigation was heading and what evidence the subcommittee had. Turner clammed up.

"How does your book fit into *our* investigation?" he asked.

139

"Part of it involves Kerr-McGee and Silkwood," she said.

Turner suddenly became very interested in Srouji. He asked her if she had any information or leads she would share with sub-committee investigators.

"Oh, yes. I've been to Oklahoma and I've seen the FBI files."

Turner was stunned. He had been whipping the Justice Department with letters and memos, and all he had seen were summaries or summaries of summaries. Was it possible that Srouji had actually read the FBI documents?

Srouji explained that the FBI agent who had directed the Silkwood investigation was an old friend whom she had known when he was a special agent in Nashville. He had let her read the file, she said. She'd copied it.

"Do you have any [documents] with you?" Turner asked. "I'd like to see them."

"They are all in Nashville," she said.

Turner was getting angry, not with Srouji, but with the Justice Department. If she was telling the truth — and he had no reason to believe she wasn't — Justice had a lot of explaining to do. Why could a journalist get copies of FBI documents when a sub-committee of the United States Congress could not? Turner had two choices. He could try to get copies of Srouji's FBI documents without telling the Bureau. Or he could confront the FBI with Srouji and embarrass the Bureau into letting him into the files. Turner decided to try to get Srouji's documents. He called his investigator, Peter Stockton.

Stockton was just as intrigued. He had tramped all over Crescent for two months with Barbara Newman a year earlier and had found few clues. And the Justice Department's stonewalling was frustrating him, too. He called Srouji, who had returned to Nashville. Yes, she had FBI documents, she told him. She'd be happy to share them, if he wanted to come to Tennessee. He certainly did.

Turner and Stockton knew it was important to see Srouji before she changed her mind or before someone else got to her. And that created a problem. Government Operations Committee rules forbid investigators to travel without telling the minority where they are going and why. The minority then has the option to veto the trip. Stockton and Turner felt that even if subcommittee minority counsel Lyle Ryter, who opposed the Silkwood investigation, didn't try to kill the Nashville trip, the plan to wheedle FBI documents from Srouji would somehow get back to the Bureau.

Stockton and Turner pulled an end run. Stockton asked his House boss, Representative Dingell, to send him to Nashville on *House* rather than Senate business. Dingell thought the trip was important and paid for it out of his personal staff budget.

Peter Stockton was a pro — thorough, dedicated, curious, and cynical. He had been a fiscal economist in the Executive Office of the President, then had moved to Capitol Hill in 1969. He worked for Senators William Proxmire and Hubert Humphrey before joining Dingell's subcommittee and his personal staff in 1973.

Stockton was an open and friendly man if he liked someone, but tough and scrappy if he didn't. He had never bought the official Silkwood story. Not only were there too many unanswered questions, but he felt that the attitude of the government toward Silkwood was somewhat sick. He had interviewed top AEC officials only one month after her death. Some of them giggled like kids and told him she was "a real flake and abused herself with bologna," suggesting she had smuggled plutonium from the Cimarron plant in her vagina, masturbated with some bologna, and then put the contaminated meat into her refrigerator. Stockton had heard that line many times. He hoped that Srouji wouldn't turn into another Pennington affair.

Joe Pennington was a newsman at radio station KTOK in Oklahoma City. He had told Stockton that Larry Olson, Sr., called him and asked to meet in a bar. Olson was confused and angry. He said he had developed a theory that Silkwood had been smuggling plutonium from the Kerr-McGee plant, and had wanted to investigate the theory. He told Pennington that he had written a letterhead memo (LHM) to FBI headquarters in Washington, asking for permission to go ahead. Washington had fired back an LHM telling Olson that missing plutonium had nothing to do with the Silkwood case. Pennington then went on the air, suggesting Olson's smuggling theory.

Pennington had told Stockton and others that he had a copy of Olson's LHMs. He promised to give Stockton copies, but never did. Stockton concluded that the radio reporter didn't have the LHMs, but he kept an open mind on the Olson-to-Washington correspondence. Stockton was hoping Jacque Srouji would give him something more tangible.

□ □ □

Srouji's book editor, Dominic de Lorenzo, met Stockton at the Nashville airport and drove him to the Aurora Publishers. De Lorenzo had been an editor at the University of Notre Dame Press before John Seigenthaler, publisher of the Nashville *Tennesseean,* hired him away to run the fledgling Aurora company, which published everything from T. S. Eliot to Chinese cookbooks.

Stockton and de Lorenzo sat in the editor's temporary office cubicle and chatted, waiting for Srouji to come over from the *Tennesseean.* The meeting was strained. De Lorenzo asked Stockton what he wanted to talk to Srouji about and why he wanted copies of her documents. Stockton told him about the subcommittee's problem in getting cooperation from the Justice Department and the FBI.

De Lorenzo assured him Srouji had letterhead memos, interview summaries, information about missing plutonium or materials unaccounted for, and 302 forms — FBI raw interview notes dictated to a secretary and typed. De Lorenzo's secretary was present during the discussion; sometimes she took notes.

When Srouji arrived, she gave Stockton a stack of papers to leaf through. Some he had seen before in the files of the Oklahoma Highway Patrol. There were summaries of interviews with Harold Smith, the wrecker driver; James Mullins, the first person at the scene of the accident; Rick Fagen, the Oklahoma Highway Patrolman who wrote the accident report; and what appeared to be an interview with Sherri Ellis. There were Pinkerton papers on A. O. Pipkin and a copy of Pipkin's accident report, with the words *FBI Evidence* taped on it.

Stockton wasn't impressed. Most of what Srouji showed him appeared to be Kerr-McGee documents. Srouji had told him earlier that she met with Jim Reading, K-M's security director, and that Olson had arranged the interview for her. She said she had a letter written by Olson to Reading to prove it.

The *FBI Evidence* taped on the Pipkin report wasn't conclu-

142

sive, either. Anyone familiar with FBI documents could forge the tape. However, a few of Srouji's documents *could* have been FBI reports. Stockton wasn't sure because the tops had been cut off.

"Look," he told Srouji. "You know these aren't FBI documents."

"Oh, yes, they are," she said. "I got them from the FBI files."

Srouji said she had at least three meetings with Larry Olson. During one meeting, she told Stockton, Olson gave her his files to copy. She showed Stockton two handwritten letters from Olson to her. Stockton had seen Olson's signature before; the letters appeared genuine. Stockton concluded that Srouji probably had some kind of relationship with Olson, but he still wasn't convinced she had actual FBI documents. It was possible that Olson had let her copy the Kerr-McGee documents in his FBI file. If so, that was another interesting question: What were K-M documents doing in Olson's files?

Stockton pressed. Could he see documents that were *clearly* FBI, he asked.

Sensing his skepticism, Srouji began to get defensive. The FBI documents were at home, she said. In a safe. She was afraid to give them to him because she didn't want to get ripped off without proper credit. She had worked very hard to get her Silkwood documents, and if she just handed them over to the subcommittee, the media would get them before her book was even published. Furthermore, she said, she had to protect her news sources. If she gave the subcommittee her records, she would compromise the confidentiality she had promised them. She told Stockton that the only way the subcommittee would ever get her documents would be to force them out of her with a subpoena.

Stockton was cautious. All he needed to see now was a headline screaming SUBCOMMITTEE SUBPOENAS JOURNALIST'S DOCU-MENTS. The First Amendment people would be all over him. Besides, he wouldn't even *consider* the possibility of a subpoena until he had some concrete proof that Srouji had actual FBI documents.

"It was obvious Srouji and de Lorenzo were trying to peddle her book, *Critical Mass,*" Stockton recalls. "You're never sure about the credibility of the information and the motive of people when you're in a situation like that."

143

When they broke for lunch, Srouji tried a new tack. She talked about her very close relationship with the Nashville police and the FBI. She said she'd known a lot of agents over the last six or seven years, and she talked about her friendship with Olson, casual and professional. That's why Olson had shared his files with her, she said.

Srouji also talked about Jim Reading, whom she claimed to have met three times, and quoted Olson as saying, "That son-of-a-bitch has lifted all my files."

Stockton was still skeptical and said he had no authority to promise a subpoena. So Srouji and de Lorenzo said they'd think about releasing some documents without a subpoena and that they'd send Stockton a copy of the *Critical Mass* galleys.

De Lorenzo asked his secretary to drive Stockton to the airport. She wanted to stop for a drink. They did, but Stockton was wary. "She was something," he recalls. "A very friendly person, but it was one of those things where you're not sure."

On the plane back to Washington, Stockton had serious misgivings. "Jacque and her documents are worth obtaining," he jotted in his notes. "Very close relationship between Olson and Reading . . . Both the FBI and K-M had stiffed the subcommittee so far. Why would Reading and Olson help Srouji?"

But there was more gnawing on Stockton than Srouji. For a publisher, de Lorenzo didn't seem very bright or articulate. Everything at Aurora looked temporary. There were no real offices to speak of; just makeshift cubicles. The space wasn't cluttered with books and manuscripts and paper like most editorial offices. In fact, de Lorenzo couldn't even come up with a copy of one of Aurora's basic books when Stockton asked to see it.

"The whole business, the whole setup there in Nashville seemed suspicious," Stockton recalls. "It seemed terribly suspicious that a guy like this could have that kind of a job. The only thing that gave me any confidence in de Lorenzo at all was the fact that Seigenthaler had hired him away from Notre Dame Press."

Stockton felt foolish. You've been had, he told himself.

Chapter 15

Soon after Stockton got back to Washington, Srouji called to say that an out-of-town FBI agent had visited her. "He was telling me I would have a lot of heartache if I went before the subcommittee," she told Stockton. "His tone toward me was the same tone that a father would have toward a child that is about to go out and get chopped up."

Srouji told Stockton that the FBI agent warned her she would be in "jeopardy" if she publicly revealed the name of her Silkwood source (Olson) in the Bureau. She was scared and needed more time to think about cooperating with the subcommittee in any way, she said.

Turner and Stockton could not wait for Jacque Srouji's documents, and it was clear that the Bureau was not going to help them. They felt they had no choice but to conduct their own field investigation to prepare for joint Metcalf-Dingell congressional hearings. Once the hearings opened, Justice Department officials could be subpoenaed and forced to answer questions under oath. After the hearings, Turner and Stockton could decide on the next step.

In early February 1976, four months after Turner and Stockton began their tug of war with the Justice Department and the FBI, Stockton reported to Dingell in a memorandum:

> The significance of the K-M/Silkwood affair to Congress at this point in time is not so much an attempt to find the smoking gun — but as a microcosm of what the nation faces as we move toward a

145

plutonium economy. Most of the major nuclear issues are involved: illegal diversion of plutonium, worker health and safety problems, quality control, falsification, etc.

The important factor is to determine, in this case study, how seriously the Federal investigating and regulatory agencies viewed these matters and whether these agencies are, in fact, equipped to handle these central issues. From the interviews and documents obtained by the subcommittee staff there are some serious questions raised about the vigor and objectivity with which the Justice Department, FBI and the AEC (NRC and ERDA [Energy Research and Development Administration]) pursued these matters. It appears from inter-agency documents that regulatory vigor is directly related to media pressure — which has serious implications.

Stockton then tore into the Justice Department and the FBI.

It appears that the K-M investigations and their earlier interviews were the original source of most of the FBI investigation. It appears that a vast number of K-M investigative documents appear in the FBI files and that FBI files also appear in K-M files . . . It appears that this was a joint investigation from the beginning — however, Justice claims that K-M was indeed a target of the investigation.

There was something else bothering him, Stockton told Dingell. If Lawrence Olson had indeed conducted a joint investigation with Kerr-McGee and had somehow cooperated with Jacque Srouji, "did he do it on his own or with the knowledge and possibly under the direction of his superiors?"

While Stockton was reporting to Dingell in the House, Turner was dealing with Metcalf in the Senate. The lawyer outlined the scope of the preliminary field investigation and emphasized the importance of getting Srouji to state under oath during the proposed joint hearings that she had FBI Silkwood documents. "If we do not get this verification," he said, "the FBI may deny it, or explain it away. We would lose the initiative."

Metcalf gave his blessing to the investigation and the joint hearings. Unknowingly, he also unleashed a series of events that would plague both his and Dingell's subcommittees for the next four months.

Stockton needed to travel to Chicago to cross-examine the

Nuclear Regulatory Commission on its investigation into the diversion of plutonium at Kerr-McGee, Silkwood's health and safety violation charges, her contamination, and K-M's security problems. Stockton then needed to go to Oklahoma to interview K-M officials and witnesses. He planned to take along congressional investigator Walter "Bud" Fialkewicz. Stockton planned on a nine-day trip, and he needed committee approval for travel funds.

Turner approached Chairman Ribicoff, who told him to get approval first from Senator Bill Brock of Tennessee, the ranking Republican member of Metcalf's subcommittee. After that, Ribicoff said, Turner would need the approval of Senator Charles Percy of Illinois, ranking minority member of the full committee.

Brock was a probusiness, antilabor conservative hawk who had voted against the Civil Rights Bill, the Voting Rights Act, the $1.00 minimum wage for farm workers, and just about all of Kennedy's and Johnson's social legislation. He was the darling of the American Security Council, the most powerful right-wing lobby in the country. The ASC generally gave Brock a 100 percent rating each year, based on his voting record in Congress.

Lyle Ryter, the subcommittee minority counsel, was Senator Brock's man, and relations between Ryter and Turner were bad. Ryter argued that the field investigation outlined by Turner was beyond the scope of the subcommittee. Others should do it, he said. The subcommittee was not used to this kind of in-depth investigation. It was too expensive, it would take too long. Stockton didn't even belong to the subcommittee. He was Dingell's man, Ryter said.

Stockton didn't get very far with Ryter, either. "What the hell do you want to go there for?" the minority counsel snapped at him. "I know what you're after — the MUF [material unaccounted for] figures."

Stockton had heard it all before. "Ryter had obviously been prepped by the NRC and the FBI," Stockton recalls. "No doubt about that."

Ryter, an economist, like Stockton, would leave the Hill soon and become a lobbyist for the American Security Council, which supported nuclear power and increased military spending. Founded by a former FBI agent and stuffed with former Bureau

147

investigators, former military intelligence officers, and retired generals and admirals, the ASC has a library of 6 million dossiers on "subversives" and "subversive organizations" on its 683–acre estate outside Washington at the foot of the Blue Ridge Mountains near the Airlie House. The 2000 corporations that belong to the ASC can ask for security checks and dossiers on their employees or prospective employees.

The Government Operations Committee minority would not allow Stockton to travel, and the investigator was angry. The FBI wouldn't talk; the Justice Department wouldn't open its files; and now he, the only investigator on the case, couldn't travel. "What kind of fucking investigation can you have without travel?" he says, still angry.

In effect, the minority Republicans on the Government Operations Committee had killed the Silkwood investigation. Turner and Stockton felt someone somewhere had used pressure, but they weren't sure who. They were discouraged, but they still had a trump card. They asked Metcalf to threaten the Justice Department with a subpoena for the FBI documents. While Metcalf was waffling, Dean McGee dropped in to chat. Senator Dewey Bartlett, a Tulsa oilman and a Kerr-McGee defender, had asked Metcalf to see McGee, and, by way of professional courtesy, Metcalf had agreed.

Turner was suspicious. He wasn't sure what the planned meeting was all about, but he would feel more comfortable if Metcalf did not meet McGee alone. "I'd like to be present," Turner told Metcalf. "We should have someone there."

It is standard practice on the Hill for one senator to ask another to meet with a constituent, but it is also common practice on matters relating to an investigation to have counsel or staff present. It was especially important in Metcalf's case. Because he was involved in so much legislation, he was not well informed about his own subcommittee's investigation.

But Metcalf brushed aside Turner's suggestion and was alone in his office when Dean McGee paid a visit. McGee brought along Philip Bennett, the manager of the Kerr-McGee Washington office. After the conference, Metcalf was tight-lipped. He told Turner that McGee claimed K-M had found all its missing plutonium, that K-M had cleaned up the mess at the Cimarron plant, and that the plant was closing down soon. Metcalf also told Turner that McGee said the OCAW had accepted the Oklahoma Highway Patrol's conclusion that there was no foul play in the death of Karen Silkwood. Turner was surprised. He had been in contact with the union during preparations for the hearing, and no one had ever told him that.

Turner tried to coax more about the McGee meeting from Metcalf, but the senator clammed up, as he usually did when there was something he didn't want to talk about. Soon after the McGee visit, Metcalf pulled out of the joint hearings. Barbara Newman interviewed him for National Public Radio. Given the confusion this interview has caused and the suspicion that Metcalf bowed out of the hearings under pressure from McGee, it is worth quoting Newman's interview at length.

After explaining how Metcalf and Dingell had planned joint hearings, Newman said: "But last week the senator pulled out of the hearings. Sources in the House told us that the reason was pressure on Senator Metcalf from the nuclear industry and the Republican minority in the Senate Government Operations Committee. But Senator Metcalf himself insists that it was something else."

"I didn't bow out of the hearings because of . . . any pressure," Metcalf told Newman.

"This morning you said that the decisive reason, in your opinion, was the fact that the president of OCAW, Al Grospiron, told you that the union did not want to have a hearing," Newman said. "I wonder if you . . ."

"He didn't tell me that," Metcalf cut in. "I . . . I . . . I'm sorry."

"I thought you said that the decisive reason was when the union told you that they really didn't see any purpose in having a hearing."

"That's correct," Metcalf said. "I . . . if . . . if the union had told me that they wanted to continue the hearing, I would have cooperated with the House . . . continued to cooperate with the House, in pursuing the inquiry. But they said that they were convinced that there wasn't anything wrong with . . . the investigation that was made by the highway patrol, and the Atomic Energy Committee . . . and . . . Commission . . . and others insofar as Miss Silkwood's death was concerned. Now, I'm *still* concerned, and I told you this morning, I'm still concerned about . . . about the loss of plutonium that's been revealed by the GAO."

Metcalf made no attempt to hide the fact that McGee visited him. The senator explained to Newman that there were forty pounds or kilograms (he wasn't sure which) of plutonium still missing from the Cimarron plant. He said that McGee claimed the company had accounted for the MUF to the satisfaction of the GAO. "A self-serving statement that Mr. McGee had made," Metcalf told Newman.

Metcalf's staff was shocked. They were not surprised when Lee bowed out of the hearings. He was sick, and he had no stomach for a long, tough investigation. Besides, he was personally guiding through the Senate a strip-mining bill important to Montana. He was up to his ears in legislation. In fact, the staff was frankly pleased that Metcalf had tossed the ball to Dingell, a younger and more aggressive man. The Silkwood case needed time, energy, and leadership. Metcalf was providing little, if any. But to say on the air that the OCAW was *happy* with the she-fell-asleep-at-the-wheel theory was clearly false and embarrassing.

Victor Reinemer, Metcalf's staff director, called the OCAW. He learned that the union was not satisfied with the Oklahoma Highway Patrol report, that Al Grospiron denied talking to Metcalf, and that Grospiron was going to say just that during an afternoon press conference. Reinemer was worried. Metcalf's credibility was at stake; so was the integrity of the Silkwood probe. He encouraged Metcalf to call Grospiron.

Reinemer was with Metcalf when the senator spoke to the union president. Metcalf told Grospiron that he was sorry about the confusion, but that "McGee had misled and misinformed" him about the union.

Dean McGee had a different story. He later said in a sworn deposition that he went to see Metcalf "to make sure that when he [Metcalf] had the investigation, he had all the facts." McGee denied he had asked for the meeting or that he had told Metcalf the union was satisfied with the Oklahoma Highway Patrol investigation. He emphasized that he had gone to see Metcalf just to give him the facts.

With Metcalf out, Dingell was now in the catbird seat, and he loved it. He was a tall, trim, balding, fifty-year-old lawyer of Polish descent from the automobile town of Dearborn, Michigan. With twenty years of experience in the House, he was a professional politician down to his conservative necktie.

Dingell had earned a reputation as a conscientious legislator who, unlike so many of his colleagues, had an outstanding record for voting and attending committee meetings. He consistently won high marks from watchdog groups representing labor, consumer interests, farmers, and conservationists.

Dingell was a legislator. In one year alone, he had introduced or cosponsored 186 bills. He was hoping that the Silkwood hearings would spark some legislation to regulate the nuclear industry. And with a safe seat in Congress, Dingell wasn't afraid to spar with the FBI. Stockton had told him that before he'd get to the bottom of the Olson-Srouji mystery, there would be an ugly fight.

Stockton was right. But before Dingell could open the first round, Jacque Srouji came to town again. This time, she was wearing the blue and white of the United States Navy.

151

Chapter 16

Srouji stepped off the plane from Nashville on April 18, dressed in a light blue Navy uniform and cap, and carrying an olive-green duffel bag filled with books. Kirk Loggins, a *Tennesseean* Washington correspondent, was there to meet her. She had called him earlier that week and asked him to pick her up.

Srouji had joined the Naval Reserve. She was assigned to the Pentagon as a Special Assistant for Manpower and Reserve Appeals with the rank of petty officer second class. She had told Mike Tate, her editor at the *Tennesseean,* that she got the high rank because of her previous Army experience. Tate found that odd, for she had once told him that she had been a WAC for only a few weeks.

Loggins drove Srouji to Military Towers, an apartment hotel in Alexandria, Virginia, within sight of the Pentagon. He asked her how she had stumbled into the Navy. As usual, she was evasive. She had joined a couple of months earlier, she said, and was surprised by the ease with which she'd got security clearance. She told Loggins she hoped to get some stories from her Pentagon assignment, but asked him not to tell anyone who called the *Tennesseean* for her where she was working or living.

The next day, Srouji dug in at the Pentagon, in the Office of the Chief of Information. Her job was to evaluate the public relations program for Project Seafarer, a top-secret, extremely low-frequency (ELF) communications system for submarines. ELF allows subs to send messages during crises without being detected.

A week later, Loggins took Srouji to a jazz concert at the Kennedy Center. Afterward, they went for a nightcap at C. R. Higgins, a Georgetown pub. Srouji told Loggins she had recently had dinner with a diplomat from the Soviet Embassy and that someone had followed them as they left the restaurant. Her Russian friend shook the tail, she said. She was going to have lunch with him the next day.

Loggins was mildly amused. "No wonder people are following you if you're going straight from the Pentagon to the Russian Embassy," he told her.

While Srouji was commuting between the Pentagon and the Soviet Embassy, Peter Stockton was still trying to get a copy of *Critical Mass*. He called de Lorenzo, but the publisher put him off once again.

"Where's Jacque these days?" Stockton asked.

"Oh, she's in the Navy," de Lorenzo said. "Navy Reserve."

"How can I get in touch with her?"

De Lorenzo gave Stockton several Washington numbers. The investigator found Srouji at the Pentagon, and she agreed to meet him at the Naval Research Laboratory (NRL) outside Washington on the Virginia side of the Potomac.

The NRL guards wouldn't let Stockton past the front gate. He flashed his congressional credentials. They weren't impressed. He waved his security clearance card. They called Srouji to the gate. She was in uniform.

Stockton was peeved. Srouji was waltzing in and out of the NRL, but here he was, with impeccable credentials, and he couldn't get past the guardhouse.

Srouji and Stockton drove to a nearby restaurant to talk about the House hearings on Silkwood, scheduled to begin the next day. Stockton already had twenty of Srouji's documents. Earlier in April, Turner had told Srouji that Dingell wanted to see her. She was reluctant to meet with the congressman, but de Lorenzo told her any publicity she could get would be good for her book. De Lorenzo went to Washington with her, and, once in Dingell's office, Srouji agreed to a compromise. She'd give the subcommittee the twenty documents she quoted in her book but keep the rest to protect her sources. The twenty documents turned out

153

to be either Kerr-McGee investigative reports or public documents, like Pipkin's accident report. There was one exception — a Nuclear Regulatory Commission report that contained the material unaccounted for figures for the Kerr-McGee Cimarron plant.

At the restaurant, Stockton and Srouji chatted about Kerr-McGee's surveillance of Silkwood and the OCAW. "Didn't you notice that some of those documents seemed to have verbatim discussions in Silkwood's apartment?" she asked Stockton.

"Yes, I noticed that." Stockton had already concluded that Kerr-McGee or the FBI (probably K-M) had done some bugging.

"Well, in the files that I looked through — both the K-M and the Bureau files," Srouji told Stockton, "there were what appeared to be transcripts of conversations."

She said she had asked Jim Reading about the apparent transcripts. Reading admitted "his group" had spied on Silkwood and probably some other OCAW members, she said. Among her documents were copies of transcripts of phone taps and bugs.

They talked some more about the OCAW and Silkwood's personal life. To Peter Stockton, two things were very clear by this time. Srouji was trying to discredit Silkwood by making constant references to sexual promiscuity, drinking, the smoking of marijuana, and mental instability. And she was trying to show him that the FBI had done a thorough investigation and that the blame for Silkwood's death lay squarely on the OCAW, which had used her.

Did the FBI feed her this line? Was the Bureau using her to spread that story? Stockton offered Srouji a deal. No subpoena for the rest of her documents, he told her, but a closed session in which she could talk about them in private and not be done out of her scoop by the press.

Srouji agreed.

"I had no idea what the fuck she was going to say," Stockton recalls. "I wasn't even sure she actually had those FBI documents."

□ □ □

The congressional hearings into the case of Karen Silkwood opened on April 26 with a statement by Representative John Dingell:

The Subcommittee is deeply distressed, indeed, appalled at the evidence which has been supplied to the Subcommittee regarding the manner by which the Justice Department and the FBI pursued the investigation.

We are also disturbed by the Justice Department's refusal to allow the staff of this Subcommittee and the Senate Government Operations Committee, with which we were originally jointly investigating this matter, complete access to their files. However, given what we have learned regarding the Department's handling of this matter, the motivation for their refusal seems evident.

Dr. Karl Z. Morgan was Dingell's first witness. For thirty years, Morgan had been director of health and safety at the Oak Ridge National Laboratory in Tennessee and was frequently called the "father" of health physics. He was currently professor of nuclear engineering at the Georgia Institute of Technology.

Dr. Morgan believed in the use of nuclear power. When he had agreed to testify, the subcommittee gave him all the AEC/NRC documents in its file relating to health and safety conditions at the Kerr-McGee Cimarron plant. He studied them carefully.

"I have never known of an operation in this industry that was so poorly operated from the standpoint of radiation protection as the Cimarron facility," Morgan testified. "It is difficult for me to comprehend and appreciate why the AEC, and more recently the NRC, permitted this facility to continue to operate for such a long time."

Morgan said he was most disturbed by the frequent contamination with airborne plutonium, the serious lack of security, a health physics program without one single qualified health physicist, and the poor criticality control.

Criticality occurs when too much plutonium is put in one place. The material then releases massive radiation, and persons in the immediate vicinity can be subjected to lethal doses. When Dr. Morgan testified, there had already been four criticality deaths in the nuclear industry.

Morgan concluded, "I consider this plant an example of how not to run the nuclear energy industry."

Sara Nelson, Kitty Tucker, and Tony Mazzocchi also testified, reviewing for the subcommittee their attempts to get some answers from the Justice Department. "The Silkwood case points up a new kind of Washington cover-up," Mazzocchi said. "In Watergate, we saw people actively attempting to cover up their tracks. In the Silkwood case, we see a pattern of nonresponse by the federal agencies in that by not looking, the scandal will never be uncovered."

Srouji took the afternoon off from the Pentagon to appear before the subcommittee. Dressed in a pink pants suit, she sat in the rear of the hearing room, awaiting her turn. Several times she stepped into the corridor for a brief conference with Chairman Dingell.

Michael Ward, majority counsel to the Dingell subcommittee, and Stockton had been working hard to get a quorum for a closed hearing. Srouji could then talk about her documents and Silkwood, protected from the press and the public. And the subcommittee wouldn't have to worry about her hyping her book.

"We don't want to hurt you," Dingell told her in the corridor. "That's why we called for a closed meeting." But Srouji had changed her mind. She wanted to testify right out in the open, she said. According to congressional rules, Dingell couldn't refuse to let her do so. It was either open testimony or no testimony.

Srouji read a prepared statement in a timid voice:

"My instant deduction of Karen Silkwood, the Kerr-McGee case, prior to any investigation on my own part, prior to going out in the field and talking with the individuals concerned, was that she appeared to be a sort of female Don Quixote, who had perhaps gotten mangled by jostling one windmill too many — namely, Kerr-McGee."

Srouji said that once she dug into the facts, however, she had changed her mind. Karen was tired the night she was killed. She had spent a draining weekend at Los Alamos, partied late the night before, had tough Kerr-McGee negotiations the next day, popped Quaaludes to settle her nerves, and was emotionally upset.

The more she studied the contamination and accident, Srouji said, the more she needed to know about Silkwood herself. "Numerous associates of the woman who were interviewed indicated — through again official and documented evidence," Srouji testified, "that she used marijuana and had the reputation of being a marijuana user, and had attempted suicide on at least two occasions, attributed to drug overdose. Now, I'm not going into this, I've got it here, the documentation on her marijuana use and suicide cases."

Dingell broke in: "We're not interested in the personal affairs of Karen Silkwood."

Srouji agreed. "I don't see any point in knocking someone who is dead," she said. "That's over with."

Kitty Tucker, sitting up front, close to the reporters' table, couldn't restrain herself any longer. "You just *did,*" she called to Srouji.

□ □ □

The "documentation" Srouji referred to was an unverified three-page summary, with attachments, prepared by Kerr-McGee investigators. The language of the K-M report was identical to the language Srouji used in her testimony.

The Kerr-McGee document said Sherri Ellis saw Silkwood smoke marijuana several times, that K-M found grass in her apartment during decontamination, and that the Oklahoma Highway Patrol found two marijuana cigarettes in her purse after her accident. The documentation also said that K-M found a handwritten note in Silkwood's apartment that said "$300 dope."

Drew Stephens later testified that Silkwood always used the word "dope" to mean marijuana, and that Karen had once bought a pound or two of grass to sell to their friends, hoping to make a few bucks. The scheme failed, he said. She never tried it again.

The Kerr-McGee report said the company also found a hypodermic kit in the apartment during decontamination. K-M assumed Silkwood had used it to shoot heroin. The report even had a picture of the kit and of the fingerprints on it. James Reading sent the kit, as well as the contents of Silkwood's purse, to the Oklahoma State Bureau of Investigation for analysis. The analysis turned up no traces of drugs on the needle, and the autopsy showed no needle tracks on Silkwood's body.

The Kerr-McGee report pointed out that records of the Gilliam Prescription Shop showed that Silkwood bought 180 Quaaludes during a ninety-two-day period. The assumption in the report was that Silkwood had swallowed all of them. It failed to point out that if she had, she might have built up a drug tolerance.

The Kerr-McGee report alleged that Silkwood attempted suicide twice — once before she began working for Kerr-McGee, and a second time in 1973, when she called Connie Edwards and said she had overdosed.

□ □ □

Srouji testified that she didn't think Karen Silkwood had deliberately contaminated herself. Who had done it to her? Srouji suggested the union had the best motive.

"Was the union . . . deliberately contaminating her in an effort to strengthen their own case, a case that was weakening by the hour?" she asked. "In effect was Karen Silkwood malleable clay for a programmer?"

Nor did Srouji believe Karen Silkwood committed suicide. She told the subcommittee that she was puzzled about the union's asking Karen to drive thirty miles to Oklahoma City when it knew she was tired from her Los Alamos trip, the Kerr-McGee–OCAW negotiations, and the AEC interviews. "This is especially unusual since the alleged documents which she was carrying with her were said to contain the facts needed to nail Kerr-McGee to the wall," Srouji testified. "The evidence was needed, and she supposedly had them. It was without a doubt the most important mission of this girl's life. And yet, the union suddenly decided that she should go to Oklahoma City, rather than picking her up or meeting her, even clandestinely, in Crescent."

Who, then, killed Karen Silkwood? "In the end," stated Srouji, "we will have to ask ourselves the question: Wasn't OCAW the only agency that has thus far benefited . . . from the death of what has become to them a tailormade martyr?"

Michael Ward led the cross-examination. He was a thirty-two-year-old former criminal defense attorney with wavy red hair and a thin red beard and mustache. He had talked briefly to Srouji before the hearings, and he intended to use her testimony to pry open the FBI files. If he could get her to say under oath what she had been telling the subcommittee all along — that she had FBI Silkwood documents — then the subcommittee could demand to see them as well. Ward was pleased that Srouji asked to testify in an open session, for the media stories about her testimony would put extra pressure on the Bureau.

Srouji had just testified that the Pinkerton Detective Agency had done some background investigations for Kerr-McGee on the Silkwood case. So Ward began to pick at the Pinkertons.

"What was Pinkerton investigating for Kerr-McGee?" asked Ward.

"Pipkin."

"Did they do this for Kerr-McGee?"

"Yes, but they didn't do it for the Justice Department . . . I have that document with me, too."

"Do you know why Kerr-McGee hired Pinkerton to investigate Pipkin?" Ward pressed.

"I think it's obvious really. Pinkerton is a very reputable firm. Perhaps they wanted an independent evaluation. They were doing a background check on somebody who had been presented to them as an expert. I think if his credentials had been verified as a nationwide expert, they [Kerr-McGee] would probably have taken him quite a bit more seriously than they did."

"You mentioned that Kerr-McGee itself had investigators other than Pinkerton. Were these company employees, or private?" Ward asked.

"They have their own, from what I could tell, pretty elaborate security section, as far as Kerr-McGee investigators go, that is separate and distinct from the guards."

"I'm kind of curious, since the Nuclear Regulatory Commission doesn't require security clearances, what did all these people do?"

"They were certainly involved in watching union activities, you know — that's probably a bad word to use," Srouji said. "I imagine in any plant, Du Pont or General Electric, you have a security force."

"Are these investigators working in the plant, do you know, as employees?"

"As far as I could tell, they had no undercover agents working in the plant, trapping the workers. The ones that I saw were strictly suit-and-tie-type people."

159

Ward changed the topic to Srouji's FBI documents. "Did you talk to the FBI?" he asked.

"Yes, I did."

"Were they cooperative?"

"In an unofficial way, yes."

Ward pressed. "Well, how cooperative were they?"

"Let me see, how can I say this without violating my confidence?" she said. "I was able to see unofficially the substance of this case as they investigated it. It was a ton of material that indicated to me quite an in-depth investigation."

"Like 1000 pages?" Ward asked. "One hundred pages?"

"I would say closer to 1000."

"You said if she [Silkwood] were a good investigator, she would have made Xerox copies of all the information," Ward said, commenting on Srouji's prepared statement. "Did you make Xerox copies of all [your] information?"

"Certainly, several. I don't want to end up in a ditch on a dark road outside of Nashville, Tennessee."

"Did you tape record the discussion you had with the FBI?" Ward asked.

"Oh, no," she said, tongue in cheek. "I would never have done that. They were probably taping me. We would have a Sony against a Wollensack."

Srouji then admitted under oath that she had got her material from a "chief investigator" of the FBI whom she had known for ten years. She refused to give his name, but she had said enough to point a finger at Larry Olson, Sr.

Srouji was the last witness that day, and reporters swallowed her up after her testimony. Unable to control her anger, Sara Nelson broke through the ring of reporters. "How could you do that?" she shouted at Srouji. "Portraying Karen like that. Don't you know about the conditions at Kerr-McGee? Don't you know what was happening there?"

"You don't know what you're talking about," Srouji snapped back. "Get out of my way."

A friend whisked Nelson out of the hearing room. "You shouldn't be talking to her," the friend said. "You have no idea what you're dealing with."

Kirk Loggins covered the hearing for the *Tennesseean,* and he invited Srouji out to dinner that night at Harvey's. She had seemed rattled by Ward's cross-examination, and she asked Loggins several times how she had come across during the hearings. She said she was worried about what the Bureau might think about her. Loggins told her she probably had blown whatever FBI sources she ever had.

He asked Srouji for Olson's phone number so that he could try to get an on-the-record comment. Srouji told him to be sure to let Larry know she had testified that the "FBI did a good job" investigating the Silkwood case.

Dingell didn't think it had. Now that Srouji had testified under oath that she had FBI documents, he demanded equal time from the Bureau. Dingell had the FBI trapped. After six months of negotiating, they agreed to a compromise: the subcommittee could not have the Silkwood documents, but it could interview the agent who wrote them.

Chapter 17

Michael Ward and Peter Stockton interviewed Olson in a windowless room in the J. Edgar Hoover Building. An FBI agent guarded the door; another, with huge biceps he enjoyed flexing under his short-sleeved shirt, stayed in the conference room. Olson placed his service revolver on the table in front of him.

Olson was a six-foot, three-inch, all-American G-man — smart, clean-cut, and very dedicated to J. Edgar Hoover. At one point during the forty hours of prehearing interviews with Ward and Stockton, Ward asked him if he would work over the weekend. "Well, I don't want to do that," Olson said. "But if Mr. Hoover says so, I will." Hoover had been dead for two years.

The ground rules for the Olson interviews were as tight as the security. Not only did the FBI refuse Ward and Stockton permission to copy the Bureau's Silkwood documents; the congressional investigators couldn't even see them. They had to sit and listen to Olson read line by line. They could quiz him, but not about Srouji.

It was clear to Ward and Stockton that they would never find out what Olson was *really* thinking unless they could get him alone. They tried to distract "Muscles," as Stockton called him, at the end of the table, but he just sat there quietly and watched. Every time Olson went to the bathroom, Ward or Stockton would tag along. But Muscles or another agent would shadow Olson right to the urinal.

"We'd stand there and try to divert the other guy's attention," Stockton recalls. "We'd try to whisper to Olson." Nothing worked.

Ward and Stockton found out that Olson liked basketball, so they asked him if he wanted to see the Washington Bullets some night. Muscles said no, Olson would be busy; there would be meetings to attend. Even Olson felt the strain. "Now I know what it would be like to be captured by the Gestapo," he said later.

Eventually, Stockton whispered a message. "God, if you have something to tell me, Larry, get ahold of me."

Olson called Stockton at night, several times. The conversations were strained and strange. "He'd try to get across that I'd *never* get to the bottom of this," Stockton remembers. "That I don't understand how to deal with the Bureau."

Whatever the case, Dingell wanted the FBI game played right out in the open. No deals. No trades. Straight talk or nothing.

The forty hours of listening to Olson read and a half-day of subcommittee testimony in closed session were not a waste of time. Olson clarified some issues and confirmed even more suspicions.

He admitted that Kerr-McGee had given him documents, but denied that he ever saw "any actual interviews" K-M held with its own employees. What they gave him, he said, was "lead information."

He said he didn't impound Silkwood's car as evidence, because he couldn't account for its exact whereabouts between the time of the accident and its appearance at Drew Stephens' hideout.

What did he conclude about the dents in Silkwood's car, the subcommittee wanted to know.

"The Oklahoma Highway Patrol," Olson said, "subjected the entire accident investigation to a review process, even going back to the previous October 31, 1974, accident . . . It indicated that Pipkin did not have any tangible evidence supporting his theory of foul play . . .

"Our investigation would concur with the Oklahoma Highway Patrol's investigation, noting that their investigation was timely . . . contemporaneous with the accident. Any findings that they observed at that time would certainly be more valid than Mr. Pipkin's conclusions."

Olson had taken paint scrapings from Silkwood's car. What did the FBI find, the subcommittee's attorneys wanted to know.

Honda paint and primer, Olson said. Nothing more.

Did the FBI lab find any cement particles in the dents?

No, they didn't, Olson said.

Well, didn't that prove the dents were not made at the accident site when the Honda was hoisted out of the ditch?

Not exactly, Olson replied. The dents could have been made at the scene without getting concrete in them. Besides, rain might have washed the cement out later. Or maybe Drew Stephens made the dents. A possibility, though there's no evidence he did.

Did Olson interview Pipkin?

No, he didn't. The Oklahoma Highway Patrol had demolished Pipkin's theory. An interview would have been a waste of time.

The Justice Department's Death Fact Memorandum said the wind was strong the night of November 13. How did the FBI know that?

The Dallas Bureau got a call from a lady who said the wind around Crescent was gusting up to seventy miles an hour. Dallas forwarded the report to the Oklahoma City Bureau.

Did he check out the wind factor?

No, he didn't, Olson said. No one else ever suggested Silkwood might have blown off the road, so he didn't think it important to check weather reports. Did the subcommittee want him to?

Martin's wrecker was called to the accident scene, the subcommittee counsel said. Then it was told to return, and Sebring was dispatched. Wasn't that strange?

No, Olson said. Two people frequently report accidents in two different police jurisdictions. Two wreckers are sent, he explained. Besides, Martin was fifteen or sixteen miles away from Silkwood's crash. Sebring had to drive only 7.3 miles.

Did he ever contact FBI or independent toxicologists to determine whether .35 mg of methaqualone was enough to put Silkwood to sleep?

No. Dr. Chapman's finding looked good to him, and no one had ever told him the medical examiner's opinion was controversial.

What about the .02 percent trace of alcohol found in Silkwood's bloodstream?

Olson said that at first he thought union members were pulling his leg when they said there was no drinking at the Hub Cafe union meetings. Later, he found out they were right. The trace of alcohol must have been a residue from the night before, he said.

What about Silkwood's documents?

There were none.

How did he know?

The AEC and a blue ribbon committee checked the Kerr-McGee fuel pins, Olson answered. The rods were acceptable.

Did he find any materials unaccounted for at the Cimarron plant?

No significant MUF to his knowledge, Olson said. And no one had ever told him there was an MUF problem at Kerr-McGee. He read about it in the papers.

Was he aware of Pinkerton's shadowing of union members? Of phone taps or bugs?

No.

Was he ever pressed to wind down the investigation?

Never. Not by the Bureau or Kerr-McGee or the OCAW.

Was he ever interviewed by Thomas Goldstein, the Justice Department lawyer who wrote the Death Fact Memorandum?

No, he couldn't recall talking to Goldstein.

What did Olson conclude about Silkwood's death?

He had ruled out suicide, Olson said. It would be unreasonable for Silkwood to kill herself to save a union. And he had ruled out murder by the OCAW. The Oklahoma Highway Patrol theory was the simplest and most logical explanation.

Could he provide the subcommittee with a list of names of the persons the Bureau had interviewed and the dates?

If that didn't violate the Privacy Act or any other federal regulations or laws, Olson said.

□ □ □

The FBI never furnished the list, but the hours of interviews with Olson had demonstrated to Ward and Stockton just how the agent had outlined his investigation. Like a good investigator, Olson operated from a set of theories, and he collected evidence to prove or disprove them. Theory one: Silkwood contaminated herself. Theory two: the union contaminated her. Theory three: Silkwood had no documents. Theory four: she committed suicide. Theory five: the union killed her. Theory six: she fell asleep at the wheel.

It was obvious to Ward and Stockton that at no time did Olson seriously ask: Did someone contaminate her to frighten her? Did she have documents? Did someone try to get those documents before she gave them to *The New York Times?* Were Pipkin's dents real?

The hours with Olson were useful for another reason. Ward and Stockton's digging had turned up a nugget. All during the tug of war with Olson, Srouji was very much on their minds. She was their best lead; she seemed to confirm that the Bureau was misleading the subcommittee and tinkering with the legislative process. Somehow they had to badger Olson into a blunder.

Olson would read a paragraph from a report. "Did you show that to Jacque?" they'd ask.

He'd read another paragraph. "Did Srouji copy that when she was in Oklahoma?" they'd ask.

They'd interrupt a sentence. "Is that one of the documents in Jacque's files?"

Most of the time Olson would let the question slip by unnoticed, or he'd say, "You know I can't talk about that." But late one afternoon, he slammed the table, jumped up, and said, "That woman has a special relationship with the Bureau and I can't answer that question!"

Ward and Stockton smiled. They finally had the Bureau over the barrel. It was time to take the offensive.

Dingell demanded an explanation, and the Bureau sent James Adams, the deputy associate director of the FBI, to soothe the chairman. Adams was an old Hooverite, and he tried to play on Dingell's sense of loyalty.

"The Srouji matter is delicate," he said. "Leave it alone. Trust us. We know what we are doing."

Dingell wanted answers. "Who is this Jacque Srouji?" he demanded. "Why does she have your documents? Did you send her to spy on the subcommittee?"

"No, sir," Adams said. "She was acting on her own."

Adams reluctantly began to sketch the FBI–Srouji connection. He told Dingell that Srouji had been an FBI source — he wouldn't be more specific — from 1964 to 1968. The FBI had a "special relationship" with the Nashville *Banner,* where Srouji had worked as a cub reporter.

According to Adams, the *Banner* owner had agreed to pay Srouji's salary, and the Bureau would reimburse the paper for out-of-town expenses. Srouji infiltrated mostly "Communist" groups. Adams said he couldn't give the names of the groups because some of them were violent and they could harm Mrs. Srouji. The Bureau had had no relationship with Srouji in the 1970s, Adams said.

The Bureau's confession was startling. Ward and Stockton had suspected that Srouji's relationship with Olson was more than the usual FBI–reporter kind. And they knew Srouji had worked for the conservative *Banner.* But they had never dreamed the Bureau had a formal agreement with Srouji and the *Banner.* That would make her the first known undercover FBI informant planted in the media. It was a dirty piece of linen. No wonder the Bureau was nervous.

Adams leaned back in his chair and put on his national security face. "Mr. Chairman," he said in hushed tones, "we're very concerned about Srouji. We have her under investigation now as a potential double agent."

Stockton almost choked on his next question as Adams explained how Srouji had been photographed going in and out of the Soviet Embassy.

"I thought you didn't have a relationship with Srouji anymore," Stockton said. Srouji had mentioned to Stockton in passing that she had visited the Russians while doing research for *Critical Mass.* But Srouji had said so many things, Stockton hadn't been sure what to believe.

"Well . . ." Adams sputtered.

"She must have had some information to trade with the Russians," Stockton said. "What?"

167

Adams sputtered some more.

It was clear to Ward and Stockton that they would never squeeze more out of the Bureau. Maybe they could find some answers back in Nashville.

Chapter 18

John Seigenthaler, the publisher of the Nashville *Tennesseean,* picked up the phone. It was Mike Ward, majority counsel to the Dingell subcommittee. How well did Seigenthaler know Srouji? Was she credible? It was May 3, and Ward had just learned about the FBI–Srouji connection.

Seigenthaler said he hardly knew Srouji personally, but she was a good copy editor.

Did Seigenthaler know she had had, and still had, a "special relationship" with the FBI?

No, Seigenthaler didn't.

Well, would Seigenthaler talk to her? He certainly would.

With the exception of a year and a half in Washington, Seigenthaler had been with the *Tennesseean* since 1949, working his way from correspondent to editor to publisher. He wasn't sure what to think about the FBI.

His father had been a cop, and he himself had worked for the Justice Department in 1961 and 1962 as an administrative assistant to Attorney General Robert Kennedy. The Washington tour turned out to be no Camelot for Seigenthaler. As a Justice Department representative on the Freedom Rides through the deep South, he had been beaten and stomped into unconsciousness by a mob attacking the bus on which he rode. The FBI stood by and took notes.

Seigenthaler didn't seem to be angry with the FBI. He explained to his son after he got out of the hospital that the Bureau just couldn't jump in and make arrests. They had no jurisdiction.

"Well, why didn't they come over and pick you up?" his son, John Michael, had asked.

"I don't know." Seigenthaler smiled. "I was unconscious."

He wasn't smiling when he interviewed Srouji over the next three days. He was incredulous and angry and emotionally drained from what he learned.

□ □ □

When Jacque von Stubbel graduated from high school, she joined the Counter-Intelligence Corps of the WACs. Several weeks later, she was honorably discharged. No one really knows why. She returned to Nashville, took a job as a South Central Bell Telephone operator by day, and studied at the Nashville extension of the University of Tennessee by night. One of her courses was newswriting and editing, taught by an editor from the Nashville *Banner,* a conservative afternoon daily with editorial offices in the same building as the liberal *Tennesseean.* The two papers had been jointly printed since 1937, but were editorially separate.

In 1963, Jacque von Stubbel applied for a job at the *Banner* as a secretary. Soon she became a general-assignment reporter and developed her own beat — the student, civil rights, and antiwar movements. She liked to ride the police patrol at night on her own time. Gradually, she developed excellent sources in such groups as Students for a Democratic Society (SDS), the University of Tennessee Student Organizing Committee, and the United States–China Friendship Association.

Jacque's stories caught the attention of the *Banner*'s executive editor, the late Charlie Moss, who liked her ambition and guts. "Tiger," he called her, and he encouraged her to attend all the radical meetings she could. She should report back to him, he told her.

Jacque bought a 35 mm camera, slipped into meetings, sometimes under the alias of Leila Hassam, took notes, and shot pictures of as many faces as she could. One morning she handed Moss her notes and three rolls of film. On her day off, she explained, she had joined the Freedom Riders in Mississippi. She thought he might like the notes and pictures.

That afternoon, *Banner* owner James G. Stahlman called the cub reporter into his office. He was as conservative as the deep South and liked to brag that J. Edgar Hoover was his personal friend. To Jacque, Stahlman was a Captain Nemo — he gave orders, ran a tight ship, but she never saw him.

Stahlman had Jacque's notes and pictures in front of him. Next to him stood an FBI agent from the Nashville office. The agent complimented Jacque on her work. It was valuable, he said. It was a good job. Would she continue to take notes for the Bureau, he asked.

She was nineteen years old, eager, impressionable, and flattered. Of course she would, especially since Mr. Stahlman thought it was a good idea.

The FBI agent warned Jacque not to tell anyone else about their meeting. He stressed she would be working for the *Banner,* not for the Bureau. The *Banner* would pay her, he explained. The arrangement was an experiment; the FBI had never worked with a reporter like this.

So Srouji continued infiltrating the movements and reporting back to the Bureau — who the movement leaders were, how the organizations were run, and what role was played by the university professors the FBI had dossiers on. She reported to one of three Nashville agents — Nick Norwood, Hank Hillen, or Larry Olson, Sr.

Sometimes Srouji would file more extensive reports with the Bureau than she would with the *Banner,* and the paper would print more pictures for the FBI than it would for itself. Once she made a trip to northern Michigan to spy on an SDS conference. The FBI paid her way. She didn't write a word for the *Banner,* but she filed a fifty-page report for the Bureau.

Jacque met Suheil Srouji in 1967. He was a Lebanese immigrant and a Nashville civil engineer. They were married in the Holy Land by the archbishop of Jerusalem. The following year, 1968, Jacque quit the *Banner* to have her first child.

Suheil was a radical Arab and a Catholic charismatic, and

Jacque learned to share his interests. She attended PLO meetings in Nashville with him. Some of the Arab leaders at the meetings were suspected terrorists.

The Bureau wouldn't let Srouji alone. It threatened to tell her husband about her "special relationship" if she didn't spy. Whether she worked for the Bureau from 1968 to 1970 is not clear.

In 1970, Srouji joined the Nashville *Tennesseean* as a part-time copy editor. She left after a year and a half to have another child. For the next four years, she free-lanced for the *Nashville* magazine, later called *Nashville!* In 1974, the Nashville Business and Professional Women's Club voted her Woman of the Year.

Early in 1975, Dominic de Lorenzo, head of Aurora Publishers, called Srouji. He had read her *Nashville!* series on nuclear power, he said, and asked her whether she would like to expand it into a book.

One of the chapters Srouji researched for *Critical Mass* was on Karen Silkwood. In April 1975, she wrote to de Lorenzo about a new Silkwood lead. "There is a possibility that Karen Silkwood may have been onto a nuclear black market and deliberately contaminated . . . a good possibility. Question: Should I spend more time on this and follow it up or do you think that we have enough on her and Kerr-McGee?"

"I don't think it is worth the expense of going back," de Lorenzo answered. "She was obviously being used by everyone, including OCAW. Concerning the nuclear black market — stay away from that because it would really cause a blow-up."

In the course of her research, Srouji visited the Oak Ridge nuclear facility. She learned that scientists there were using a piece of Russian equipment, and decided she ought to talk to the Russians. So on the way back from a trip to Los Alamos — with a stop in Oklahoma to visit Olson — she flew to Washington.

Srouji knocked on the Soviet Embassy door. She spoke some Russian, she said, and wanted to talk to a scientist about nuclear energy. Colonel Sergei Zaitsev appeared. With his wild white hair, he reminded her of Ben-Gurion. He was a kind, fatherly, charming KGB agent.

He crossed the room and sat down in a leather chair. "Now,

172

Jacque," he said, correctly pronouncing her name. "What is your problem and how can we help you?"

Srouji offered her press credentials. Zaitsev shrugged them off.

"I believe you," he said. "There are no need for these."

Srouji and Zaitsev chatted about nuclear power, fusion, his family, her family. Zaitsev was interested in her book. Maybe he'd have it translated into Russian, he said.

Srouji also visited the Liaison Office of the People's Republic of China. The FBI was more than curious. They called her in and showed her movies starring Jacque Srouji entering and leaving the Soviet Embassy. What was she doing there? She explained about her book. The FBI asked if she would be willing to continue seeing Zaitsev and report back to the Bureau. Why not, she said.

Five months later, after Srouji had finished a rough draft of her book, she was back sipping vodka with Zaitsev. He casually quizzed her about the Arnold Engineering Center in Tullahoma, Tennessee, sixty miles from Nashville. He showed her pictures. He needed someone to get information from Tullahoma. Wouldn't she help?

The Russians had been sending two "illegals" to pick up information from a spy they had planted in the Arnold Center. The spy had been leaving the goodies in a dead-drop near Pulaski. It would be easier for Srouji to bring the information to the embassy, Zaitsev told her, as she apparently had FBI and CIA clearance to enter. Zaitsev gave her $400 for expenses. She reported back to the FBI and gave the Bureau the money. The Bureau told her to report to Special Agent Larry Thomas in Columbia, Tennessee; he was experienced in counterespionage.

Srouji attended a nuclear energy conference in Washington for Zaitsev, toured the Arnold Center in Tullahoma, was hostess to visiting Russians in Nashville, and guided a Russian "journalist," who had accompanied a Siberian scientific exhibit, to an experimental farm near Summertown, Tennessee.

Srouji rejoined the *Tennesseean* staff in 1975 as a copy editor. The Bureau approached her again. Informally, this time. Could she fill them in on two *Tennesseean* reporters, Dolph Honicker and Jerry Hornsby? Honicker had written columns highly critical of nuclear power; Hornsby had once been a member of the Socialist Party.

Early in 1976, the FBI also tried to use Srouji to bag Special Agent Joseph Trimbach, the boss of the Bureau's Memphis office. Trimbach, a former Marine, had led the FBI charge on the American Indian Movement leaders who were holding eleven hostages on the Oglala Sioux Reservation at Wounded Knee, South Dakota, in 1973. Two FBI agents were killed, and the Nashville agents blamed Trimbach.

Srouji told Seigenthaler she had a tip that certain South Dakota court records would embarrass Trimbach.

Seigenthaler knew by that time that Srouji had good FBI contacts. One night in December 1975, for example, word filtered back to the *Tennesseean* newsroom that a big gambling raid was going on. Srouji was at the copy desk. Wayne Witt, who coordinated the coverage, sent out several reporters. They came back empty.

Could she help, Srouji had asked Witt. She had contacts in the Bureau, she told him. In a few hours, she had the names of places raided and people arrested.

That and other FBI bones Srouji fetched for the *Tennesseean* had impressed Seigenthaler. He told her she could go to South Dakota. She came back with the story, but Seigenthaler suspected the FBI Nashville office was trying to use the paper, so he never ran it.

□ □ □

When Srouji finished telling Seigenthaler about her FBI escapades, she pulled out two FBI letterhead memoranda to prove her "special relationship." Seigenthaler knew an LHM when he saw one, and these two were hot. So hot, he has refused to show them to anyone or to say what was in them. All he will say is that they were not about Project Seafarer or Zaitsev, and that if they were made public, contracts would go out on three people.

Seigenthaler told Srouji she needed a lawyer and then fired her. She left his office in tears and later that night wrote him an emotional letter. "If my stream of life becomes infested with pir-

174

anhas and a sacrifice is in order," she said, "then I can certainly take some of Mr. Hoover's finest along for the swim."

□ □ □

John Seigenthaler volunteered to testify before the Dingell subcommittee on May 20. Shortly before he appeared, the FBI's congressional liaison, Paul V. Daly, stopped in to see his old friend Representative William Cohen of Maine, a Republican. Cohen was a member both of Dingell's Small Business Subcommittee on Energy and Environment and of the prestigious House Judiciary Committee. Daly asked Cohen to stop the Silkwood investigation, suggesting that the congressman try to get the other three minority members of the Dingell subcommittee to protest further hearings and to recommend that the Silkwood investigation be turned over to the Judiciary Committee, where it would be sure to die on the vine.

The night before Seigenthaler was to testify, the minority members met; they needed little convincing. Besides Cohen, two other Republicans were from the Northeast: Silvio Conte of Massachusetts and Hamilton Fish of New York. The fourth minority member, Caldwell Butler, was from Virginia. None of them had supported the Silkwood probe, which was draining the subcommittee's limited resources away from hearings on such "safe" East Coast issues as the oil shortage.

The Republicans were also upset with Peter Stockton, whom they considered an antinuke using the subcommittee to bring down the nuclear industry; the last thing they wanted was to get embroiled in a no-win nuclear power controversy. Ranking Republican Conte was particularly sensitive, for there was a proposal for a nuclear plant in his congressional district. He had asked the subcommittee assistant minority counsel, Paul Kritzer, to help him "stay low on the nuclear stuff."

Furthermore, the minority were convinced that the Silkwood hearings had already gone beyond the subcommittee's original concern about nuclear health, safety, and security and were mired in Silkwood's death, contamination, and sex life. Jacque Srouji had been the last straw. The minority had not been told she was

175

going to testify; they had no idea why she was allowed to testify; and they were not sure where she was leading the subcommittee. And now John Seigenthaler was going to testify the next day about the FBI–Srouji connection.

The minority agreed to stand up to Dingell, even though they were outnumbered two to one. The next day, after Dingell had introduced John Seigenthaler, Conte said:

> I have a statement, Mr. Chairman. I am concerned about the direction and focus of these hearings. I've spoken to you several times about this. This is a Committee on Small Business. We are charged by the House of Representatives to focus on issues of importance to small business . . .
>
> It is hard for me to justify how this committee has jurisdiction over safety standards of a large plutonium plant or the adequacy of a criminal investigation by the Federal Bureau of Investigation.
>
> But we are now investigating how Karen Silkwood died, how she was contaminated, whether an FBI agent leaked official documents, whether a former witness before this committee testified at the direction or under the control of the FBI, and whether the FBI spies on antinuclear activists.
>
> It's my understanding of the House rules that questions of murder, spying, and other criminal conduct do not belong in the jurisdiction of this committee . . .
>
> We've become bogged down in an issue that is no longer focused on the interests of small business, is outside the jurisdiction of this committee, is beyond the competence of the staff, and will probably never be resolved because of the masses of circumstantial and contradictory information.
>
> I hope this segment of this so-called small business hearing can be wound up today so we can get back to the real business of this panel.

Representative Cohen jumped in. "Will the gentleman yield?"
"Yes," said Conte.
Cohen went on:

> I thank the gentleman for yielding. And I want to say that I support his remarks, but I also have serious concern about the wayward force of these hearings. The subcommittee, I think, is going far afield from the intended purpose of its investigation . . .

While I understand the popular interest in this case, I agree with Mr. Conte that it's well beyond the interest and competence of this subcommittee. If any congressional panel is going to get involved with allegations of murder, domestic spying, or smear tactics, it should be the Judiciary Committee.

I am appalled as anybody else in this room at the suggestion that the FBI maintains any sort of files or compilation of names for those people who oppose the nuclear fuel development . . .

There's one final concern I have about these hearings, Mr. Chairman. The subcommittee has been advised that the FBI is now conducting an internal investigation involving matters brought up before the subcommittee that might very well lead to criminal indictment [how Srouji got her FBI documents]. So I am concerned that further subcommittee pursuit of this matter might endanger or prejudice constitutional and civil rights of people who might stand accused of criminal conduct.

For those reasons, Mr. Chairman, I would support the gentleman from Massachusetts in his urging the subcommittee to conclude these hearings as soon as possible.

Dingell would not budge. He said he would return to the basic Silkwood issue as soon as he could. "But when questions relate to the very core and the very fundamental nature of the investigation, then it becomes necessary that the subcommittee pursue those questions . . ." he said. "The Chair thanks the gentleman for his comments. Mr. Seigenthaler, would you rise, please?"

Seigenthaler outlined with an appropriate amount of detail all he felt he could reveal about the FBI–Srouji connection. He said he wasn't sure how much of what Srouji had told him was true, but he felt the *Tennesseean* had been snookered. "I resent the Bureau doing this," he testified.

Seigenthaler also filed a complaint with the Justice Department against what he called "improper conduct on the part of agents of the Bureau."

John Seigenthaler would pay for embarrassing the FBI. So would Peter Stockton. So would John Dingell.

Chapter 19

The FBI smeared Peter Stockton. He believes they did it deliberately.

Months after the 1976 congressional hearings on Silkwood were over, Stockton found a copy of an FBI report about his 1975 trip to Oklahoma with Barbara Newman. The document was tucked in among hundreds of pages of Silkwood-related reports subpoenaed from the Bureau. In all the documents it released, the FBI had been careful to protect the innocent and its own sources, innocent or not. It had deleted all the names in the Stockton report — except Peter Stockton's.

The report dealt with alleged sex hanky-panky. It said, in part:

> [Name deleted] had several brief conversations with Pete Stockton and [name deleted], both of whom said they were affiliated with National Public Radio. Again [name deleted] declined to talk with the reporters, but he did learn from certain of his sources that they didn't accomplish too much because Stockton spent most of his time visiting local prostitutes and not working . . .
>
> The last time Stockton called was in the Spring of this year. He told [name deleted] that he was on "journalistic leave" with some House Committee but [name deleted] still refused to talk to him.

It was a typical FBI raw report. The information from the source — Kerr-McGee's James Reading — had never been independently verified. And the FBI agent who interviewed Reading hadn't bothered to find out who Reading's "sources" were. The agent would have, if he had been directed to do so. Thus, the Bureau's report on Stockton's alleged tumbles with prostitutes was rumor at best and character assassination at worst.

To Peter Stockton, it was no innocent clerical error. He is suing the Bureau.

The FBI smeared John Seigenthaler with even less finesse. While the publisher was interviewing Jacque Srouji, on May 3 through 5, 1976, the FBI offices in Memphis and Nashville were sending telexes to Director Clarence Kelley. ALLEGATIONS OF SEIGENTHALER HAVING ILLICIT RELATIONS WITH YOUNG GIRLS, one message said.

On May 8, the *Tennesseean* published a story that said Srouji had been fired. Two days later, the FBI Nashville office sent another telex to Kelley. Seigenthaler eventually got a copy. The FBI had deleted everything on the page but the date.

On May 13, Seigenthaler and his lawyer, William Willis, visited the Justice Department to file a formal complaint about the way the FBI had used the *Tennesseean*. Then they visited Dingell and told him what they knew about Srouji.

Would Seigenthaler testify, Dingell asked.

Yes, but he would not show the subcommittee any of the documents Srouji had given the *Tennesseean,* Seigenthaler answered. "She got those documents from a source she wants to keep confidential, and we will protect that."

Seigenthaler had to give a speech the next day to a group of sociologists in North Carolina. He drove Willis to Washington's National Airport and returned to his room at the Jefferson Hotel, where he found a message to call Bill Kovach, news editor in the *New York Times* Washington bureau and a former *Tennesseean* reporter.

Seigenthaler reached Kovach at home; the *Times* editor was concerned. About ten o'clock that morning, Homer Boynton and T. J. Harrington, two FBI agents, had dropped into the *Times* office to discuss with Kovach and one of his colleagues, John Crewdson, the paper's "poor play" of a Kelley speech in Missouri. They'd drifted onto the Srouji mess. "Seigenthaler and the *Tennesseean* are not entirely pure," Boynton had told Crewdson.

When the agents left, Kovach asked Crewdson to call Boynton and find out what he had meant. Crewdson did. "John, you're a good reporter," Boynton said. "Go down to Nashville and make an investigation of Seigenthaler."

Seigenthaler was stunned. "Investigate me for what?" he asked Kovach, who had once worked for Seigenthaler at the *Tennesseean*.

"I think they're sending you a message," Kovach said, adding that Crewdson was going to Nashville to investigate. "If we find anything, we'll put it in the paper."

Seigenthaler urged Kovach to report in the *Times* what Boynton had said and then to investigate him. "Write it. Don't let them get away with it." Kovach said Crewdson, who regularly covered the Justice Department and the FBI, preferred to investigate first, then write.

Seigenthaler called home. John Michael answered. "Something has happened here and I need your help," Seigenthaler told his son. "I am convinced our phones are tapped. This phone in the hotel may be tapped. Go to a pay phone and call me here at the hotel. I'll go into the lobby and phone you back on a pay phone."

John Michael drove three miles to a pay phone. "I want you to talk to Bill Willis the minute he gets home from Washington," Seigenthaler said when he called back. "You contact him, but not on our home phone. Have him go to a neighbor's house and call me here. I'll come back into the lobby and call him."

The next day Seigenthaler called an old friend in the Justice Department who had connections in the Hoover Building. Could the friend find out what was behind Boynton's "not entirely pure" comment, Seigenthaler asked.

"I haven't seen anything in your file," the friend called back later. "But I understand that there are two exchanges of telexes recently between Washington and Tennessee that mention your name. It has something to do with white slavery."

Seigenthaler dashed off a letter to Clarence Kelley, demanding the two telexes. In the meantime, John Crewdson was investigating Seigenthaler. Crewdson spent a week in Nashville, but came

180

up empty. He was reluctant to mention Boynton's name in his story, but Kovach ordered him to. The reporter gave in. "I've just ruined a good man's career," Crewdson told Seigenthaler. "They'll fire Boynton."

Seigenthaler tried to reach Boynton for an explanation. "Not available . . . out of town," the agent's secretary said. Worried that the FBI might leak the "white slavery" smear to the press, Seigenthaler called a family council and told them about the telexes.

"I don't want you children reading something like that in the *Banner,*" he said, "or in my own paper, if we should have it leaked to us, and not have you aware in advance. It's stupid, but I wanted you all to know what might come."

Was John absolutely sure he wouldn't get into trouble with the FBI? Seigenthaler's mother wanted to know.

"Except for the fact I'm in a fight with the most powerful police agency in the world, can't sleep, can't eat, I'm fine," he told her.

Eighteen months later, Seigenthaler finally got copies of the FBI telexes. The first communiqué was two pages long. Page one contained the following words: "coded . . . nitel . . . Director . . . Memphis . . ." The rest of the page had been deleted by Kelley or his censors. Everything on the second page was deleted except "allegations of Seigenthaler having illicit relations with young girls, which information source obtained from an unnamed source."

The second telex was three pages long. Page one said: "Director . . . Memphis . . . immediate." Everything else had been censored. The second page said: "volunteered . . . heard rumors that . . . Seigenthaler . . ." The rest was blank. The third page had a little more: ". . . involved in having illicit relations with young females . . . Inasmuch as the information furnished by the source was unfounded rumor since it could not be corroborated [deleted] no record of this information was made in Memphis files."

The Seigenthaler story had a happy ending. He wrote an article himself for the *Tennesseean,* candidly quoting the "young girls" allegation. Other papers around the country picked up the first-person account. He won the Sidney Hillman Award for courage in publishing; the Justice Department apologized and purged his files.

□ □ □

The FBI smear of John Dingell was even clumsier. The congressman wrote a searing letter to Attorney General Levi on June 16, 1976, after the hearing. It was a good summary of Justice Department stonewalling and is worth quoting at length:

> To date, you have defied this Subcommittee's April 5th request that a specifically named employee [Olson] appear before this Subcommittee. You have refused to comply with this Subcommittee's April 28th, May 5th, May 13th, and May 25th requests for information. You have ignored our May 25th request that you submit certain reports to this Subcommittee by specified dates. You have initiated investigations of two individuals [Olson and Srouji] who provided information to this Subcommittee, and a high-ranking FBI official has allegedly made derogatory remarks to a member of the press about a third individual [Seigenthaler] who appeared as a witness before this Subcommittee in the course of this inquiry. Furthermore, you have yet to respond to my April 28th inquiry wherein I ask that you provide a statutory basis for your continuing refusal to honor this Subcommittee's requests. Now, your Office of Legislative Affairs is not returning the Subcommittee Counsel's phone calls. I believe this pattern of behavior clearly indicates your feelings towards this Subcommittee, and I hope that your apparent contempt for us is not reflective of your attitude towards Congress as a whole.
>
> I must say that I find your reluctance to supply this Subcommittee with requested information extremely curious in light of the willingness of some of your employees to discuss these same issues with members of the press. For instance, Mr. Boynton personally called a nationally syndicated columnist to explain the meaning of his remarks about Mr. Seigenthaler. Additionally, Mr. Michael Shaheen, who is the Director of your Office of Professional Responsibility and who is conducting your Department's inquiry into this matter, has declined to discuss any aspect of his investigation with the Subcommittee Counsel, yet he has found it appropriate to talk at length about his inquiry with various reporters. On the basis of

such information I can only conclude that it is your policy to with-hold information from this Subcommittee alone.

I must say that I am disturbed by this pattern of behavior and I cannot adequately express my personal indignation at the manner by which you have chosen to proceed. I again call upon you to voluntarily comply with our previous requests.

Four days later, on June 20, a thirty-three-year-old Detroit-area call girl told reporters Dingell was a customer. Her name was Lois Herman. On the street, she was known as Teri Cole, and she wasn't cheap — $100 a night.

Cole had been arrested on April 25, 1976, after she allegedly had approached an undercover Southfield police officer. She was charged with keeping and maintaining a house of prostitution, gross indecency, and accepting the earnings of a prostitute. All that added up to a possible thirty years.

The police released her on bail, but before long the Royal Oak police arrested her for accosting and soliciting in the Holiday Motel. If the police wanted to throw the book at her, she would be in deep trouble.

Somebody leaked Dingell's name and Cole's phone number to the press. Two Detroit *News* reporters picked up the bait. They grilled Teri Cole for two hours. She told them she had been working with another call girl, Lynn Stevens, who wanted to retire. Cole bought her out. Cole said she kept the 200 to 300 names of her own customers in a shoe box and that Dingell — "tall, balding . . . nice, sweet . . . marital difficulties" — had been a customer. She and Dingell had romped in the Northland Inn in Southfield, she said.

Cole explained that Dingell's name was not in her shoe box, but on a more complete list she kept. The police had the shoe box and a list, she told the *News* reporters. But not the *complete* list. Only she had that, she said. And Dingell's name was only on the "complete list."

Cole also said she joined a "Mafia, Inc., party" while working with Lynn Stevens. Among the organized crime figures she had entertained was "Sam-the-Mustache" Norber, who dated back to the old days of the Detroit Purple Gang.

183

The *News* ran the story with a banner headline in the Sunday edition: CALL GIRL SAYS KELLEY, DINGELL WERE CUSTOMERS. (Frank J. Kelley was the attorney general of Michigan.) Cole later said during a TV interview that the *News* reporters had told her their chat was "off the record and that everything discussed was just to help me." But Cole stuck to her Dingell story and told the TV reporter she'd be willing to take a polygraph test to prove she wasn't lying. It all sounded very convincing.

Dingell denied he had ever slept with Cole. He said he might have met her, but that he wouldn't recognize her "if she fell on top of me."

The courts went easy on Lois Herman, alias Teri Cole. For a guilty plea to one felony count, the prosecutors dropped the other two charges. All she got was probation and a lecture about street corners, motels, and shoe boxes.

It was all an FBI put-up job, according to columnist Jack Anderson, in an article about how the Bureau used sex to smear Stockton, Seigenthaler, and Dingell.

NOVEMBER 1976 TO FEBRUARY 1977

Dingell had never lost an election, and he won in 1976 despite the smear. On November 1, he announced there would be more Silkwood hearings during the first week of December. The Justice Department would be the star witness. It would be "called on to explain how Mrs. Srouji obtained access to [FBI] files, to define her 'special relationship' with the Bureau, and to explain what connection that relationship may have had with her appearance before the Subcommittee [in April]."

Under pressure from Dingell, the FBI finally allowed the subcommittee to read the Bureau's Silkwood reports. Majority Counsel Mike Ward, Peter Stockton, and Assistant Minority Counsel Paul Kritzer pored over more than a thousand pages of FBI interviews. The reports were disappointing, and the subcommittee staff found no new significant facts. Ward was suspicious. Silkwood had been proven right. She had alleged that the Kerr-McGee plant was unhealthful and unsafe. The AEC investigations and the testimony of Dr. Morgan had confirmed that without a doubt. She had also alleged that Kerr-McGee had cheated on quality control, and the AEC had found more than forty deliberately fogged quality-control negatives. His mind kept going back to the crash on the night of November 13. The FBI reports did not answer even the most elementary question: Did someone tamper with Karen's brakes or steering? Ward was convinced that she had been awake when she hit the concrete wingwall. Why hadn't she applied her brakes? The FBI reports didn't even raise the question. Either the Bureau had more documents, or it had conducted a very superficial investigation.

Before the Silkwood hearings could resume in December as scheduled, the House passed a rule, at the urging of its young Democrats, that no member could chair two subcommittees. Dingell was the chairman of the Energy and Power Subcommittee as well as of the Energy and Environment Subcommittee, which was holding the Silkwood hearings. Since Energy and Power had more legislative teeth, Dingell gave up Energy and Environment.

With Dingell out, there was no longer support in the subcommittee for the Silkwood probe. December came and went without the promised second set of hearings. Then, on January 13, 1977 — just over one year after Tucker and Nelson had met with Senator Ribicoff — Mike Ward and Paul Kritzer recommended that the Silkwood investigation be iced. Bill Silkwood had already filed a civil suit against Kerr-McGee and the FBI, and the subcommittee had failed to find cause.

"Due to litigation recently brought against parties and witnesses involved in the Subcommittee's investigation by the Silkwood estate," Ward and Kritzer advised in a memo, "the Subcommittee staff believes it would be improper to continue with this study in that such activity may jeopardize the legal rights of all parties. For this reason the Subcommittee was compelled to indefinitely postpone hearings scheduled for December 2 and 3, 1976."

What sparked the memo was another *Rolling Stone* article by Howard Kohn. Ward and Kritzer were worried about the "inac-

185

curate impressions" they thought Kohn had created, and they devoted most of their memo to picking at the reporter.

The subcommittee investigation into the Silkwood case had not been a murder investigation, as Kohn seemed to imply, Ward and Kritzer stressed. However, the interrelationship of events made it difficult to separate Silkwood's death from the unauthorized diversion of plutonium, which did interest the subcommittee. "In considering these issues," they wrote, "the staff could find no credible evidence to challenge the conclusion that Silkwood died in a one-car accident."

Ward and Kritzer said in their memo that the subcommittee had found no evidence that a plutonium theft-and-smuggling ring existed at the Kerr-McGee plant. (Kohn had suggested that there was one.) The MUF evidence at the Cimarron plant, however, was "inconsistent or contradictory," they said. There were still twenty pounds of plutonium missing, despite Kerr-McGee's claim that all the plutonium had been accounted for.

Ward and Kritzer argued that the evidence indicating Silkwood had documents the night she was killed was "inconsistent and contradictory, second-hand hearsay and inconclusive."

They summed it up this way: "We have no documents. We have no witnesses who can identify what documents Silkwood had, if any, or describe their content. We have no evidence that documents were surreptitiously taken from Silkwood's car at the accident scene."

Ward and Kritzer were tough on the FBI. They said the Bureau's investigation of Silkwood's contamination was one of the "major deficiencies in the FBI's investigation." Kerr-McGee security was poor, they argued. Plutonium *had* been diverted. And the Bureau had not responsibly investigated the theft. In fact, they said, they didn't think the FBI had conducted a very good investigation, and they thought the Justice Department had done "an embarrassingly poor job of reviewing and summarizing the FBI's investigation."

Ward and Kritzer wanted the subcommittee members to know all this so that they could better respond "to constituent inquiries" about Kohn's *Rolling Stone* article.

It was over. The congressional investigation had ground to a halt. It went out, as Ward and Kritzer quoted, "not with a bang, but a whimper."

It had smoked out Jacque Srouji. It had exposed the FBI and the Justice Department as incompetent. Through the testimony of Dr. Morgan, it had established the Kerr-McGee Cimarron facility as an unhealthful and unsafe place to work. It had raised doubts about the whole MUF and security system. But it hadn't learned much about Karen Gay Silkwood's contamination and death.

Sara Nelson, who had worked so hard with Kitty Tucker to launch the congressional investigation, had mixed feelings. She wasn't surprised Congress had dropped Silkwood, for, as the hearings began to unfold, she came to recognize that Congress was not the powerful independent body she'd once thought it was. Yet Nelson was angry — not at John Dingell, but at the FBI, the Justice Department, and the political pressures that had finally killed the investigation and the hearings.

On another level, Sara Nelson was pleased. During the year and a half just before and during the congressional investigation, she and Tucker had built up a national grassroots network that would keep the Silkwood issue alive in the months ahead. And because of the media's coverage of the investigation and hearings, both Congress and the public had a better understanding of Karen Silkwood and the abuses she had struggled to correct.

It was up to the courts now — that agonizingly slow and frequently unfair system of justice. And where Congress had ended with a whimper, the courts would enter with a roar.

The Courts

Chapter 20

Daniel Sheehan was watching television at Herman Thunder-hawk's on the Oglala Sioux Reservation at Wounded Knee when he first heard of Karen Silkwood. Thunderhawk's home in Porcu-pine was one of the American Indian Movement's nerve centers; it was just past the church and trading post where the 1973 Indian siege began.

A small black and white TV droned in the dark shack as Sheehan and friends chatted, torn between "The Reasoner Re-port" and the election of the new Sioux tribal chief. On the screen, the ABC camera panned lonely Oklahoma Highway 74 as David Shoumacher explained: "That night she was on her way to meet with a Washington official of her union and a reporter from *The New York Times* to deliver the documents she had been gathering. A mile from the plant, she was killed. The documents have dis-appeared."

"Fuck, can you believe that?" Sheehan said. "Somebody's ass is in a sling. Somebody is going to get them for doing that." Little did Sheehan realize that he would be the somebody to try.

Sheehan was a thirty-year-old attorney who didn't smoke, drink, or care about money. With curly black hair hanging over his ears and forehead, he looked like a French poodle in need of a trim. But he had a terrible bite. Self-confident and blunt to the point of rudeness, he was driven by a passion to defend the Con-stitution and by a huge appetite for important cases.

After finishing Harvard Law School in 1970, Sheehan joined

191

the Wall Street firm of Cahill, Gordon, and Reindel, one of the best litigation law firms in the country. Cahill promised Sheehan the world. He wouldn't have to be a "desk jock"; he'd get plenty of trial experience; he could work on First Amendment and civil rights cases; and he could do *pro bono* (free) work outside the law firm, as long as he didn't act in behalf of Cahill or litigate against its clients.

Sheehan was one of five Cahill attorneys who represented Earl Caldwell, a reporter who had refused to give his notes to the police; *The New York Times,* which printed the Pentagon Papers; and Daniel Ellsberg, who leaked them. But it was Sheehan's *pro bono* work that got him into trouble. He defended Black Panthers accused of a plot to blow up Macy's and the New York Police Department; conscientious objectors to the war in Vietnam; and prisoners in Manhattan's Tombs who had rioted against prison conditions.

Cahill sacked Sheehan from his $21,000-a-year job, saying that the firm's and Sheehan's interests clashed. Sheehan was ready to move on anyway, for he had learned an important lesson on Wall Street: big law firms have a conflict of interest between what is good for the public and what is good for their rich powerful clients, and they won't bite the hand that feeds them.

Sheehan then joined flamboyant trial attorney F. Lee Bailey for $35,000 a year, but quit after twelve months. He didn't fit into the hard-drinking, macho Bailey crowd, which was more interested in winning than in human rights or the Constitution.

In the fall of 1973, Sheehan enrolled at Harvard Divinity School for a doctorate in social ethics. He was struggling with questions that had plagued philosophers and theologians for centuries: What are human rights? Where do they come from? The State? Or a law higher than the State? But before he could delve deeply into these issues, Joseph de Raismes, an old Harvard chum, called from the Rocky Mountain regional office of the American Civil Liberties Union in Denver. The Wounded Knee trials were about to begin. Would Sheehan like to defend Russell Means and the other American Indian Movement Sioux?

Sheehan spent a year as ACLU's Rocky Mountain litigation director. Besides Russell Means, he defended the movie *Last*

Tango in Paris against a pornography charge by the State of Idaho, and Wyoming inmates against the Wyoming State Prison. When the ACLU Rocky Mountain region ran out of money, Sheehan went back east, where he met William Davis, a short, gutsy Jesuit priest from Montana who walked on the edge of Catholic social theology. Davis believed religious leaders of all denominations ought to take a public stand on such critical social issues as war, human rights, military spending, and nuclear power. Davis was the director of the Jesuits' new Office of Social Ministries, but he had no staff, no program, and no budget. He offered Sheehan free room and board plus one half of his meager Jesuit stipend if he'd help build the Social Ministries program.

Sheehan's first case with Davis was to defend Fathers Daniel and Philip Berrigan and comedian Dick Gregory, arrested for a sit-in on the White House lawn to protest President Gerald Ford's limited amnesty program for Vietnam draft evaders and deserters. The demonstrators were acquitted, and, after the trial, one of the defendants told Sheehan that NOW needed legal advice on how to deal with the Karen Silkwood case. Sheehan learned that Lee Metcalf was about to conduct a congressional investigation into the Silkwood matter but that NOW, not sure Congress would get anywhere, was searching for some kind of legal alternative.

Sheehan met with Sara Nelson, Kitty Tucker, Steve Wodka, and several others interested in the Silkwood case. They told him that Bill and Merle Silkwood had already approached Michael Kennedy, a civil rights attorney, about filing a complaint against Kerr-McGee, but that Kennedy wanted $25,000 up front. Bill Silkwood didn't have that kind of money, they said. Would Sheehan talk to Kennedy?

Sheehan asked Kennedy on which legal theory he planned to build his case against Kerr-McGee. Kennedy told him he hadn't given the Silkwood matter much thought, and wouldn't until he got paid. Sheehan volunteered to put some ideas together for Kennedy without charge.

Sheehan believed that Silkwood had been killed, but, as far as he knew, there was absolutely no evidence to prove that. He concluded that all Bill Silkwood could do was sue Kerr-McGee for negligence in safeguarding the plutonium that had contaminated

Karen, and then hope some clues about the possible violation of Karen's civil rights would pop up during the investigation in preparation for the negligence trial.

Sheehan attended the Dingell hearings in April and May of 1976, and was puzzled by Jacque Srouji's testimony. She seemed to know a lot more than she was telling Dingell. What were those thousand pages of FBI documents she had? What was in them? Sheehan cornered Peter Stockton and asked him about the Nashville journalist, but the congressional investigator just smiled. Sheehan suspected there was more to Srouji than Stockton or anyone else on the subcommittee was willing to say.

After the hearings, Bill Silkwood asked Sheehan if he'd take on Kerr-McGee. "I know you've done more work on Karen's case than Kennedy has," Bill said. "I don't have twenty-five thousand dollars. I want to know who killed my daughter. How much will it cost?"

Sheehan told Bill that if there was a case, he'd do it for nothing. "The main thing right now," he said, "is to file before the statute of limitations runs out. We can worry about money later."

Sheehan explained to Bill Silkwood that he would have to file a *civil* suit — to try to recover damages for some alleged injury to Karen — and that he could not file a *criminal* suit alleging someone had committed a crime against Karen. Only city, state, and federal prosecutors did that. However, if, in the course of a civil suit, Bill uncovered a crime such as murder, then there would be a good chance that a government prosecutor would file criminal charges.

Sheehan further explained that the law requires civil suits to be filed within two years after the alleged injury, and that in Karen's case there were only two realistic civil suit possibilities: to sue for damages resulting from employer negligence (a personal injury case), or to sue for damages resulting from a violation of Karen's civil rights. If Bill could prove that Karen had been deliberately murdered or contaminated, he could sue on the grounds of a violation of her civil rights. But, Sheehan cautioned, there was no evidence that she had been murdered, deliberately contaminated, or deprived of her civil rights in any other way. Even if there had been a crime, Sheehan said, there was no evidence pointing to who may have committed it. If Bill charged into court alleging a civil rights violation, he would destroy his credibility and run the risk of losing the personal injury case, which was solid.

Bill wasn't interested in the fine points of the law. Someone had murdered his daughter, he said, and he wanted to know who.

"I agree to file for you, Bill, based on whatever evidence there is to support the case," Sheehan promised. He had a lot of work to do, for the statute of limitations would run out on November 13, just seven months away.

Sheehan rarely made an important decision on the scope of a case without consulting William Taylor, a private investigator whose integrity and instincts Sheehan had learned to trust. Sheehan spent thirty-six hours with Taylor in Florida, reviewing the Silkwood case and trying to talk the investigator into heading up the probe into Kerr-McGee. But Taylor wasn't interested. He told Sheehan that he was a criminal investigator, didn't like working on civil suits, and was not going to risk his life just to make somebody a millionaire.

What if Karen had been murdered, Sheehan asked. What if the civil suit was not for a lot of money? Taylor said he'd think it over.

Sheehan needed Taylor, because he had already concluded that to prove even the negligence case against Kerr-McGee, he would need a thorough investigation. If he wanted Taylor, he'd have to convince him that a crime had been committed.

Sheehan's best lead was Peter Stockton, and he leveled with the investigator in his subcommittee office. Sheehan said he wanted to take the case for Bill Silkwood and that he knew there was more to Jacque Srouji than had come out in the hearings. What was going on?

Stockton opened up. He told Sheehan that he suspected the congressional hearings and investigation would soon grind to a halt and was glad Sheehan wanted to pick up the pieces. He told Sheehan everything Srouji had said but had not testified to: that Karen Silkwood's phone had been tapped and her apartment bugged; that Srouji had seen the transcripts of bugged or taped conversations; and that there were FBI documents in the Kerr-McGee files and K-M documents in Bureau files.

Stockton grew excited as he talked, punctuating his sentences with exclamations of "Jee-sus *Christ*" for emphasis. He told Sheehan that he wasn't sure which documents Srouji actually had in her safe-deposit box because he had not seen them, or how much of what Srouji had said was true. But, Stockton said, he had concluded that there had been some wiretapping, bugging, and surveillance, and that the FBI and Kerr-McGee had conducted some kind of joint investigation into the Karen Silkwood case.

That was the first time Sheehan heard of the wiretapping, bugging, and surveillance allegations. If Srouji's allegations were true, they would constitute a violation of Karen's constitutional rights, and Bill Silkwood could file a complaint that his daughter's civil rights had been violated.

Sheehan invited Bill Taylor to Washington. He told the private investigator what Stockton had said and introduced Taylor to Sara Nelson and Kitty Tucker. All three tried to convince Taylor to take on not only Kerr-McGee, but the FBI as well. Sheehan argued that Taylor *would* be investigating a crime, even though he would be working on a civil suit, and that he was just the man to find out who had killed Karen. He told Taylor that he wouldn't be putting his life on the line to make someone a millionaire because Bill Silkwood was going to ask for only $160,000 in damages.

Then Nelson and Tucker went to work on Taylor. They told him there was a conspiracy to cover up the Silkwood case and that the FBI was in on it. They gave him Howard Kohn's *Rolling Stone* articles and a stack of documents to read. Taylor told Sheehan, Nelson, and Tucker that he'd look over the material and think about it.

Taylor was not the kind of investigator who wanted to be confused by secondhand reports in newspapers and magazines, so he studied two primary documents — the Oklahoma Highway Patrol accident report and the autopsy report. After probing hundreds of accidents as a criminal investigator, Taylor knew what he was looking for. And to him, nothing looked right in the Silkwood reports.

On the one hand, the facts indicated that Karen was awake behind the wheel. Her Honda hit the shoulder at about a 45 degree angle, somehow straightened out, and then somehow traveled in a straight line for 240 feet along a grassy shoulder parallel with the highway. She didn't slip into the ditch, drift into the pasture, or bounce back onto the road. To Taylor, that seemed impossible unless someone was steering the car.

On the other hand, if Karen had been awake, why hadn't she slammed on her brakes before hitting the wingwall, or at least tried to scoot back onto the highway? She was going only forty-five miles an hour, according to the OHP report. If she had hit her brakes, she would have skidded into the wingwall broadside, not head on, and would probably have come out alive. Furthermore, if she was awake, she would have instinctively tensed before impact, and braced her legs. If she had done that, her legs would have splintered like dry pine branches. The autopsy report, however, didn't describe those kinds of injuries, indicating she had not been in a fully braced position at the time of impact.

Finally, Oklahoma Highway Patrolman Rick Fagen reported that Silkwood was asleep and under the influence of alcohol and drugs. There was no way he could have known that when he wrote his report unless someone had fed the story to him.

The whole thing smelled suspicious. Taylor told Sheehan he'd do a preliminary check, and if he then felt a crime had been committed, he'd join the team.

Taylor had sources everywhere in the law enforcement world — FBI, CIA, National Security Agency, Immigration and Naturalization, Secret Service, and police departments around the country. He called "Echo," a source in the FBI's Washington headquarters. "The Silkwood case is a doozy," Echo said. "The Bureau is covering so much shit on Karen Silkwood, you wouldn't believe me."

"Try me," Taylor said. "Was she murdered?"

"I don't know," Echo said. "Everyone is tight-lipped. A lot of agents have been called on the carpet. One [of them] threatened the Bureau. 'If I'm nailed,' he said, 'I'll tell all.' "

Echo told Taylor that field reports on Silkwood were still coming in — two years after her death — and that she had been under heavy surveillance. Echo said he had seen what looked like transcripts of taped conversations in the Silkwood files.

"There's no damn way you'll uncover that shit," Echo said. "It's buried too deep."

All Taylor needed to hear was that the Karen Silkwood case was impossible to crack. He told Echo to keep his ears open, and then called Sheehan. "Let's meet," Taylor said, meaning he didn't want to talk over the phone.

Taylor told Sheehan he was now convinced that Karen had been murdered and that the Silkwood case smelled like a cover-up. The FBI had backed off. Why? The accident data were all wrong. Why? The Pinkertons had been on Pipkin. Why? Lieutenant Fagen had concluded, using erroneous and secondhand information, that Karen had fallen asleep at the wheel. Why?

Taylor told Sheehan what Echo and other sources had said. "She was bugged and placed under surveillance," Taylor said. "It was a national security tap. They suspected she might try to smuggle out some plutonium."

What Taylor had learned from his sources only confirmed what Srouji had told Stockton and others. With that information, Sheehan drafted three possible complaints against Kerr-McGee and members of the FBI, ranging from weak to strong. Bill Silkwood chose the middle one.

First, the complaint charged Kerr-McGee with *negligence* in handling the plutonium that contaminated Karen, arguing that the plutonium belonged to Kerr-McGee and not to someone else, that plutonium is extraordinarily toxic, and that Kerr-McGee not only failed to establish and maintain safety at the Cimarron plant but had refused to do so.

Second, the complaint charged twenty-three members of the Kerr-McGee Corporation with a *conspiracy* to deprive Karen of her civil rights by, for example:

□ trying to prevent Karen and others from improving the working conditions at the Cimarron plant between November 1972, when the OCAW struck, and 1976, when the plant shut down for good;

□ punishing Silkwood, Tice, and Brewer for complaining to the Atomic Energy Commission about working conditions at the Cimarron plant;

□ approving a plan to collect dossiers on Silkwood and others illegally; break into their lockers, desks, cars, and homes; tap their telephones and bug their apartments; hire unknown "operation agents" to spy on them and harass them on the highway in order to frighten them; fire and transfer them as punishment for union activities; and interfere with their rights to talk to the media, submitting them to polygraph tests if they did.

Third, the complaint charged the FBI special agent Lawrence Olson, Sr., his Oklahoma City office boss, Theodore Rosack, and an unidentified FBI agent in Washington had joined the Kerr-McGee officials in a *conspiracy* to cover up the illegal activity. It also charged that Jacque Srouji joined the conspiracy cover-up in 1975 when she learned, while doing research for *Critical Mass,* that Silkwood's civil rights had been violated.

Bill Silkwood asked for $160,000 for Karen's loss of property, personal injury, and mental anguish, and as a punishment to Kerr-McGee and the others named in the complaint.

Sheehan could legally file Bill Silkwood's complaint in one of three places: Texas, where Bill, the administrator of the Karen Silkwood estate, lived; Washington, where the FBI was headquartered; or Oklahoma City, the home of the Kerr-McGee Corporation. Sheehan chose Oklahoma City. "Hit them in their strongest place," his old high school football coach used to tell him. Kerr-McGee's strongest place was Oklahoma City, where the Kerr-McGee Towers dominated the downtown skyline, casting their shadows on the courthouse, three blocks away; where K-M was a household word, with Kerr-McGee Plaza, Robert S. Kerr Boulevard, the Dean A. McGee Eye Institute, and the Kerr-McGee logo atop dozens of city gas stations.

Besides, Sheehan believed that Silkwood was not just a civil rights case. It was also a nuclear power issue, which ultimately, he believed, had to be settled in the village square. What was closer to the village square than the heartland of Oklahoma City?

Sheehan finished drafting the Silkwood complaint late at night on November 4, 1976. Three NOW volunteers were using the Jesuit Social Ministries Office on Washington's Embassy Row to type, proof, and Xerograph copies of the complaint. The Jesuits were not comfortable about having so many lay persons in the office and had instructed Sheehan not to leave the women alone.

But Sheehan thought their concern was bullshit, and left for a good night's sleep. He had a press conference the next morning to announce the complaint, and he needed to be fresh. At 1:00 A.M., one of the NOW volunteers opened the Jesuits' back door and set off the burglar alarms. The police dragged a sleepy Jesuit out of bed to come down to the office and vouch for the NOW women. The Silkwood case was off to a bad start.

The next day, November 5 — just eight days before the statute of limitations ran out — Sheehan sent the complaint against Kerr-McGee and the FBI by airplane to Oklahoma City, where an associate filed it in the federal courthouse.

The Kerr-McGee Corporation denied it had injured Karen Silkwood, caused her to be injured, or was negligent in any way. It argued that any personal injury to her because of her contamination was caused "solely by the negligence of Karen Silkwood, her failure to use ordinary care, or by her failure to follow the safety practices, rules and regulations prescribed by the Kerr-McGee Corporation or the Kerr-McGee Nuclear Corporation for Cimarron employees."

Srouji, Olson, and Rosack also denied the cover-up charges against them and asked the court to dismiss the case. Olson further argued that his "investigation was made in a competent, reasonable, and workmanlike fashion, and that he is without fault."

The twenty-three Kerr-McGee defendants individually denied the charges against them. Each eventually would give the court a sworn affidavit saying: "It is alleged . . . that I was an active member in the devising and execution of an illegal conspiracy to deprive Karen Silkwood and other workers at the Cimarron facility of the equal protection of the law, and to conceal this alleged wrongdoing by acts other than in the normal course of my corporate duties. This allegation is false. I deny it without qualification. I have neither devised, executed nor participated in any conspiracy to deprive Karen Silkwood or other workers at the Cimarron facility of the equal protection of the law. Further, I am unaware of any facts indicating that any of the other defendants in this action have devised, executed or participated in any conspiracy to deprive Karen Silkwood and other workers at the Cimarron facility of equal protection of the law."

□ □ □

Taylor began to line up his sources across the country. "I think this may get hairy," he told each one. "If I find out there has been some plutonium smuggling going on, I might get blown away. Keep your ear to the track so at least I'll hear the train. If you can't get me information, then give me security."

Echo called back. "The Bureau has gathered all the Silkwood files and moved them," he said.

"How many?" Taylor asked.

"You have no idea how big this case is," Echo said. "You'd need a truck to cart the boxes around. Thousands and thousands of papers. That damn case has files from here to Union Station."

Chapter 21

Bill Taylor is a big man with red hair, a neatly trimmed beard, and a handshake like a vise. He prefers to work alone. He remembers names, faces, phone numbers, dates, and times. He checks behind mirrors, pictures, and curtains for bugs, and calls from "safe" telephones and "safe houses." He is always looking for pieces that don't fit — the grease monkey with manicured fingernails, the wet stones on the dry path, the darting eyes over a relaxed smile. He approaches his work like an animal, wary and disciplined, always assuming his life is in danger. Not without reason. Taylor has been shot at more times than he likes to recall, and he has been stabbed several times, once through the stomach and out the back. The knife is stuck in the huge United States Marines Coat of Arms that hangs on his den wall, a reminder of the best of times and the worst of times.

Criminal investigations are what Taylor knows and does best. When he was fifteen — a big kid for his age — the Boston police used him to set up barkeeps who served drinks to minors, and prostitutes who needed a VD check. Taylor would buy a bourbon or a blonde with marked money, flash the signal, and the cops would move in for the collar. Taylor would get $20.

At the age of sixteen, Taylor enlisted in the Marine Corps and spent three years in Vietnam, fourteen months of that undercover. For the rest of his nine years in the corps, he did criminal investigations.

When Taylor left the corps in 1971, he became a private eye.

One of his first clients was F. Lee Bailey, who was indicted in Orlando, Florida, in the Turner mail fraud case; one of Taylor's first assignments from Bailey was to find out about Daniel Sheehan. Bailey respected Sheehan's legal mind but couldn't figure out the young attorney. He didn't smoke or drink and didn't seem interested in money or power inside Bailey's firm. Bailey had no idea where Sheehan was coming from, and that made the famous attorney uneasy. He assigned Taylor to work for Sheehan and to report back to him. He wanted to know what Sheehan liked, what he wanted from life, what his politics were.

Bailey's plan backfired, and Sheehan and Taylor became close friends, even though they were worlds apart. Taylor was still a United States Marine at heart — fiercely loyal to the flag, ready to go back to Nam, not afraid to fight or to kill. Sheehan was a liberal Easterner — fiercely loyal to the Constitution and human rights, unwilling to go to war or to kill for anybody. But Sheehan and Taylor had two things in common: they formed strong loyalties, and they had a passion for getting to the bottom of a case.

□ □ □

Soon after Sheehan filed the Silkwood complaint, Taylor slipped into jeans and cowboy boots and headed for Oklahoma. The Kerr-McGee workers around Crescent were still hostile toward Karen Silkwood. "We had a house and a nice car," they told Taylor. "Now we're living in a trailer. We got nothing. It's all *her* fault. Sure we'll help you out, you asshole."

Like Fagen and Pipkin, Taylor studied the accident site. One of the first things he did was to ride down the left shoulder toward the concrete culvert, just as Karen had done, but much more slowly. It was like riding down a washboard, and there was no way Taylor's car could keep a straight line for 240 feet without help from him. Two things were clear, and they confirmed the investigator's earlier reading of the accident report: no one could be asleep in a car that bounced around so badly; and without someone steering, any car would bound out of control into the ditch and pasture or back onto the highway long before it reached the cement culvert.

Taylor knew, from her friends, that Karen had been an excel-

203

lent driver, and he concluded she'd been handling the Honda as any racing driver would have done — driving with hands off, letting the car take the bumps, gently guiding it, not forcing the steering wheel this way or that. But why was she on the shoulder in the first place? And why didn't she slam on her brakes before she hit the wingwall? To Taylor, there was only one logical explanation: Karen Silkwood had been forced off the road at a 45 degree angle, as the tracks in the grass had indicated; she had straightened out the Honda and sped along the shoulder, hoping to swing back onto the highway. But she couldn't, because another car was running parallel to her, keeping her off the road. She didn't see the culvert because she was looking over her right shoulder at the car chasing her. Then, just before she hit the wingwall, she turned the car slightly to the right — as the Honda tracks had shown — not to get back onto the highway, but to move a little higher onto the shoulder, where the bank wasn't so steep. She flew over the first wingwall without realizing what was happening. Probably she saw the other wingwall in the split second before she crashed. In either case, it was too late to brake.

The scenario fit all the known facts and answered all Taylor's questions:

☐ How did Silkwood get onto the shoulder? Someone pushed or chased her there.

☐ Why did the car hit the grass shoulder at a 45 degree angle and then straighten itself out? Karen straightened the Honda out herself rather than run into the pasture. She was trying to escape from the car chasing her.

☐ How could the Honda travel 240 feet in a straight line on a bumpy, slightly inclined shoulder? Karen was steering the car.

☐ Why didn't the Honda wheels chew up the grass, indicating that Karen was trying to get back onto the road? Karen was driving hands-off, and, like a good racing driver, did not apply her brakes.

☐ Why didn't Karen try to avoid hitting the wingwall head on? She didn't see it because she was looking over her right shoulder at the car chasing her.

☐ Why did the autopsy report describe injuries consistent with a driver asleep at the wheel? Karen was not braced, and the crash took her by surprise. She only had time to tighten her grip on the steering wheel.

Taylor kept returning to the scene of the accident, mostly to sit, think, and sort over notes, looking for clues in what people had said, or didn't say, or how they had reacted to him. He'd park in a field away from Highway 74, where his car wouldn't be noticed, and walk back to the accident site along the little creek that ran through the concrete wingwalls into the culvert under the highway. Things always seemed clearer as he sat on a rock, close to the spot where Karen had died. Sometimes, in his frustration, he'd talk to her. "Come on, woman," he'd say. "If I'm going to get them, you have to give me some ideas."

Taylor's mind usually drifted back to an inconsistency in the Silkwood story that stuck in his craw. From interviews with her friends, enemies, and fellow workers, Taylor had concluded that Karen was smart, enjoyed playing mouse to the Kerr-McGee cat, and knew exactly what she was doing. She was passionate about "kicking their ass," her friends had told him. But if she was so smart, why was she carrying such important documents around in her car the day she was killed? Except for a short lunch break with Frank Murch, she had been in contract negotiations or in conference with AEC investigators from nine until five, when she left for a union meeting at the Hub Cafe. Anyone could have broken into her car in the Kerr-McGee parking lot and taken her documents. It was a careless risk to take. Unless . . . unless Karen didn't *have* any documents and was too embarrassed to say so. That was a possibility, and Taylor gave it a lot of thought.

One idea that came to Taylor as he sat by the accident site, pondering the document problem, was that he should search the creek bed and banks for some sign of Karen's evidence. There had been a light east wind the night she was killed, and it was possible that an x ray of a defective weld had blown into the water and floated downstream. Things like that happened all the time, and even after two years, the evidence could still be trapped in a shrub or between the roots of a river birch along the bank.

Taylor followed the creek back into the fields. He found a 1974 New Jersey license plate close to the culvert. He had a source check the tag later. The plate was not renewed in 1975, and there was no record of its ever being issued. Taylor concluded the license belonged to a federal agent of some kind; it is standard practice for each state to issue a certain number of tags to agencies like the FBI without keeping a public record of them. Since Kerr-McGee was handling 700 pounds of plutonium in 1974, it would not have been unusual for federal agents to be nosing around.

Other than the license plate, which was interesting but not helpful, Taylor found nothing. As he was climbing over a barbed wire fence along the oil road that fed into Highway 74 just north of the spot where Karen was killed, a truck with an oil-drilling rig on the back came by. The driver lumbered to a stop.

"Hi! Can I help you?" he asked.

"Nope," Taylor said. "Just looking for some property to buy. I'm from back east. Never saw how these work." Taylor pointed to the oil wells pumping quietly in the fields.

"All this property belongs to Ellis — Ellis in Guthrie," the driver said. "This and most of the wheat fields up to Seventy-four."

The driver offered to take Taylor along and show him how the crude was pumped, but the investigator thanked him, walked back to his car, and took off for Guthrie. He had a suspicion.

The property deeds showed that the land did belong to Ellis, and that there was a farmhouse and two barns a mile back down the road from Highway 74. A discreet check showed the land and house belonged to Sherri Ellis' father and that Sherri had grown up on that farm. Taylor was disturbed by both Sherri and Drew Stephens. Drew had been out to the accident site with him and had never told him about the Ellis farm. And why hadn't Sherri said something? Why hadn't she told him, "I grew up there and my roommate died there, at the foot of my road"?

Taylor decided to check out the farm for himself. Something just didn't fit right, and he had a feeling. He drove past Kerr-McGee on the knoll, crossed over Highway 33, took the first right turn off Highway 74 onto the oil road, and passed the spot where he had met the oil roughneck. The farmhouse was locked up, and it looked as if no one had lived in it for a long time. There was a horse in one barn; Taylor assumed it belonged to Sherri. He checked the potato cellar, not even sure what he was looking for, and saw nothing that interested him. Then he went into the old tractor barn. It was empty. But in the west corner, Taylor found a shelf. Old bottles and junk had been pushed to the left side as if to make room for something.

Taylor poked around carefully. At the back of the shelf, he found a brown manila envelope covered with dust and pigeon droppings. On the barn floor lay a soiled, white, number ten business envelope. Both envelopes were empty and unaddressed. But both carried the Kerr-McGee logo in the corner.

Karen must have stashed her documents in the barn, Taylor reasoned. She knew that Sherri kept her horse there, and she probably had gone to the farm with her roommate. Just a mile from the Cimarron plant, the abandoned barn was an ideal hideout. The farmhouse was empty; she could come and go unnoticed. And she could easily drive there after work if she had something to get or hide.

Taylor reasoned that either the barn had been Karen's permanent hiding place for the documents or that she had kept the evidence in her apartment and then, when the apartment was quarantined on November 7, had taken the documents and hidden them in the barn until the thirteenth. After the union meeting at the Hub, she had stopped off at the farm to get the rest of her documents before heading for the Holiday Inn in Oklahoma City.

There was no other logical explanation. Why else would empty Kerr-McGee envelopes be sitting on a shelf in a barn that had been abandoned well before Kerr-McGee had opened the Cimarron facility? Mr. Ellis did not work for Kerr-McGee, and it was very unlikely that he would have access to Kerr-McGee stationery. Sherri could have put the envelopes on the shelf, but why? She had no reason to do so.

To Taylor, the empty envelopes pointed to Karen Silkwood, and the scenario solved another basic Silkwood problem. Indeed, Karen *had* been too smart to carry all her documents around in her car on November 13. Instead, she had found the perfect hiding place, the cloak-and-dagger kind of niche Taylor would have expected her to find.

DECEMBER 1976

It was snowing in Oklahoma City, almost Christmas, and Taylor had been on the Silkwood case off and on for two months. He had talked to more than 200 people, watched the Kerr-McGee Cimarron plant (which had a skeleton crew on duty), and checked on former K-M management personnel. He had even hired a helicopter to fly him over the plant, the death site, and Ellis' farm, hoping to see a new angle, something he had overlooked. He felt he understood who Karen Silkwood was and why she was trying to kick Kerr-McGee's ass. But everywhere he went, he found closed doors and tight lips, anger and fear.

Taylor and "Fairy Godfather," one of his best Oklahoma sources, were eating in an Italian restaurant. There was something unusual about the dining room — something Taylor sensed but couldn't put his finger on. After dinner, he and Fairy Godfather went to the Prairie Lady bar for a drink, and Taylor saw something familiar in the crowd, a striped shirt he had noticed earlier in the restaurant. Taylor nudged Fairy Godfather; they casually left the bar. A Karmann Ghia was blocking Fairy Godfather's car, so Taylor lifted the back and then the front of the sports car and slid it over in the snow so that Fairy Godfather could back out. A few blocks from the Prairie Lady, Taylor noticed headlights behind them. "Someone's tailing us," he said. "Lose him."

Fairy Godfather drove a mile of turns, around corners and through alleys and parking lots, until he was certain he had lost the shadow. He and Taylor ended up at Chadwick's, a new club close to the Northwest Holiday Inn, where Karen had been supposed to meet David Burnham, and where the AEC stayed whenever they were in town. Chadwick's had security guards for its opening night to make sure that no one walked off with the glasses and that there were no drunken fights. After a couple of drinks, Taylor made his way toward the john. Out of the corner of his eye, he caught the striped shirt following him and decided it was time to find out who the tail was. Picking his way between the tables, Taylor crossed the crowded room to the front door, glass in hand.

"Hold this for me," he told the security guard. "I'll be right back. I have to catch a friend before he takes off."

Outside, Taylor took two rights into an alley next to a small parking lot and crouched behind a low wall. In less than a minute, he heard footsteps scraping in the snow, then saw the striped shirt. The man was taller than Taylor, but not as hefty. Pound for pound, Taylor had the advantage. He grabbed the spook, pulled him to the ground, and dragged him into the snow behind the wall.

Taylor sat on the man's chest, pinned his arms to the ground with his knees, and began slapping him.

"What the fuck are you?" Taylor asked.

The man said nothing.

Taylor hit him harder. "Who the fuck are you working for?"

Taylor rifled the man's pockets, but found no identification. Then he heard more footsteps in the alley. He gave the man in the striped shirt one more good punch, got up, and staggering, to make it look as if he'd been jumped, went over to the two Chadwick security guards.

"Thank God it's you," he said breathlessly to the guard he had asked to hold his drink. "That guy over there followed me and jumped me."

The man in the striped shirt was rolling on the ground, his face bloody. The other guard cuffed him.

"You all right?" the first guard asked Taylor.

"Shaken up," Taylor said. "Otherwise fine."

"I knew something was up," the guard said. "I saw him follow you out the door."

"Thanks," Taylor said. "I appreciate it."

Taylor followed one of the guards back into Chadwick's. When the guard went to phone the police, Taylor nodded to Fairy Godfather, and the two left the bar. By the time the patrol car skidded to a halt in front of Chadwick's, Taylor and Fairy Godfather were gone.

Taylor knew the man in the striped shirt was a pro. Fairy Godfather hadn't been able to shake him, and, as Taylor discovered when he'd quickly frisked him, the spook carried no I.D.

Taylor figured he was a federal agent of some kind — CIA, FBI, National Security Agency. Whatever he was, he was mighty curious about what Bill Taylor was up to. The private investigator now knew the Karen Silkwood case was as hot as Echo had told him.

Chapter 22

Jacque Srouji was the key to Dan Sheehan's conspiracy case. Bill Taylor hadn't found any evidence to support the allegations of the case — that Kerr-McGee had wiretapped, bugged, and conducted illegal surveillance on Karen Silkwood, or that the FBI had found out about the illegal snooping and had covered it up. If Srouji would say under oath what she'd told Peter Stockton and others, and, even better, if she would produce FBI documents to substantiate her allegations, Sheehan would be well on his way to cracking the Silkwood case. So he subpoenaed Srouji to appear in the federal courthouse in Oklahoma City for a sworn deposition, and he subpoenaed all her Silkwood-related documents.

In a civil suit, the court sets aside a certain number of months for "discovery," during which attorneys for both sides develop the facts needed to support their cases. During discovery, each side has the right to subpoena relevant witnesses to give depositions, taken under oath and transcribed by a court recorder. The depositions themselves are normally not used in any trial resulting from the complaint; rather, they are tools for uncovering facts or investigative leads. If a deposed person proves to be a valuable witness, he or she may be called later to testify at the trial, if there is one. If a deposed person lies in the deposition, he or she may be charged with perjury.

Dan Sheehan had no idea what Srouji would say in a sworn deposition, and he was quite willing to strike a deal with the jour-

211

nalist. If she would tell all she knew about the Silkwood case and hand over the documents, and if her testimony and documents indicated she was not part of a conspiracy to cover up the violation of Silkwood's civil rights, he would drop her from the case. So Sheehan called one of Srouji's attorneys and said he'd like to talk to her before taking the deposition. In the long run, he argued, a predeposition conference would save Mrs. Srouji a lot of time and emotional energy. But Jacque refused to meet Sheehan. She told her lawyer that the FBI was following her everywhere and that she was scared.

Thus, Sheehan was surprised when he got a message on February 13 from Sister Caroline, of the convent of the Little Sisters of Jesus, that Jacque Srouji was trying to reach him. Srouji knew the sisters because they had given her lodgings for a few nights when she had been in Washington.

Sister Caroline was waiting when Sheehan knocked on the convent door. She told him Jacque had called her in tears. "I don't belong in this case," Sister Caroline said Srouji had told her. "I don't have anything I can help with. Please tell them to stop persecuting me. Why are the Jesuits after me?"

"Don't cry, Jacque," Sister Caroline said she had told Srouji. "Just relax. They are going to *help* you."

"Would you call and talk to her?" Sister Caroline asked Sheehan.

Sheehan told the nun that court rules forbade his speaking to Mrs. Srouji without her lawyer on the line because she was a defendant and he was an attorney for the plaintiff. "If *you* want to talk to her, Sister, go ahead," Sheehan said. "But tell her I can't. I'll meet with her and her attorney if she wants. I'll be in Nashville on Saturday, February nineteenth. Tell her to name the time and place."

Sister Caroline called Srouji's neighbor, since Jacque had told the nun she thought her phone was tapped and that it would be safer to call next door. Sister Caroline told Srouji what Sheehan had said, then covered the receiver with her hand. "She doesn't want to tell her lawyer about meeting with you," Sister Caroline told Sheehan.

212

"Tell her no meeting without a lawyer," Sheehan said. "Tell her it would be good to bring her husband along, too."

Srouji finally agreed to meet Sheehan at St. Henry's Catholic Church on the afternoon of February 19. She explained to Sister Caroline that her husband, Suheil, was a deacon in the church and would be assisting at the celebration of Mass that evening.

Sheehan was cautious when he arrived at St. Henry's with Father Bill Davis, now a full-time investigator on the Silkwood case. Sheehan had talked to Taylor, and the investigator had told him what Echo had said: "She won't make a deal. The Bureau got to her. Be careful and don't go to see her alone."

"Where's your attorney?" Sheehan asked when Jacque, Suheil, and their daughter, Claire, finally arrived.

Srouji said that Dominic de Lorenzo had promised to send Aurora Publishers' attorney to help her. She seemed upset that he hadn't arrived yet. "Everyone's always letting me down," she complained. "People are always betraying me."

They waited for half an hour, but Aurora's attorney did not arrive. Srouji insisted on starting the meeting without him. "He'll be along in a moment," she said.

"Look," Sheehan said, "there won't be any conversation without your lawyer here."

Srouji asked Suheil to call "Jack" and have him come right over. She then suggested they go into the church office to prepare for the meeting. Once inside, she began to complain again. She said she was being driven into bankruptcy. She took out some papers. "Do you think I'm kidding?" she asked. "Here, look at these bills for yourself."

Sheehan wouldn't touch them.

"Please don't subpoena me," she begged. "I'm afraid they're going to kill me if they find out I'm talking to you. Please let me out. I don't know of anything that would be of use to you."

Sheehan explained that in a civil rights conspiracy case, if the plaintiff dismisses one alleged conspirator, the court could throw out the entire case unless the plaintiff has a good reason for dropping the charges. He said he couldn't risk that, and that her lawyer would explain the legal point to her when he got there.

213

Sheehan went on to tell Srouji that his sources had told him she had some information about FBI wrongdoing and was reluctant to spill it because she was afraid the FBI would do something to her. "Is that true?"

"Do you have any reason to believe I'm in danger?" Srouji asked.

Sheehan was cautious. If he said he had some evidence that she was in danger, she could later accuse him of threatening her, and the court could dismiss the charges against her. "Do *you* have any reason to believe you're in danger?" he asked.

"Yes," she said.

She told Sheehan and Davis that someone had tried to break into her hotel room recently. She had awakened when she heard the doorknob rattling, she said. In the morning, she found the knob on the floor, outside the door.

"Look, you have nothing to worry about if you talk," Sheehan told her. "The best way to protect yourself is to talk during a deposition. It's crazy for anyone to try to do anything then."

"They told me the only thing that can get me in trouble is my mouth," she said.

"Who told you that?" Sheehan asked.

"I can't say."

"Was he from the FBI?"

Srouji nodded yes, and as soon as she did, Sheehan suspected that she was wired with a mini tape recorder and that the whole conversation so far was planned to get him to talk about the case without her attorney present and to say she was in danger. He became even more cautious and was glad that he had taken Taylor's advice and brought Davis as a witness.

Srouji kept looking at her watch. "I've got to go to the ladies' room," she said eventually and left the office.

When she returned, Suheil and Jack (Sheehan found out later that he was not a lawyer, but a family friend) were waiting. Srouji's mood had changed. "Look, I'm not going to testify," she said. "I know what you want, and I *know* what you want to know. I've got it locked in a safe-deposit box. It's not in the city. I have two tapes with everything you want to know on them."

Suheil cut in. "Jacque, why are you saying this —"

"You're going to have to dig the information out of me," Jacque said.

The meeting was over. It was clear to Sheehan and Davis that they were wasting their time. All they could do now was to depose Srouji under oath and follow her own advice — "dig the information" out of her.

"She's a lot more sophisticated than I suspected," Sheehan told Davis on the way back to the hotel. "This lady is a professional. She knows exactly what she's doing. She's no amateur."

□　□　□

Presiding over the Silkwood case in Oklahoma City was Judge Luther Eubanks, a stooped, potbellied man with a red face and bulbous nose. Srouji's attorney, Claude Love, had filed a motion arguing that, as a free-lance reporter, Jacque was protected by the Oklahoma newsperson's privilege, which said: "No newsman shall be required to disclose . . . the source of any public or unpublic information obtained in the gathering, receiving or processing of information."

The Oklahoma statute had two exceptions, however. If the attorney seeking to depose a reporter can prove that the information he asks for is absolutely vital to his case and cannot be obtained in any other way, the court can compel the reporter to answer all questions. Or if the attorney can prove the reporter officially gathered information for a government agency such as the FBI or the CIA, the court can strip the reporter of the newsperson's privilege.

On February 25, just before Srouji's deposition, Judge Eubanks heard arguments on Love's motion. Love argued that Sheehan could get 90 percent of what Srouji knew from Kerr-McGee and the FBI and had not even tried to do so. Love asked Judge Eubanks to stop Sheehan from going on a "fishing expedition" and to allow Srouji to reveal only those news sources she had documented in *Critical Mass*.

If Eubanks granted Love's motion, there would be very little Sheehan could ask Srouji, for she would duck behind the reporter's privilege every time she didn't want to answer a question.

Sheehan argued, in turn, that he supported the First Amendment right of every reporter, including Srouji, but that Srouji had forfeited her privilege because she was an informant for the FBI. He said Srouji had infiltrated the PLO for the Bureau.

"PLO? PLO?" Judge Eubanks asked. "Why, isn't that the group that kidnaped that woman in San Francisco?"

"No, Your Honor," Sheehan said. "That was Patricia Hearst and the SLA — the Symbionese Liberation Army."

"They're all the same," Eubanks said. "If this woman has anything to do with those people, she will have to tell everything she knows."

Glenn Whitaker, the Justice Department attorney representing Larry Olson, objected. He tried to point out to Judge Eubanks that he had misunderstood Sheehan's allegation. Srouji didn't *work* for the PLO; she had infiltrated the PLO, according to Sheehan, to help the FBI.

Eubanks brushed Whitaker aside and ruled that Srouji did not have a blanket privilege, but had the right to invoke the reporter's privilege on particular questions. If Sheehan objected, he, Eubanks, would review those questions she had refused to answer and then decide if she would have to address them later.

Sheehan had asked to take Srouji's deposition in Judge Eubanks' large federal courtroom to impress upon the journalist that she was under oath and that if she lied, she would be committing perjury. And over the objections of Kerr-McGee and Justice Department attorneys, Sheehan got Eubanks' permission to videotape Srouji's deposition. If for some reason she couldn't be a trial witness, Sheehan at least could play the videotape to the jury so that it could watch her face and listen to the nuances in her voice.

Sheehan had a strategy for the deposition. Besides wanting facts, evidence, and leads, he needed the opportunity to evaluate Srouji's credibility and candor under oath in case he called her as a trial witness. Thus, how she acted and how she answered his questions were important.

Sheehan would also try to establish that Srouji was an informant for the FBI so that he could have the court strip her of her newsperson's privilege at the point in the discovery process that was most advantageous to him. Finally, he would find a legal loophole entitling him to depose Srouji a second time when he had more facts and more questions.

The stage was set for a major confrontation; the clerk swore Srouji in.

"Did there ever come a time when you began to perform a service for a United States Government agency, gathering information for such an agency?" Sheehan asked, laying the groundwork to strip her of her newsperson's privilege.

"In what capacity?"

"I'm not asking in what capacity. Did you ever collect information for a United States Government agency?"

Srouji turned to Claude Love. "Can I answer that?"

"Yes, ma'am. Answer the man's question."

"I thought there was a problem here," she told her attorney.

"No, there is no problem. Answer the man's question with the truth."

Love dropped his microphone. "Hold the cameras," he said, and, as he fidgeted with the mike, he whispered to Srouji off the record.

Sheehan asked Srouji for the third time whether she had gathered information for a government agency.

"At the Nashville *Banner*," she said.

"Now, can you tell us what agencies you performed this service for?"

"It was only the FBI."

"Can you tell us what the services were that you performed for the Bureau?"

"It involved civil rights and antiwar stories that I covered. You know, it was a sharing of stories with them."

"Sharing of stories?"

"Usually in the office of the publisher of the *Banner* or in the office of the executive editor," Srouji explained. "As did other photographers . . . other reporters at the *Banner*. Their stories were sometimes given to the Bureau."

"Did you attend meetings of the Students for a Democratic Society, using an assumed name?"

"I attended the meetings but not using an assumed name."

217

"Did you tell them that you were a reporter from the *Banner?*"

"Well, they knew it," she said. "One of them told me one time that they knew what I was doing. You know, it wasn't any closely guarded secret."

"Did you confirm it when they told you that?"

"I just smiled."

Srouji went on to deny that she had served as an FBI "confidential source" after she left the *Banner* in 1967, with the exception of reporting to the Bureau about the Russian, Colonel Zaitsev, which she said she did "out of self-protection."

So far, Srouji's testimony was consistent with the known facts, but from that point on, the journalist became evasive and contradictory. She denied she had ever spied for the Bureau on anyone who had worked for the Nashville *Tennesseean,* where she had been a copy editor. Yet her publisher, John Seigenthaler, had testified before the Dingell subcommittee that she told him she reported to the FBI on two of his journalists.

Srouji denied that the FBI ever suggested she write a story on Special Agent Joseph Trimbach. Yet Seigenthaler had sworn under oath that she had told him the FBI encouraged her to do the story.

Srouji denied that the FBI or its agents had ever given her documents other than press releases or public source materials. Yet both Peter Stockton and Howard Kohn had signed sworn affidavits saying Srouji had told them Olson had given her FBI documents on the Silkwood case, and that some of those documents were sensitive letterhead memoranda.

Sheehan began to pin Srouji down. "Do you recall testifying on April 26, 1976, to the United States Congress that, in fact, an investigator of the FBI gave you documents?" he asked.

"No, sir."

Sheehan read her the Dingell hearing transcripts: "Srouji: 'Let me see, how can I say this without violating my confidence? I was able to see unofficially the substance of this case as they investigated it. It was a ton of material that indicated to me quite an in-depth investigation.'

"Ward: 'Like 1000 pages? One hundred pages?'

218

"Srouji: 'I would say closer to 1000.' "

Sheehan asked: "Do you recall that testimony?"

"Yes, I do," Srouji said.

"Do you ever recall having been put into possession of those — of any of those documents?"

"No, sir."

"Did you at any time take documents and Xerox them, documents that had been given by the FBI?"

"Some, yes."

"Were they at any time on FBI letterhead?"

"No, sir."

"Did you see summaries [of FBI interviews about Silkwood]?"

"Yes, sir, I did."

"Were you shown those summaries by an FBI agent?"

"I saw that stack of papers, and a man was sitting there, but he was not showing me the papers."

"Did he allow you to read those summaries?"

"No, he did not, sir. He did not."

"Then are you saying that you *never* read those summaries?"

"No, sir. I did not read the summaries."

Once again, Sheehan read Srouji her own sworn testimony before Congress: "But I saw tons of material, you know, interviews with the guy that drove the wrecker, comments he made. Particularly the contamination itself, that seems to be a big focal point as to what was occurring."

"How did you find that out if you didn't read them?" Sheehan asked.

Srouji did not answer.

"Couldn't it have been from reading those, Mrs. Srouji?"

"No, I saw them stacked up. I did not read them."

"If you didn't, how do you know that?" Sheehan pressed.

"I would have to invoke confidentiality of sources."

Sheehan moved on to another topic. "Were you ever told by an FBI agent or an agent of the Justice Department that the only thing that will get you in trouble is your tongue?"

"Yes, sir."

"Do you recall telephoning Mr. Peter Stockton, a congressional investigator, February 13, 1976, and telling him that you had contact with an FBI agent, and that subsequent to that contact you were having to reconsider your cooperation in giving information on the Silkwood matter?"

"I recall that, but it had nothing to do with an FBI agent."

"Can you tell me who contacted you, if it wasn't an FBI agent?"

"No, I can't . . . I just — I don't remember."

Srouji began to talk off the record to Claude Love, but Sheehan pressed. "Are you saying that you won't give that to us?"

"No, I won't give it to you," Srouji said.

"On what grounds?"

"I don't think you are after the truth." Srouji began to cry. "You told me last Saturday I was going to be killed and I want a judge to hear it."

"Who told you this last Saturday?" Love asked her. He knew about Srouji's meeting with Sheehan and Davis because Sheehan had told him.

"Mr. Sheehan," Srouji said.

"Where? Where did you see Mr. Sheehan?" Love asked, feigning indignation.

"You know perfectly well where I saw her," Sheehan interrupted.

"But I want it on record here," Love said. "Why are you afraid, Mrs. Srouji?"

"He told me last Saturday I was going to be killed."

"That's not true," Sheehan said. Then he returned to the question of who had told her not to cooperate with Dingell's subcommittee. "Are you saying at this time that you don't recall who that person was?" he asked.

"I — I recall who it was now," she said.

"Okay, can you tell us who it was?"

"I can't tell you," she said, invoking her newsperson's privilege and crying once again.

"Now," Sheehan continued, "did you at any time during [your] conversations with Mr. Seigenthaler give him any document that was confidential or in any other way classified?"

220

"I refuse to answer on the grounds it might tend to incriminate me."

"Did you obtain that in the course of your services for the Navy?"

"I refuse to answer on the grounds it might incriminate me."

"*Did* you or did you *not* give a document to Mr. Seigenthaler about which he testified before Congress?"

By that time, Sheehan was furious, not so much at Jacque Srouji but at the government. He felt Srouji was lying. In fact, he had concluded that after the first half-hour of her deposition, and he reasoned that she would never have perjured herself so brazenly if Washington had not promised to protect her. Sheehan concluded that the government had made a deal with her — she'd got it into the Silkwood mess, and if she would perjure the government out, it would protect her. He would eventually report her to the U.S. Attorney for perjury.

Sheehan stopped the deposition, telling Srouji and Love that he would ask Judge Eubanks to compel her to answer all the questions she had dodged so far. Only then, Sheehan said, would he continue taking the deposition. It would take Eubanks weeks, if not months, to rule on the twenty or so questions she had refused to answer during her 300-page deposition. That would buy Sheehan some time.

JUNE 1977

James Ikard, the Oklahoma City lawyer who had filed the complaint in person in the clerk's office for Sheehan, took the second deposition from Jacque Srouji. A former Kansas State University basketball player with prematurely gray hair and beard, Ikard looked like a thin Papa Hemingway. He had a soft, almost gentle voice, and Sheehan had asked him to continue deposing Srouji so that she could not try to tie her testimony in a phony legal knot, alleging that Sheehan had threatened her.

221

Srouji dodged questions as artfully during the second deposition as she had during the first. She swore she couldn't remember if anyone was present when she met James Reading to get information on Silkwood for her book, where and how many times they had met, and if she had taped their conversation or taken notes. She denied that Reading had ever told her that his people had placed Silkwood under surveillance, wiretapped or bugged her apartment, and she denied that she had ever told Peter Stockton and Howard Kohn that Reading had. Yet Stockton and Kohn had signed sworn affidavits stating she had told them about the alleged bugging and tapping.

She testified that she had come to believe that Karen contaminated herself but that her crash was no accident. But she took the Fifth Amendment when Ikard asked her which FBI agent had discussed the Bureau's Silkwood investigation with her.

"Do you have any information or knowledge that agents of the federal government ever wiretapped any telephone used by Karen Silkwood?" Ikard asked.

"I claim my newsman source, First Amendment rights," Srouji said.

"Have you ever been given access to documents not available to the public?"

"I would claim First Amendment rights on that question."

"Have you ever received any FBI documents not available to the public in an unofficial capacity?"

"I refuse to answer on the grounds it may incriminate me."

"Were you allowed access to the Kerr-McGee investigative files in the Silkwood matter?"

"First Amendment privilege."

"Did you ever meet Larry Olson in an Oklahoma City motel and examine his file on the Silkwood documents?"

"No, sir."

"Did you copy the documents that Larry Olson had in his FBI file?"

"I'm going to invoke the Fifth on that."

"Is it true you made copies of about fifty pages from the file?"

"I'm going to take the Fifth."

"Did Mr. Olson take you step by step through his investigation?"

"I'm going to take the newsman privilege."

"Did he ever discuss the FBI letterhead memo?"

"Fifth and First."

"And did he tell you his theory about the missing plutonium and the OCAW?"

"I'm sure it was discussed."

"Do you recall what he thought had happened to the plutonium when you had your conversation with him?"

"I'll take the First on that."

"It was your opinion that you gave under oath that 'every individual even remotely connected with Karen Silkwood had been interviewed.' Now, what's the basis for that statement under oath?"

"I think I'll take the Fifth."

Srouji had Sheehan up a tree. It was clear that she would not crack, even if the court stripped her of her newsperson's privilege, for she could still call on the Fifth Amendment or say she couldn't remember or just lie, knowing that someone in Washington would protect her. Sheehan realized there would be no easy way to prove the conspiracy to deprive Karen Silkwood of her civil rights and the conspiracy to cover up the crime. His investigators, Taylor and Davis, would have to dig in Washington, in Oklahoma City, in Crescent — anywhere the leads took them — because Sheehan suspected that the Silkwood case had the tentacles of an octopus.

Chapter 23

A friend from ACLU's Atlanta office called Dan Sheehan with a lead. The Atlanta *Journal* had just published an exposé on the Georgia Power Company's secret security force, the friend said. There might be a link to Karen Silkwood.

Like others who had followed the activities of the intelligence community, Sheehan had concluded that there was a private, national network that collected dossiers on antinuclear activists for the FBI, the CIA, and the police. He had heard names like *Information Digest* and Law Enforcement Intelligence Unit (LEIU) bandied about, but didn't know what they were. If either or both spied on antinukes, Sheehan reasoned, they might have been watching Karen Silkwood. It was a thin thread, but Sheehan didn't have much else, so he sent Taylor and Davis to Atlanta.

□ □ □

The Georgia Power Company began keeping files on "subversives" and "sexual deviates, sickos and pinkos" in 1973, the year before Karen Silkwood was killed, according to former Georgia Power investigators. The Georgia Power nine-man undercover intelligence unit, with an annual budget of $750,000, was made up of veterans from Army Intelligence; the FBI; the Federal Bureau of Alcohol, Tobacco and Firearms; the CIA; and the Georgia Bureau of Investigation. Each Georgia Power surveillance car had a pistol, shotgun, radio, camera for nighttime photography, and a flip switch that changed headlight configurations to confuse the driver being tailed. Georgia Power's surveillance equipment was so good that the FBI borrowed it occasionally.

The Georgia Power Company was plugged into a national network that circulated secret information on so-called dissidents. One link in that network was *Information Digest,* a thirty-page biweekly newsletter that New York State Legislature investigators described as a cover "to develop dossiers on thousands of patriotic and decent Americans who had committed no crime and were not suspected of committing a crime . . . the string that held together a network of hidden informants whose information was recorded by police departments throughout the nation without the individual knowing the process and without independent checking by police as to the validity and source of the derogatory information."

The National Security Agency, the CIA, FBI, IRS, Drug Enforcement Agency, and United States Customs had all used *Information Digest,* edited by John Rees, a known countersubversive who worked out of the congressional office of Georgia representative Larry P. McDonald. *Information Digest* was protected from public disclosure. If questioned about the newsletter, Rees (alias John O'Connor and the Reverend John Seeley) could claim the newsperson's privilege, and McDonald could hide behind congressional immunity.

Tom Baxter, author of the Atlanta *Journal* exposé, led Bill Taylor and Father Davis to two former Georgia Power Company investigators — John Taylor and William Lovin. John Taylor, who had been an intelligence officer in the United States Army for nine years, told the Silkwood investigators that Georgia Power probably had a dossier on Karen Silkwood. He recalled how the security force had laughed and winked at each other when they heard about Karen's accident. "I wonder what happened to her?" they said, according to John Taylor.

Bill Lovin said he actually had seen a Georgia Power Company file on Karen Silkwood, taken either from *Information Digest* or a Law Enforcement Intelligence Unit report. Lovin, who was unemployed and almost broke, talked about a national computer with dossiers on so-called dissidents. "I could get anything I wanted on your background — by going directly to a sheriff or possibly a chief of police in this state — or anything that's been fed into the national computer," he said. "They'd say, 'When we give you the computer printout . . . after you've transcribed it, burn it.' "

Lovin told the Silkwood investigators the man they *really* wanted to talk to was Art Benson, chief of Georgia Power security and a good friend of John Rees's. Georgia Power had hired Rees to infiltrate an Atlanta citizens' group opposed to nuclear power. Lovin said Benson floated in and out of CIA ventures, including the Bay of Pigs, and that Benson had once been a cop in Boynton Beach, Florida, along with two other Georgia Power investigators. Jack Holcomb, who ran a wiretapping school in Fort Lauderdale, was a close friend of Benson's, Lovin said.

Lovin was a good break. All Sheehan had to do was get the Georgia Power Company file on Karen Silkwood. He subpoenaed the file, along with Art Benson and Bill Lovin. But when Bill Taylor went to serve Lovin weeks later, the former Georgia Power investigator was gone. Most of his furniture was still in his house; it looked as if he had packed his toothbrush and fled in the night. Lovin had left no forwarding address and apparently hadn't told anyone where he was going.

Taylor remembered that Lovin's wife was German, so he asked "Sierra," a source with connections throughout the intelligence community who also had access to Immigration records, to trace the couple. Sierra called Taylor back in a few days. The Lovins had moved suddenly to Simmerfeld in West Germany, Sierra said.

When Art Benson got his subpoena, he sent up a legal smoke screen to avoid appearing in court. When that didn't work, he requested the court to quash the subpoena, claiming that he knew nothing about Silkwood. That left Sheehan, once again, holding an empty bag. There was no doubt in his mind that someone had bought off Bill Lovin. Sheehan felt his investigators were on to something, so he sent Bill Taylor to dig in Fort Lauderdale and Boynton Beach, and Bill Davis to chase down the Law Enforcement Intelligence Unit.

□   □   □

Davis learned that LEIU was an intelligence system that collected information on known criminals and "terrorist suspects," linking the intelligence squads of major police forces in the United States and Canada. As a private club, LEIU was not subject to federal, state, or citizen control. And hardly anyone knew it existed.

LEIU was founded by policemen in 1956 in an attempt to break the FBI's monopoly on the files of criminals and suspects. By 1975, LEIU had grown into a mini-Bureau, with 225 law enforcement agency members. New applicants had to be sponsored by one LEIU member and endorsed by at least three more. Before voting the LEIU board of directors conducted an intensive investigation to find out if the applicant could be trusted and was leakproof. Once accepted, *one* person in the law enforcement agency received LEIU secret suspect files — 5-by-8-inch index cards with background information on the suspect and a source to contact for more details.

When the LEIU representative in the law enforcement agency left, membership was suspended until another representative could be appointed and approved. And if any LEIU member organization either gave the secret files to non-LEIU members or could not maintain secrecy, it could be suspended.

Several LEIU members had been charged with illegal wiretapping and spying, among them the Michigan State Police and the Chicago, Baltimore, Houston, and Fairfax, Virginia, police departments. The Fairfax Police Department had close ties with the CIA, headquartered down the highway in Langley, Virginia. And LEIU had contacts within the Department of Labor and in Air Force and Coast Guard Intelligence.

Even more important, LEIU was secretly receiving government funds. The Law Enforcement Assistance Administration, a Justice Department program, gave the California Department of Justice more than $1.7 million to put LEIU files on a central computer at the Michigan State Police headquarters in East Lansing. The national network, with thirty on-line computer terminals, was called Project Search.

In his hunt for a Silkwood link, Bill Davis learned that the Oklahoma State Bureau of Investigation in Oklahoma City was an LEIU member. And it was the OSBI that, at the request of

James Reading, Kerr-McGee's security director, had analyzed the cigarettes and medications found in Karen Silkwood's purse the night she was killed, as well as the pills, leafy substance, and hypodermic kit taken from her apartment during decontamination.

☐ ☐ ☐

Back in Fort Lauderdale, Bill Taylor was checking out Jack Holcomb's connection with Art Benson. Until 1968, it turned out, Holcomb had been a private detective and free-lance wiretapper who bragged that he had tapped and bugged for more than 400 law enforcement agencies, including the FBI, that gave him the jobs they "wouldn't touch."

Holcomb founded the Solar Research Corporation to manufacture and sell electronic surveillance equipment, but the company was destroyed by fire under mysterious circumstances. Somehow, Holcomb got himself appointed councilman, fire commissioner, and police commissioner in Sea Ranch Lakes, Florida, close to Boynton Beach. But he was fired by the mayor because of his "background."

In 1969, Scotland Yard and the British paratroopers booted Holcomb out of the tiny Caribbean island of Anguilla, claiming he was helping the underworld to get a toe hold there under Acting President Ronald Webster. Holcomb denied the charge, saying he was just Webster's "economic" adviser, and that he wanted to build a construction company on the island.

Then, in 1970, just two weeks before the Haitian Coast Guard revolted, Haiti also kicked Holcomb out, accusing him of being an agent of the United States Government. Holcomb denied the charge, saying he was in Haiti only to buy native black art for resale in the United States.

Sometime between 1970 and 1971, Holcomb became friends with Leo Goodwin, Jr., a former paratrooper, former Army Intelligence officer, former Texas Ranger, and millionaire police buff. Goodwin convinced his father, the founder of the GEICO Insurance Company, to create the Leo Goodwin Foundation, which would, among other things, help the nation maintain law and order. The old man put up $100 million, then died four months later, making Junior one of the twenty-five richest men in the country and the president of a foundation.

Goodwin helped Holcomb build the Audio Intelligence Devices Corporation (AID) in Fort Lauderdale; it became the largest private company to design and sell high-grade wiretapping, bugging, tracking, and other surveillance gear. In 1973, Holcomb and Goodwin founded the National Intelligence Academy (NIA). The NIA was housed in a separate wing of the AID building, next to the Fort Lauderdale Executive Airport, where NIA/AID owned two heliports and a private airstrip long enough to land a 727. Goodwin owned the building. The NIA got free rent; AID had to pay. Goodwin and his foundation poured more than $3 million into NIA/AID.

The NIA offered two-week courses on the state-of-the-art in electronic surveillance. The students, who lived at the Tradewinds Hotel (owned by Goodwin), learned how to bug a room and tap a phone in five minutes in the NIA's secret Classroom D — a fully equipped telephone city with nearly every kind of indoor and outdoor terminal used by phone companies. In the middle of the room was a rigged telephone pole, which the students used for practice. The NIA records showed that one quarter of its students came from states like Oklahoma, where wiretapping was illegal without a court order.

A sign at the NIA's entrance read: U.S. GOVERNMENT REGULATIONS PROHIBIT ANY DISCUSSION OF THIS FACILITY. SORRY, THE RECEPTIONIST IS INSTRUCTED NOT TO ANSWER RELATED INQUIRIES. And Holcomb bragged that both the NIA and AID had "secret" clearance from the Department of Defense.

When Bill Taylor cased NIA/AID, security was tight. There were armed guards at the building and airstrip, television cameras, and roving patrols who watched the streets around the building. Taylor saw a lot of dark-skinned people, both men and women, who looked and dressed like foreigners, going in and out of the school. He noted that CIA-owned airlines, like Air America, used the airstrip and that there were thirteen flights a day to and from Andros Island, in the British Bahamas. Taylor called Sheehan and suggested he check out Andros.

229

A helicopter dropped Taylor into the island's jungles, and he spent four and a half days snooping. He noted the United States destroyer 974 in the bay and saw that the island was full of underground facilities. He learned that the CIA, United States Naval Intelligence, and British Intelligence had command posts there. And he found out the jungles and bases were used for specialized training of some kind.

Taylor came back to Florida. As far as he could tell, NIA/AID students were being sent to Andros Island for more training; the setup in Fort Lauderdale might well be a red herring to draw attention away from Andros.

It made sense. In the 1960s, the United States International Police Academy (IPA) trained foreign police, intelligence, military, and security officers in surveillance skills, gave them field training in counterinsurgency, and recruited promising students as CIA informants and operatives. Congress closed the IPA when the public learned that it had trained men who turned out to be assassins and torturers. Holcomb's National Intelligence Academy had taken over the first IPA duty — to train foreign students in surveillance. Was Andros Island the school for teaching them counterinsurgency and screening for the CIA?

Taylor bought a source inside NIA/AID who told him that Oklahoma police officers on the state and local level had been trained at the NIA, that the Oklahoma City Police Department had bought equipment from AID, and that Jack Holcomb had a file on Oklahoma.

When Taylor returned to his motel room one afternoon, a journalist from the Miami *Herald* was waiting. The reporter said that he had heard Taylor was conducting some kind of investigation in the Fort Lauderdale area, and he wondered what it was. Maybe he could help Taylor, since he knew Fort Lauderdale well. Maybe they could swap information.

Taylor was suspicious. He got rid of the reporter and called the *Herald*. The newspaper's personnel department had no record of the man. Taylor then called his source, Sierra. "Put a check out on him."

Sierra called back two hours later. "He works for Intertel."

International Intelligence, Incorporated, is a private intelligence-agency-for-hire, founded by veterans of the Naval Intelligence Agency and the National Security Agency. Intertel's staff consists of former FBI, NSA, and Secret Service men, and the company specializes in working for conglomerates and multinational corporations. Among Intertel's clients were billionaire Howard Hughes, who had CIA links; financier Robert Vesco, who fled the country before he could be tried for fraud; and ITT, which was tied to Richard Nixon and Watergate as well as to the CIA's efforts to dump Chilean Marxist president Salvador Allende.

Bill Taylor felt like a novice gumshoe for blowing his cover. "I fucked up," he reported to Sheehan.

Sheehan wasn't worried. That Intertel shadowed Taylor only confirmed that his investigator was moving in the right direction. Sheehan made two important decisions. Because the discovery period was limited and money tight, he couldn't afford to investigate thoroughly both the conspiracies and the negligence counts. Since Sheehan felt the negligence charge against Kerr-McGee was relatively easy to prove, he decided to concentrate almost exclusively on the conspiracies. Given enough time and a lot of luck, he might even find out who had killed Karen and why.

Second, the conspiracy investigation could either begin with the big picture outside Crescent and then narrow down to Oklahoma, or it could start in Oklahoma and then expand as the investigation found links to Karen Silkwood outside the state.

Sheehan chose the broad approach. For one thing, Sheehan believed Bill Taylor's life was at stake, and the investigator would be safer if he knew what and whom he was stalking. For another, Sheehan was convinced his team would learn more, make fewer mistakes, and same more time and money if they knew *what* they were looking for in Oklahoma.

□ □ □

One of Taylor's best Oklahoma sources was James V. Smith, a former division manager at the Kerr-McGee plutonium plant. Smith and Taylor had hit it off from the start, and Taylor told Sheehan that Smith would make an excellent trial witness. As part of the Kerr-McGee management staff, he was an insider; as a manager, he didn't belong to the union; he had worked at the plutonium plant from the day it opened to the day it shut down; he was cooperative, didn't seem to have an ax to grind, and came across as open, honest, and sure of himself.

Jim Ikard put Howard Kohn of *Rolling Stone* on to Smith, and Kohn interviewed him in depth. Soon after Kohn's article appeared in the magazine, attorneys for Kerr-McGee subpoenaed Smith. This came as no surprise, for all of Kerr-McGee's legal maneuvers so far had been to try to get the case against the corporation dropped without a trial:

□ They had subpoenaed Bill Silkwood in an apparent attempt to find out if Sheehan had prompted Karen's father to file the complaint. If Sheehan had, they could accuse him of "fomenting litigation," grounds for disbarment in the State of Oklahoma. Bill Silkwood couldn't be trapped.

□ They had subpoenaed Bill Davis to see if there was a Jesuit plot to discredit Kerr-McGee. But Father Davis represented himself, not the powerful order.

□ They subpoenaed Kitty Tucker and Sara Nelson to find out if the Silkwood case was secretly financed by a left-wing antinuke group. It wasn't.

Now it looked as if Kerr-McGee attorneys were trying to catch Howard Kohn fabricating quotations or stretching the truth. If he were, Sheehan suspected that Kerr-McGee would sue *Rolling Stone* for libel to take the heat off itself. If that was Kerr-McGee's strategy, it didn't work.

From 1952 to 1969, Jim Smith worked at the Rocky Flats, Colorado, plutonium plant run by Dow Chemical. He had scratched his way up from fireman to plutonium operations supervisor, and in 1969, when he heard that Kerr-McGee was building a plutonium plant and was looking for workers, he applied. Kerr-McGee called him to Oklahoma City for an interview. Smith toured the uranium plant in Crescent, inspected the plutonium plant under construction, signed a contract to work in the plutonium plant once it was ready, and went home. A few months later, Kerr-McGee asked him to report for work. To his utter amazement, K-M assigned him to the uranium plant.

"That place wasn't fit for man or beast to work in," he said during the deposition to William Paul, one of the attorneys Kerr-

McGee hired to defend it. "One hundred twenty, thirty degree temperatures . . . Why you never saw such a pigpen in your life . . . Uranium everywhere. You could wade in it, go by and see piles of it on the floor."

Smith said that in 1969 Kerr-McGee did not require its uranium workers to wear protective booties, so they picked up contamination on their shoes and brought it home. "You could see the powder on the floor," he said. "You could look at it on the people's clothes . . . powder, black smudges, green smudges — depending on the form."

Bill Paul soon changed the subject to the plutonium plant. He asked Smith if he knew whether union workers had deliberately contaminated air filters, as Kerr-McGee health physics director Wayne Norwood had alleged.

"Not intentionally," Smith said. "I don't know why anybody would be stupid enough to get one of them contaminated. All *that* does is just put you in a full-face respirator. If you've ever wore one of them, you don't want to wear them."

Paul then asked Smith if he knew of any attempt to intimidate Karen Silkwood or other union members.

"I feel there was," Smith said. "Jerry Brewer was a plutonium worker who is a union official and he was transferred, I might say, to the uranium plant under conditions he opposed greatly . . . And Ray Stillwell was another union official. He was transferred out . . . Almost every union official ended up on what we call the rock pile."

"What's the rock pile?" Paul asked.

"That's when they pulled them all off the process area and put them out there to sweeping sidewalks, and had a guard with them all the time like a bunch of convict labor."

"Any armed guards?"

"Not armed," Smith said. "He was a foreman, and his only job was not to let them out of his sight. Well, it would be no different from going down here and taking a work gang out of McAlester [State Prison]. Only difference is, he didn't have a gun."

"How did you know his only job was not to let them out of his sight?" Paul asked.

"Because he was *my* foreman. I had to give him up and send him over there to watch them."

"Did he tell you that was his job?"

"Yes, sir," Smith said. "Almost anybody that was a union official ended up on the thing. Now, that is an odd coincidence."

"Do you know who picked them out?" Paul asked.

"No, sir. We were just told. The foremen were all called and asked, 'Send so and so and so up.' That was it."

"What was the company's reason for having a rock pile?"

"They didn't tell me," Smith said. "I assumed to get them out of the production areas . . . They were watching them very close."

"Is there anything else you could say about intimidation?"

"The intimidation was at all of the union officials," Smith said. "They were given a bad time."

"During Karen's lifetime," Paul asked, trying to narrow Smith's answers, "what bad times were given to them?"

"She was harassed," Smith said. "Her boss told me so, Jerry Schreiber." (Gerald Schreiber had supervised the Emission Spec Lab.)

"How was she harassed?"

"I don't know. Schreiber told me one time which I'll quote. He says: 'I don't give a damn what they say about her. She's still a good tech, and she does her job, and she doesn't give me any trouble.' They was putting the heat on him to put the heat on her."

"Were you aware of any unlawful management activity in connection with efforts to learn what the union was up to?" Paul asked.

"They never consulted me," Smith said. "But they always seemed to know."

"Who is they?"

"Morgan."

"Morgan Moore?" Paul asked. Moore was the Cimarron facility manager.

"He always seemed to know what was going on," Smith said.

"What was going on at the *plant?*"

"No," Smith said. "At the union meetings and so forth. I

234

didn't know what his pipeline was . . . He would come to the [management] meeting in the morning and mention something that had taken place at the union meeting, or something that was said at the meeting."

Bill Paul directed the deposition to Karen Silkwood's contamination the week before she was killed. Smith testified that he couldn't see any reason for Wayne Norwood and his crew to rip up Karen's apartment; he thought a lot of her belongings could have been decontaminated easily and saved. Then Smith made a startling statement.

"The barrels, after they were sealed and inventoried, were brought back into the plutonium plant and reopened for the inspection of other people," Smith testified.

"For whom were they reopened?" Paul asked.

"I don't know."

"And you were *told* they were reopened?"

"I *know* they were reopened," Smith said. "I was told to go out and remove the Silkwood barrels from storage and bring them in to be opened for inspection, of whom I don't know, on the night shift."

Smith testified that the next morning he was ordered to put the barrels back into storage.

Bill Paul moved on to allegations that Kerr-McGee had tried to sneak around AEC regulations. "Do you know of any time that the company did not properly report to the AEC?" he asked.

"On a lot of spills, I would say."

"Contamination incidents?"

"Yes," Smith said. "For instance, the reporting stated that any time the plant went down for twenty-four hours, this was a reportable incident . . . But to by-pass this reporting, people were kept in respirators and continued to keep it operating which made it nonreportable. Do you see what I am saying?"

"In what respect were Kerr-McGee and the AEC putting on rose-colored glasses?" Paul asked, referring to a *Rolling Stone* allegation.

"For instance, when they ordered the shutdown to find the [missing] plutonium," Smith said, explaining that the AEC had ordered Kerr-McGee to shut down because the corporation had reported some missing plutonium. Then, suddenly, the AEC changed its mind and told K-M it could reopen the plant.

"If the goddamn plutonium was missing, and missing in that plant," Smith testified, "it had a chance of having a criticality and killing all of us. They should have shut it down until they found it . . . It could have been all in one little pile and just kept accumulating until it went off. They didn't know where it was. Why was it allowed to operate, not knowing where that plutonium was?"

"Well, how much information do you have about the AEC inspection in that instance?" Paul asked.

"I know that they came in and we were on a shutdown notice, and it wasn't but a couple of days later until we were released when the MUF was at 1.8 kilograms roughly. And Kerr-McGee got permission to go to 5 kilograms [11 pounds]."

"Are you charging there has been impropriety by the AEC?"

"Somebody did something."

"But are you confident that the numbers didn't come out right?" Paul asked. "And they started up the plant when they shouldn't have?"

"Well, I have been working on inventories for twenty some years. I think I would have a pretty good knowledge when something was wrong."

"Well — "

"We took an inventory and started up and were ordered to shut right back down again," Smith said. "We were told what it was for. We were short."

"Who told you that?"

"The plant manager . . . We were continually short every time we took an inventory. It just growed from one inventory to the next."

Smith went on to say that the managers were told in their morning staff meetings when the AEC inspectors were coming to the plant.

"Well, is there something improper about that?" Paul asked. "Do you charge there was?"

"It was supposed to be *unannounced* inspections, so somebody — well, there was a leak somewhere."

"How much lead time would you have?"

"Two or three days, maybe a week. We always had plenty of time."

Smith then told Paul about the public address–system operator who forgot to turn off the outside speaker one day when the "unannounced" AEC inspectors came. " 'All right, fellows,' he told the workers over the P.A. 'Shape up. Here they come down the hill now.' Them guys come in the door laughing and said, 'Thank you for the greetings,' " Smith testified.

Bill Paul began to probe allegations that Kerr-McGee security was not up to snuff.

"For the first few years," Smith testified, "there was zero."

"Zero security?"

"Zero security, yes."

"No fence out there?" Paul asked.

"Not originally, no."

"Anybody could walk into the plant?"

"Anybody that wanted to."

Paul asked Smith whether K-M security was lax in 1974 after the fence had been built and security guards hired. Smith told him security was so loose, someone stole $5000 to $10,000 worth of platinum used in making fuel rods and all of the hand-held calculators used in the office. Someone even carried off a huge stereo past the security guards.

"So if they could steal all of that," Smith said, "I guess they could steal about anything."

Smith ended his deposition by commenting on the contaminations in the plutonium plant. "We just continued to clean while we operated," he said. "Basically, you should shut down and clean up. In fact, there is *still* contamination there from previous leaks, right *now* . . . We always seemed to clean enough just to get by, to keep operational . . . But I don't think that anything was ever completely cleaned up from an incident."

Chapter 24

Sheehan and Davis moved from Washington to a cheap motel outside Oklahoma City to try to figure out what to do next. The Silkwood case had reached a crossroad. There was less than $1000 in the kitty — hardly enough to take one more deposition — and the investigation was just getting under way. Judge Luther Eubanks would not keep discovery open much longer; and without more facts, the case against Kerr-McGee would never hold up in court.

The Jesuits had asked Father Davis to resign as Social Ministries director, and he had done so. The Jesuits were upset because Davis had concentrated most of his resources on social projects outside the religious order, such as the Silkwood case, when they expected him to improve the ministries within the congregation. Furthermore, the Jesuits had taken a strong dislike to Dan Sheehan, whom they considered disrespectful and pushy.

The Quixote Center, a small peace-and-justice institute in Washington, adopted Davis, Sheehan, and Karen. Like Davis, the center's codirector, William Callahan, was a Jesuit and an outspoken critic of the Catholic Church's fear of tackling touchy social issues. Callahan felt that if no one else would adopt Karen Silkwood, the Quixote Center could at least keep her alive by paying subsistence salaries — less than $700 a month — to Sheehan, Davis, and, eventually, Sara Nelson. Even that would almost break Quixote's fragile financial spine.

Sheehan called Sara Nelson. "Unless I get twenty-five thousand dollars by the end of the month I'll withdraw two-thirds of the case," Sheehan told her, threatening to drop the two conspiracy counts against Kerr-McGee and the FBI. "I must, because I can't go out on the limb with the court without losing my credibility as a lawyer and the credibility of the case. We're at a turning point with the court. If we can't back up our complaint with facts, the court will throw it out."

"All right," Nelson said. "Proceed as if you had the money."

Nelson was in a bind. She had helped raise the initial $15,000 that had kept the case afloat so far. The $2500 needed to develop and file the complaint had come from the National Emergency Civil Liberties Foundation; the LARAS Fund had kicked in another $7000; and private individuals like Maryanne Mott Meynet and Marion Edey had made up the difference. If Nelson were to take the responsibility for raising the money needed to see the case through trial, it would be a full-time job. What about her NOW Labor Task Force programs?

Nelson did what she always did when she faced a big decision: she prayed, asking for a sign. If she could raise $35,000 within a month, she would know that she should give up NOW and join Silkwood. She talked to David Hunter, of the Stern Fund, and Jon Wenner, publisher of *Rolling Stone*. She leaned on friends, and friends of friends; then she called Sheehan.

"So far, I have pledges for thirty-five thousand," she said.

The investigation was on again, and Sara Nelson became a full-time Silkwood fund raiser. By the end of October, she would raise a total of $48,000.

Bill Davis did his share to relieve the financial pressure, too. He was driving past Corpus Christi Catholic Church on the northeast side of Oklahoma City one Saturday afternoon. Since he hadn't said Mass, he stopped to join the congregation there. Dressed in faded jeans and sporting a beard, he looked more like an oil-rig foreman than a Jesuit priest. When no one began the Mass, Davis checked at the rectory next door. He was told the pastor, Father Paul Gallatin, was out of town and that Gallatin's substitute apparently had forgotten to show up. Davis vested and said the Mass. Later, he called Father Gallatin to introduce him-

239

self so that the pastor wouldn't wonder if an impostor had donned the cloth.

Father Gallatin, it turned out, knew all about the Silkwood case, and when he heard that Davis, Sheehan, and Taylor were sharing a cheap motel room to cut costs, he invited them to live at the rectory. He was the only person in a big house, he said, had plenty of room, and would enjoy the company. Davis could help him with the parish work, time permitting, and that would be all the room-and-board fee he'd ask.

The Karen Silkwood case had a new home.

☐ ☐ ☐

The FBI finally released 2000 pages of the Silkwood documents Sheehan had subpoenaed. They were disappointing and intriguing at the same time, and so full of deletions — dates, names, words, sentences, paragraphs, pages — that Sheehan, Davis, and Bill Taylor spent days trying to piece them together. Over 200 of those pages dealt with the Larry Olson–Jacque Srouji connection and, according to the documents, the story went like this:

Between 1970 and 1975, after she had quit the Nashville *Banner,* Srouji worked as an undercover agent for a federal intelligence agency. The FBI blacked out the name of that agency ("Srouji claimed to be a ——— for the ———") but it was clear from the context that the mystery agency was not the FBI. It could have been the National Security Agency, the Naval Intelligence Agency, or, most likely, the CIA. The agency had suggested that Srouji "write a book on the nuclear industry in order to make contacts in that area."

As a counterespionage agent, Srouji kept up her contacts with the FBI. Thus, in March 1975, Larry Olson got a call from a Bureau friend in Nashville telling him that Jacque Srouji said she would be stopping off to visit Olson on her return trip from the West Coast, where she was researching nuclear plants for her book. The Nashville agent told Olson that Srouji had national security information he needed for one of his cases, but that she wouldn't give it to him because she didn't trust him. He asked Olson either to get the information from her or to put in a good word for him. Olson said he'd do what he could.

Olson met Srouji at the Will Rogers World Airport in early March 1975 and took her out to lunch. She told him she wanted to do a chapter of *Critical Mass* on Karen Silkwood, so they talked briefly about the FBI investigation. Olson asked Srouji about the national security matter; she told him she would discuss it when she returned to Oklahoma in April. All references to the specific national security matter they had discussed were blacked out in the FBI documents.

On April 13, 1975, Srouji returned to Oklahoma, as she had promised, and once again Larry Olson met her at the airport. She told him she had "some stuff as hot as hell," and that from then on she would work with the FBI on "current" contacts but would not discuss her past activities. Olson convinced her to stay overnight.

At that point in the story, the FBI documents became contradictory. Srouji said Olson made a reservation for her at the Northwest Holiday Inn, where Wodka, Burnham, and the AEC had stayed the night Karen was killed, exactly six months earlier. (Olson denied he had made the reservation.) Srouji paid $14.50 for a single room "even though the registration card reflected that two persons occupied the room." The FBI documents did not say who the second person was. That night, Olson took Srouji to dinner at Hoe Sai Gai's at his own expense.

The next day, Olson and his Bureau superior (name deleted, though probably Rosack) interviewed Srouji about the "hot" national security matter. She gave them "only a moderate amount of information and sort of clammed up." After the interview, Olson filed a report to Washington, but the FBI documents deleted all references to the substance of the Olson-Srouji interview.

Olson said he did not give Srouji FBI documents at any time, but that he did mail her some news articles and press releases "to keep her sweet" because she seemed "to be a good friend of the Bureau," had "unusual potential for further development," and showed "no signs of unreliability or instability." Srouji, however, said Olson gave her a thick, black ledger that had holes in the top and was held together with an Acco fastener. She said Olson left the report with her at the motel and that she photographed the cover and copied about fifty pages containing "all the details of

241

the [Silkwood] case." The photocopying, she said, cost her $30. But when Bureau agents checked the Holiday Inn's 3-M Copier records for April 1975, they found that no one had paid more than $4.00 for copies on the days Srouji had been a motel guest.

Srouji said that the next day Olson arranged for her to meet James Reading, to whom she talked for several hours. Olson, however, said he had merely given Srouji Reading's phone numbers, had never contacted Reading for her, and had never arranged for any meeting.

Reading confirmed Srouji's story, according to the FBI documents. Reading had known Olson professionally for "several years," and had given Olson "probably almost everything" on Kerr-McGee's Silkwood investigation. He met Srouji at Olson's request, for Olson had told him "he had a friend from Tennessee who was writing a book favoring the peaceful uses of nuclear energy and he wondered if Reading could give any assistance to the friend."

According to Reading, Olson introduced Srouji to him either in the Hilton motel lobby or in the nearby Sports Page restaurant, and then left. Reading was nervous about the meeting, because he had orders from Dean McGee not to talk to the press about the Silkwood case. But Reading saw Srouji anyway, "as a favor to Olson."

Reading said he met with Srouji for only fifteen minutes, talking about the accident, and gave her a few leads but no Kerr-McGee investigative documents.

When Srouji left Oklahoma the day after her meeting with Reading, the Memphis office, which supervised all FBI activities in Tennessee, assigned Special Agent Larry Thomas in Columbia, Tennessee, to contact Srouji as soon as she returned to Nashville. Thomas, who had been with the Bureau for twenty-four years, spent four hours debriefing Srouji on the national security matter she had brought up in Oklahoma. Over the next year, Srouji continued to feed Thomas other national security information that she picked up while researching *Critical Mass*.

Thomas found Srouji to be extremely reliable. "If she gave me information that indicated a violation of the law," he said, "I would conclude it would be sufficiently accurate and truthful that I would act upon it."

It was clear from the FBI documents that Jacque Srouji testified before the Dingell subcommittee the next year, 1976, to protect the Bureau from critics and to blast the Oil, Chemical and Atomic Workers Union:

☐ Srouji called Olson one month before she was to testify. "The subcommittee is not interested in the facts of the Silkwood matter," she told him, "but is trying to make the department look bad, embarrass the FBI and mess up the Nuclear Regulatory Commission." She discussed her congressional testimony with Olson beforehand "so they could get their stories lined up."

☐ Just hours after her April 26, 1976, testimony before the subcommittee, she called Dominic de Lorenzo to say she "thought her testimony went quite well and it had changed the thrust of the subcommittee's inquiry concerning the union and Silkwood."

☐ That same night, she called Larry Thomas and told him the subcommittee had tried to show that the Bureau had been negligent in its Silkwood investigation. She said she had testified that the Bureau's work had been thorough. Thomas reported her conversation back to Washington. "She tried to put the Bureau in the best possible light," he said.

☐ Srouji told another FBI agent that "her main concern was the union activities which she felt should be blasted, and that the committee had gone astray." And she called Olson again. "Don't worry," she said. "I didn't let you down." Olson later told his FBI superiors that he didn't know what Srouji had meant by "I didn't let you down."

Three days after the congressional hearing, Srouji called Thomas again, this time to complain that the FBI told Dingell it had never heard of her, had never given her any documents, and was going to charge her with perjury. Thomas gave her the name of an FBI agent in Washington whom she could call to find out if the perjury threat was true; then he telexed headquarters: SROUJI STATED SHE HAD VIGOROUSLY DEFENDED THE FBI AND NOW SHE FELT SHE WAS A DEFENDANT INSTEAD OF A SUPPORTER.

Four days after the hearing, Srouji walked into the J. Edgar

Hoover Building after she had finished work at the Pentagon and volunteered to be interviewed under oath. She was evasive about the exact nature of her relationship with Olson. She admitted she had taken about fifty Kerr-McGee documents and about twenty letterhead memoranda from Olson's briefcase, copied them, and was keeping the copies in a safe-deposit box. She refused to tell the FBI how she filched the documents and got them back without being caught. Several times she invoked the Fifth Amendment.

"It was pointed out to Srouji," said the FBI summary of that interview, "that if she maintained that her testimony was true and an interested party furnished a different version or denied her testimony, then she could well understand that a matter of perjury could be involved."

Srouji told the two agents who interviewed her that she would never talk about what she had done or learned in Oklahoma. "She stated that she does know that Olson had done a very thorough investigation, and she did not want any harm to come to him as he had been a dedicated investigator. She stated that she would not hesitate to perjure herself in this matter, and asked the rhetorical question as to whether she would get thirty days."

The Bureau took her seriously. "It was pointed out to her that in all instances where there is a violation of a federal statute the matter is presented by the FBI to the appropriate U.S. Attorney or the Department of Justice for a prosecutive opinion," the summary said.

If she would return her Silkwood documents, would the FBI forget the whole matter? Srouji asked. The agents said that they could not promise anything at that time, but that the Bureau would send someone to visit her once she got back to Nashville.

At the end of the interview, Srouji signed a sworn affidavit that said, "At no time did I officially receive any documents from an FBI employee." That wasn't exactly what the Bureau had in mind. It pressed her to remove the word "officially" from the affidavit, but Srouji refused to do so.

Before Srouji could fly back to Nashville, Mike Ward asked to see her. The subcommittee counsel told her that Dingell still wanted her FBI documents, and that it "might be necessary to subpoena them" if she didn't give them voluntarily. If she refused to obey the subpoena, Ward said, she could be cited for contempt of Congress and, like Watergate conspirator Howard Hunt, end up in jail.

Srouji told Ward she was thinking of giving her documents back to the FBI. Ward advised her to hold on to the papers, for if she returned them, she would have no defense against a perjury charge.

The advice seemed sound, so when the FBI visited her in Nashville two days after her conversation with Ward, Srouji refused to return the documents. She said her attorney had told her that the papers were the key to her credibility.

Soon after Srouji's voluntary appearance at FBI headquarters, the Bureau began to investigate the Olson-Srouji connection to see whether Olson had given Srouji any sensitive or classified documents, and if he had, which ones. The FBI Internal Affairs division caught Olson in a web of inconsistencies that bothered them. First, Olson had told them he had seen Srouji in Oklahoma only once. When they confronted him with evidence that it was more than once, Olson admitted it was twice. (Srouji had told Stockton it was more than twice.) Next, Olson had told Internal Affairs he did not give Srouji any documents — officially or unofficially, sensitive or classified or unclassified. When they confronted him with evidence that his statement was not accurate, Olson told them he had given her only newspaper clippings. Finally, when pressed, he admitted he had also given her press releases.

□   □   □

The subpoenaed FBI documents about Silkwood's death turned out to be superficial. There were no forensic reports on the Honda's brakes and steering system, nor was there any indication that Olson had tried to find out if anyone had followed Karen from the Hub Cafe the night she was killed. There was no critical analysis of the Oklahoma Highway Patrol's theory that Karen had fallen asleep, when it should have been clear that the Honda could not have stayed on the grassy shoulder without a conscious driver.

The FBI documents also confirmed that Olson had made no serious attempt to find any documents Silkwood might have had or to answer some obvious questions raised about them:

□ Olson interviewed Jean Jung, who had said under oath she saw Silkwood with documents in the Hub Cafe. He wrote: "Karen did not tell her what the papers were nor did she actually observe the contents of the folder. She assumes that the papers in the folder dealt with the information Karen Silkwood had acquired regarding the falsification of the fuel rod quality control records, but she does not know this to be a fact. It is merely an assumption." With that, Olson dismissed Jean Jung.

□ Olson reasoned that Silkwood could not have had any documents because Kerr-McGee had told him none was missing from its files. But there was no evidence in the FBI documents that Olson had checked out the K-M claim.

□ Jung had described in detail the folder and the notebook that Karen had clasped minutes before she was killed. They were never found. There was no evidence indicating Olson had tried to find out who may have taken them.

□ Rick Fagen of the OHP said he let the AEC into Sebring's garage after midnight on November 14. Yet Sebring and Fagen talked about Kerr-McGee's Wayne Norwood being there and reading the documents found in the car. Olson did not even allude to this contradiction.

□ Fagen found in the Honda personal letters addressed to Karen and read aloud from at least one of them. Those letters disappeared. Olson never even referred to the "missing letters."

Moreover, the Bureau would normally go crazy investigating the mere possibility that someone might have stolen some plutonium. In the Silkwood case, where plutonium was actually stolen from the Cimarron plant and was used to contaminate her apartment, the Bureau did not seem to be very concerned.

Next, the FBI contamination and death reports seemed to miss no opportunity to sprinkle in details about Silkwood's sex life, drug use, and emotional stability, while, at the same time, never saying anything derogatory about Kerr-McGee. In one instance, Olson clipped an Oklahoma City Times article, quoting Howard Kohn saying that Karen Silkwood was a young woman from a hard-working Baptist family who was trying to recapture early aspirations for a scientific career when she joined Kerr-McGee in 1972. Olson penned in the margin: "Who is he trying to kid? She was a pig!" Olson placed the clipping, with his comment, in the FBI's Silkwood file.

Finally, the FBI documents and three Kerr-McGee memoranda suggested a close relationship between Olson and Kerr-McGee. For example, someone at K-M reported to Olson that Tony Mazzocchi of the OCAW was on the board of directors of the Natural Resources Defense Council, which Kerr-McGee described as an antinuke group. Someone at K-M also sent Olson a Supporters of Silkwood news release, describing a national campaign to educate, organize, and raise funds for the Silkwood lawsuit. Kerr-McGee pointed out to Olson that donations for the Silkwood case were being accepted by the Emergency Civil Liberties Foundation, which K-M described as a communist group. Olson had the Bureau check out these (and other) Kerr-McGee leads.

But there was more to the relationship between Olson and K-M than the Bureau's chasing down Kerr-McGee leads, according to the documents. Less than two weeks after Silkwood's death, Olson visited Reading at the Kerr-McGee headquarters in downtown Oklahoma City and informed him that the corporation was an FBI suspect in the Silkwood case. Olson carefully explained the potential charges and the law, and then asked Reading to set up a meeting for him with top Kerr-McGee officials so that he could explain the problem to them. Reading arranged for Olson to see Dean McGee and the corporation's attorney, Derrill Cody.

Three months later, while Kerr-McGee was still a suspect in the case, Reading picked up Olson at the FBI headquarters in Oklahoma City and drove him to Crescent for interviews Reading had arranged. In a March 4 memorandum to Dean McGee, Reading explained that the new special agent in charge of the Oklahoma City FBI office had just arrived from Miami and had introduced himself to Reading. The new agent "was very friendly and expressed a desire to meet you, possibly for lunch, at your convenience," Reading told McGee.

Reading also told his boss that the current thrust of Olson's investigation did not involve Kerr-McGee, but rather "what were the means used by Silkwood to remove plutonium from the Cimarron Facility, or if in fact the plutonium she was contaminated with came from some other source." Reading also said in the memo that Olson was investigating to see if Karen Silkwood, Jerry Brewer, or Jack Tice had been involved in falsifying quality-control documents.

Reading went on to explain to McGee that Olson had interviewed three Kerr-McGee management people in Reading's presence and that the managers had given Olson a list of suspects. Then, according to Reading, Olson sat down with K-M facility manager Morgan Moore and outlined the scope of his current investigation.

There was one puzzle in the pile of subpoenaed FBI documents. Bureau Internal Affairs investigators had found on Olson's desk "photographic" copies of sixty-six pages of notes relating to the Silkwood case. Olson had not logged those papers in accordance with FBI procedures, so Washington had questioned in a report what the notes were and why Olson had them.

Karen Silkwood

Karen Silkwood's car (Photo by A. O. Pipkin)

Highway 74
(Photo by A. O. Pipkin)

Silkwood's car
showing steering wheel
bent in at the sides

Dean A. McGee

Below: the Cimarron plant

James Reading

Jacque Srouji

James V. Smith

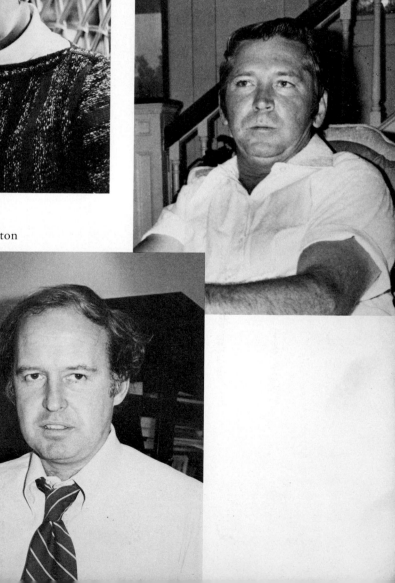

Peter Stockton

The Kerr McGee
defense lawyers, headed
by Bill Paul,
fourth from left
(Photo by Theo Wise)

The Silkwood legal team holding their daily debriefing in the YMCA Jacuzzi.
Left to right: Father Wally Kasuboski, Father Bill Davis, Jim Ikard.
Standing: Dan Sheehan, Jerry Spence, and Art Angel (Photo by Theo Wise)

Karen's children, Dawn, Michael, and Beverly, with her former husband, Bill Meadows (*Daily Oklahoman,* March 8, 1979. Copyright © 1979, the Oklahoma Publishing Co.)

Dan Sheehan and Sara Nelson in front of the Oklahoma City courthouse, after the verdict (UPI telephoto, courtesy of the *Oklahoma Journal*)

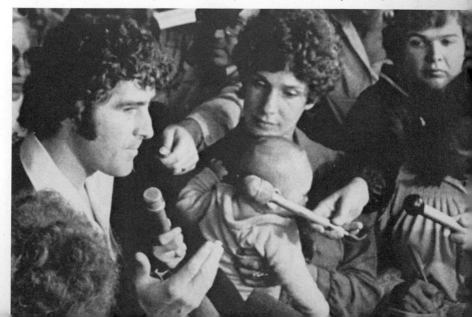

The vindication of Karen Silkwood (Photo by Dale Wittner)

Chapter 25

Echo called. "The Bureau hasn't given you sixty percent of the Silkwood file," he told Taylor. "There are at least four more file cabinets full of documents."

Echo said many Silkwood documents were "June Mail" — the FBI code for secret reports — and that June Mail was carefully guarded. He told Taylor he had seen June Mail surveillance reports on Karen Silkwood, Drew Stephens, and Bob Ivins (in whose garage Drew had stashed Karen's Honda). These surveillance reports were transmitted under a National Security Agency (NSA) code, Echo said, and were classified as top secret. Echo said that some of these surveillance reports dated back to 1973, before Karen was killed.

Echo also said that the Oklahoma City Police Department had CIA or NSA undercover personnel either in the intelligence unit or in the department's upper ranks. He told Taylor that the department had a wiretapping and bugging unit equipped with the best electronic gear in the central United States, and that its headquarters would be a good place to start looking for whoever tapped and bugged Karen Silkwood.

Echo cautioned Taylor not to rule out CIA involvement in the Silkwood case, because the CIA was directly involved in monitoring the nuclear industry, and the security director of each nuclear facility was under CIA surveillance. Echo said Kerr-McGee would have had to deal directly or indirectly with the CIA before it got the AEC contract to work with plutonium.

249

Finally, Echo confirmed that the CIA had been diverting plutonium from nuclear plants and giving it to countries friendly to the United States and that a number of CIA agents had been contaminated with the plutonium they had diverted. Several had died. But Echo did not know if the CIA had taken plutonium from the Kerr-McGee Cimarron plant; he did not have direct access to that report.

□ □ □

Wyoming attorney Gerald Spence took the deposition from Larry Olson, Sr. Sheehan had called Spence, a millionaire rancher as well as a lawyer, for a financial contribution to the Silkwood fund. Spence became fascinated with the case. He was the best personal injury attorney in the Tenth Federal Circuit, which included Oklahoma, and he enjoyed going after the big boys with a lot of money, like Kerr-McGee. One day in October, Spence called Sheehan to say he would be passing through Oklahoma City and would be delighted to take a deposition for Sheehan so that he could see firsthand what the Silkwood case was all about. Sheehan tossed him Larry Olson.

Olson began his testimony by saying he had a master's degree in geology. Spence grinned. "And so you know something about being between a rock and a hard place, don't you?" he said.

Part of Spence's job was to help Sheehan prove that Jacque Srouji was an FBI informant and therefore not entitled to hide behind the newsperson's privilege. Under Spence's probing, Olson admitted that in Nashville, between 1965 and 1970, Srouji had supplied him with information "of a sensitive security nature." Olson described Srouji as a "confidential source," cautiously avoiding the word "informant."

Olson denied that he had ever wiretapped during his more than ten years with the Bureau, or that he had ever discussed wiretapping with Jacque Srouji during the ten years he had known her. (She had told Stockton and Kohn that Olson discussed the wiretapping of Silkwood with her.) Olson admitted that Srouji was supplying information to the Bureau in 1975, but said it was not related in any way to Silkwood or Kerr-McGee.

Spence began to pick apart Olson's comment, quoted in the

FBI report released just three days earlier, that he tried to "keep her sweet . . . she appeared to be a good friend of the Bureau . . . she had unusual potential for further development."

"What caused you to feel that she was a good friend of the Bureau?" Spence asked.

"Being friendly."

"Was she friendly because, in your opinion, she provided information to the FBI?"

"That could have been a factor," Olson said.

"Well, *was* it a factor?" Spence demanded, raising his voice.

"Yes, sir."

"Was she personally friendly, Mr. Olson?" Spence asked innocently.

Olson avoided the question. "Well, as a journalist, she wrote several articles . . . that generally spoke favorably of the FBI," he said.

"Was she an informant of the FBI?"

"I object to that," said Olson's attorney, Glenn Whitaker of the Justice Department. "Mr. Olson has already testified that she was a confidential source."

"What do you mean by 'unusual potential'?" Spence asked Olson.

"Unusual circumstances."

"What do you mean by 'unusual circumstances'?"

"We object again," Whitaker said.

"And what do you mean by the words 'for further development'?" Spence asked.

"Just what I said."

"What did you mean by 'further development'?" Spence repeated.

"We'll object to these questions," Whitaker said. "I think further questions concerning that matter are not only irrelevant, but they are also privileged."

Spence ignored the objection. "What were you developing her for?"

"I repeat my objection," Whitaker said.

Spence read again from the FBI document: " 'Srouji claimed to be a (deleted) for the (deleted) . . .'

251

"Now," he asked, "what did she claim to be?"

"We object to that," Whitaker said. "The matter is of a sensitive security nature and is privileged."

"What do you do to somebody when you 'develop' them?" Spence asked a fourth time.

"I am going to instruct the witness not to answer," Whitaker said.

"What do you mean by 'keeping her sweet'?" Spence asked.

"Maintaining her good will," Olson said. "It is not an FBI term. It is an expression that I used. It seemed to be descriptive. Keeping her good will at that particular time . . . to situations regarding national security beyond the Silkwood matter."

Olson explained to Spence the difference between national and internal security. National security usually involves a crime by a foreigner, such as espionage, he said. Internal security is limited to the crimes of United States citizens.

As the cat-and-mouse questioning continued, Olson blundered, as he had done when he admitted to Ward and Stockton that the Bureau had a "special relationship" with Srouji. Under Spence's probing, Olson finally let it slip that Srouji had been an FBI informant in 1975 and 1976.

The distinction between an FBI confidential source and an FBI informant is not academic. A confidential source volunteers information; an informant is directed and controlled by the Bureau. If Srouji were just a confidential source, there would be no grounds to strip her of her newsperson's privilege. Olson and Whitaker would later try to take back the word *informant,* but they couldn't; it was on the record.

Spence began to probe Olson's Oklahoma meetings with Srouji. "Did you give her information [about Silkwood]?" he asked.

"Not beyond what was in the media," Olson said, sticking to the story he had told FBI Internal Affairs.

"Are you aware of the fact that Jacque Srouji is reportedly so friendly with the FBI that she said she would perjure herself to protect the FBI or members of it?" Spence asked, paraphrasing another FBI document just released.

"No."

"Did she ever have your file, and Xerox material from your file?" Spence asked, paraphrasing what Srouji had sworn to during her voluntary testimony at FBI headquarters in Washington.

"No."

"Did you let her see a 'ton of material'?" Spence asked, quoting Srouji's congressional testimony.

"No."

During her depositions, Jacque Srouji had changed the story she told the FBI and Congress, swearing she had merely *seen* the Silkwood reports piled on Olson's desk, but that she had never read or copied them. But during his deposition, Olson told Spence that the FBI Silkwood files were neither on his desk nor anywhere in view when Srouji was in his office.

Olson went on to deny under oath that he had made Srouji's motel reservation, as she had claimed, or that he had sent her a letter confirming that he arranged for her to see James Reading, as she claimed. (She told Stockton that she still had the letter.)

Olson testified that the first time he had ever met James Reading was in March 1974, when he was appointed liaison to the Cimarron plant. Yet Reading had told the FBI he had known Olson "for several years."

One of the twenty documents Srouji eventually gave Dingell's subcommittee was a copy of Pipkin's accident report with the words *FBI Evidence* taped on the cover. Olson denied that he or anyone else in the Bureau had placed the special FBI tape on the Pipkin report, implying that Srouji must have stolen the tape from an FBI desk somewhere and put it on the document herself. Olson never explained why she would have placed the *FBI Evidence* tape on Pipkin's report and not on other, more important documents she gave Dingell.

Srouji had also given Dingell a copy of a Silkwood handwritten household budget, listing $300 for "dope." Wayne Norwood had found the document while decontaminating Silkwood's apartment. Olson admitted Reading had given him a copy of the note and that he had placed it in the FBI Silkwood file, but he denied giving a copy to Srouji.

Srouji had given Dingell several Kerr-McGee investigative reports on Silkwood's accident written by Reading and his free-lance investigator and former K-M boss, W. Spot Gentry. Olson denied he had ever seen those reports before they appeared in the appendix of the Dingell hearing transcripts. Yet Reading had told the FBI he had given Olson "probably about everything" on K-M's Silkwood investigation.

Spence moved on to wiretapping.

"I have no knowledge," Olson said, "that the police department, the Crime Bureau, any other law enforcement agency, be it state or local or federal, engaged in any wiretapping. If they had, I would have blown the whistle on them."

Spence began to probe Olson about Srouji's book, *Critical Mass,* for if Sheehan could learn who put her up to writing it, he would be able to narrow the focus of his investigation.

"Didn't she tell you she was instructed to write a book in order to make contact with people in the nuclear area?" Spence asked, paraphrasing a subpoenaed FBI document.

"Object to that, you know," Whitaker said.

"From whom did she receive her instruction?"

"Object to that."

Olson went on to defend his investigation of the Silkwood matter. He testified that there was no "information whatsoever to indicate that some outside individuals contaminated Karen Silkwood," and that he had tried to find them but couldn't. He swore that he did not investigate missing plutonium because neither Kerr-McGee nor the AEC had told him a substantial amount was gone. In sum, Olson testified that he had not only done a thorough job investigating the Silkwood matter, but that he had taken the probe "further than most people would have."

□　□　□

Dan Sheehan deposed James Reading the next day. His investigators had dug up some interesting background on the Kerr-McGee security director. As an Oklahoma City policeman in the 1960s, Reading had had excellent rapport with the FBI. In a memorandum dated June 8, 1962, the Bureau's Oklahoma City office stated that Reading had set up an intelligence unit in the Oklahoma City Police Department and that he was willing to exchange information with major cities in the surrounding states. "It is noted, " the memo said, "that Reading is a graduate of the FBI National Academy and is extremely cooperative with this office."

After early retirement as chief of detectives, Reading took a job as head of security at the First National Bank of Oklahoma. Then, in 1973, he joined Kerr-McGee as assistant director of security under Spot Gentry, working out of the Kerr-McGee headquarters in Oklahoma City. When Gentry retired, Reading took over. He had overall responsibility for the security of all operations in the K-M empire. Bill Taylor's Oklahoma source, Fairy Godfather, had reported that during 1973 and 1974, Reading frequently visited William Vetter, then commander of the police department's intelligence unit.

Sheehan began the deposition by questioning Reading about his April 1975 meeting with Jacque Srouji. Repeating under oath what he had told the FBI previously, Reading said Olson had indeed called him and said he wanted him to meet a journalist who was writing a book favorable to nuclear power. He said Olson set up the meeting at the Hilton, introduced Srouji to him, asked him to help her so that she "could conduct her own investigation," and then left. But Reading stuck to his story that his meeting with Srouji in her motel room lasted only fifteen minutes, not several hours, as she claimed.

"Didn't you in fact tell Srouji during that time that she ought to investigate Karen Silkwood's sex habits?" Sheehan asked.

"I don't recall mentioning Silkwood's sex life habits at all," Reading said.

"Didn't *you* investigate them?"

"No, sir."

(Olson had told Stockton that the Kerr-McGee investigation was good, but had got bogged down in Silkwood's use of dope and her sex life.)

"Then you don't have any knowledge about them?" Sheehan asked.

"Yes, I do."

"Just fell out of the sky?"

"Yes, sir."

Reading went on to say that the first time he had ever heard of Karen Silkwood was on November 9, 1974, two days after

Kerr-McGee had found her apartment to be contaminated. Reading said Norwood told him someone had stolen a K-M alpha counter from the apartment, and Norwood asked him to investigate the theft. But Reading denied he knew that Silkwood was collecting documents until he had read about it in the *Daily Oklahoman* after her death.

Reading also testified that Kerr-McGee had not asked him to investigate missing plutonium, or who might have had access to pellet lot 29, from which some of the plutonium that contaminated Silkwood had come. Yet Reading admitted that finding missing plutonium was one of his responsibilities as security director.

Reading testified that he began the Kerr-McGee Silkwood investigation during the week after Karen's death because Kerr-McGee was concerned about possible lawsuits. "We was walking on eggshells," he said, denying he was assigned to protect the corporation against criminal indictment by the Justice Department.

Reading denied he ever wiretapped Karen Silkwood's apartment, bugged her phone, harassed her, or hired anyone else to do those things. He said he had spoken to Bill Vetter only three or four times in the past five years, and that was just to say "Hello, how are you." Yet Fairy Godfather had learned that Reading had met with Vetter frequently during 1973 and 1974. Reading said he had never heard that the police department ever wiretapped or bugged anyone, or that the intelligence unit he founded ever possessed tapping or bugging equipment. And he denied that Kerr-McGee used the Pinkertons or any other "suit-and-tie" investigators to snoop on Cimarron union members, as Srouji had testified.

Reading swore he never discussed with Olson the direction of the FBI investigation. Then Sheehan read Reading his own March 4, 1975, memorandum to Dean McGee: "The current thrust of the [FBI] investigation is as follows . . . What were the means used by Silkwood to remove plutonium from the Cimarron facility, or if in fact the plutonium she was contaminated with came from other sources."

Reading seemed flustered. He swore the memorandum was inaccurate and that he had made a poor choice of words. "To be fair and straighten the matter out," he testified, "I suggest in all honesty, God is my witness . . . Olson did not indicate to me in any manner that Silkwood was the one that carried that crap out of there."

Reading reluctantly admitted that on March 4, 1975, Kerr-McGee was still an FBI criminal suspect in the Silkwood case.

Sheehan began to probe K-M's putting pressure on its Cimarron workers to take polygraph tests. "Now, Mr. Reading, isn't it in fact true that you were attempting to find out, in the use of these polygraphs, whether workers at the plant had antinuclear attitudes?"

"I would say yes," Reading said. "But there's more to it than that."

"Isn't it in fact true that you were determining if a person had an antinuclear attitude, that they were one of these people you might suspect, that you might not be able to trust?"

"Yes, sir."

"Now, isn't it in fact true that you undertook other methods than simply giving polygraph examinations to try to determine whether or not any employees were antinuclear?"

"Yes, sir."

"Can you tell me what they were?"

"Interviews."

"With whom?"

"Everyone."

Sheehan gave Reading a list of the polygraph questions Kerr-McGee said it had used. The OCAW charged that K-M had departed from the "official list" and had asked questions about union membership and sexual activities. Reading admitted he had helped draw up the questions, but denied Kerr-McGee had departed from the official list.

"They remained the same," Reading said. "The relevant questions, such as their name, their age, and so forth and so on . . . And the control questions? I don't know what the polygraph people used in those. They could have asked them, 'Is a redbird red?' "

Under Sheehan's probing, Reading admitted that he'd had to admonish the polygraph testers to stay within the official list.

"What were the questions that the polygraph operators asked

that caused you to admonish them to stop?" Sheehan asked.

"One or two of the females would get off into areas of sex and so forth, and the polygrapher would pursue it," Reading said, absolving Kerr-McGee of all blame. "And then the female would gripe about it to the people outside, and I would go in there and get that polygrapher and say, 'Hey, fellow, stay within those questions. If they want to talk about something else, or gossip, avoid that part of it.' "

Sheehan read a question from the official list: " 'Since being employed by Kerr-McGee, have you participated in any antinuclear activites?"

"Can you tell me whether you drafted, and authorized them to ask, that question?" Sheehan asked.

"I assume I did."

"What did you mean by 'antinuclear activities'?"

"Just what it says."

"What did you mean when you put the word in there?"

"Just exactly what I said, sir."

"A third time," Sheehan pressed. "What did you mean by it when you put it in there?"

"Oh, demonstrations," Reading said. "It is hard to explain . . . Unlawful acts."

"Demonstrations are unlawful acts?"

"I don't know," Reading said.

"What kind of other conduct did you have in mind when you asked about whether they participated in antinuclear activities?"

"I've answered the best I can, sir."

"In fact, isn't it true, Mr. Reading, that you meant *lawful* First Amendment exercises and expression of their opinions in opposition to nuclear power? Isn't that what you mean?"

"No, sir."

"Now, Mr. Reading, I direct your attention to a document which was taken from the FBI files dated April 11, 1975." Sheehan began to read aloud: " '(Name deleted), Kerr-McGee Corporation, Oklahoma City, advised it was KMC's understanding that Anthony Mazzocchi . . . is on the board of directors of the Natural Resources Defense Council. (Name deleted) advised it was his understanding this council is considered to be an antinuclear group.'

"Are you in fact the person —"

"No, sir," Reading cut in.

"You know who it is?"

"No, sir."

"Isn't it in fact a policy of the Kerr-McGee Nuclear Corporation, as a corporation, to keep records on people they consider to be antinuclear?"

"I wouldn't know anything about that, sir."

"As chief of security, weren't you involved in collecting and disseminating that type of information?"

"No, sir."

"Have you ever seen that FBI record before?"

"No, sir."

Sheehan addressed the judge. "I would request, Your Honor, I would request that the FBI remove the blank out of this. There is no legitimate reason why we shouldn't know who within the Kerr-McGee Corporation was reporting to the FBI on the membership of Mr. Mazzocchi in a totally lawful organization."

Glenn Whitaker interrupted. "We don't have any problem giving the name, as long as Mr. Sheehan will agree to protect the privacy of third-party names that are blanked out."

Sheehan refused. "I would daresay that whoever's name is under that is in significant trouble, and I'm not going to keep anything confidential about that person."

"Mr. Sheehan," Whitaker said, "if you can't agree to protect the privacy of the third-party name, I'm not sure we can agree about *anything*."

"I'm not sure we *have* to agree," Sheehan said. "I'm asking the court to *order* you to do it."

Whitaker gave in. "Mr. Reading's name is the name that has been deleted."

□ □ □

The FBI-Reading-Srouji-Olson story was filled with contradictions and inconsistencies.

THE FBI: Washington Bureau Agent James Adams had told Representative John Dingell that the FBI had no relationship with Jacque Srouji in the 1970s. Yet the Bureau's own documents and the sworn testimony of FBI agents Larry Thomas and Larry Olson, Sr., indicated that Srouji did indeed have a special relationship with the FBI, at least during 1975 and 1976. Thomas and Olson had tried to define that relationship as a "confidential source." But Olson had let it slip under oath that she was actually an "informant" during those two years.

The FBI also had told Dingell after Srouji's congressional testimony that it had never heard of a Jacque Srouji. But the sworn testimony of Larry Thomas and the FBI documents released so far proved that Srouji was reporting regularly to Thomas on matters of a "sensitive national security nature," and that Thomas was filing regular reports back to Washington about what she had told him.

JAMES READING: The Kerr-McGee chief of security agreed with Jacque Srouji on two major points. He said he met her in a motel at the suggestion of Larry Olson to discuss the Silkwood case; she agreed. He said Larry Olson was there to introduce her to him; she did not disagree.

But Reading contradicted Srouji on even more major points. She said they had met for several hours; he said they met for fifteen minutes. She told Stockton and Kohn that Reading had told her his people wiretapped Silkwood. Reading denied that and said he hadn't even heard of Karen Silkwood until her contamination, less than a week before her death. Srouji said Reading had given her Kerr-McGee documents; Reading denied it.

LAWRENCE OLSON, SR.: It was obvious from the FBI reports released so far that Olson was defensive about his meetings with Srouji in Oklahoma City. FBI Internal Affairs had caught him in some strange inconsistencies: he said he'd met her once, then admitted it was twice; he denied he gave her documents, then admitted he gave her newspaper clips, and finally admitted he also gave her some press releases.

Olson contradicted *every* major thing Srouji had said up to her 1977 depositions. She told Peter Stockton, when he visited her in Nashville, that she had met with Olson in Oklahoma "at least three times," once at the airport. A desk clerk at the Northwest Holiday Inn in Oklahoma City told the FBI that she recognized Jacque Srouji and had orders not to let the Nashville woman stay at the motel anymore because it had had trouble collecting payment from her. The clerk also told the FBI that Srouji had registered under the name "Jacque Stubbel" (a variation of her maiden name); yet she also registered in mid-April 1975 as "Srouji," according to motel records. Therefore, unless the clerk was totally mistaken, the journalist must have met Olson in Oklahoma "at least three times," as Srouji had told Stockton.

Prior to her depositions, Srouji said she had discussed the Silkwood case at length with Olson; he said he had mentioned Silkwood only in passing. Prior to her depositions, Srouji also said Olson had given her FBI documents, including letterhead memoranda. She even described to the FBI under oath what the notebook looked like. Olson denied he gave her or let her read FBI reports and LHMs. After Srouji had changed her story during her 1977 depositions to say she only *saw* the Silkwood documents on Olson's desk, he denied that the Silkwood documents had been anywhere in sight.

Both Reading and Srouji said Olson set up their April 1975 meeting. Reading said Olson had introduced him to Srouji at the motel; Olson denied he had even talked to Reading about Srouji, much less introduced him to her. Reading said he had known Olson professionally for "several years." Olson said he had not known Reading before March 1974.

JACQUE SROUJI: The Nashville journalist was an intelligent woman. John Seigenthaler called her a good *Tennesseean* copy editor; the Nashville Business and Professional Women's Association had given her an award; and her Naval Reserve supervisor at the Pentagon had recommended her for a promotion. Furthermore, both Thomas and Olson considered Srouji reliable, emotionally stable, and accurate in reporting national security information to the Bureau.

What was important was that Srouji's story up to her 1977

depositions had been consistent. She told a congressional investigator (Stockton), a nationally known journalist (Kohn), and a congressman (Dingell) substantially the same story under different sets of circumstances. Moreover, she never contradicted herself during the many meetings she had with Stockton, Ward, Dingell, and Turner. And when she finally testified under oath before Congress, her basic story hadn't changed, a tough cross-examination by Ward notwithstanding.

Even after her congressional testimony, Srouji told the very same story to the FBI under oath. The Bureau tried to get her to sign an affidavit saying she never got documents from the FBI, but she refused to sign until she could add the word *officially*. The Bureau obviously believed she had the documents, for it sent an agent from Washington to Nashville to get them. Finally, she showed John Seigenthaler some classified and highly sensitive FBI documents unrelated to Silkwood, proving that it would not be exceptional for her to have sensitive FBI Silkwood documents as well.

Then, suddenly, Srouji changed her story. After she told the FBI she would perjure herself before she'd hurt Larry Olson, she was deposed by Dan Sheehan and Jim Ikard. During those two lengthy depositions, she actually said nothing that would hurt Larry Olson, even though that meant telling a new, contradictory story.

□   □   □

Before anyone could straighten out the Reading-Olsen-Srouji story, the Silkwood case began to run into trouble with Judge Luther Eubanks. Glenn Whitaker, who represented both Larry Olson and the Bureau in the matter of its subpoenaed documents, asked Judge Eubanks to slap a gag order on Sheehan and Ikard. Whitaker argued that the two Silkwood attorneys were telling the media the contents of the FBI documents released earlier in the month, and that the documents were not part of the "public" record. He further argued that Sheehan and Ikard "colored" his clients and created "a hostile attitude in the press."

Sheehan and Ikard, in turn, argued that Whitaker failed to show that their statements to the media posed a "clear and present danger" to the government or to a fair trial. Therefore, they reasoned, a gag order would violate the First Amendment rights of the media. What Sheehan and Ikard did not tell Judge Eubanks was that they were giving the media the facts as they uncovered them because they believed that too much about Silkwood had been covered up for too long, and that the media would keep the pressure on the court, helping them in the long run.

On October 27, Judge Luther Eubanks refused to grant Whitaker's motion to gag Sheehan and Ikard. "I cannot conclude that there is a reasonable likelihood that anyone cannot get a fair trial," Eubanks ruled. But he warned the two Silkwood attorneys that he would examine their future statements to the media. If he considered any of them out of order, he said, "I will slap a gag order on you tomorrow." When Sheehan objected, Eubanks told him to sit down.

Then the judge tore into Sheehan and Ikard. "Apparently the lawyers have taken this thing and started running with it and used the Silkwood parents as a tool," he told the open court. He added that Sheehan and Ikard's case did "not amount to a hill of beans," and that they were acting in an "unprofessional" manner.

Judge Eubanks went on to characterize the discovery phase of the Silkwood case as creating "a Roman holiday atmosphere," Ikard as a "magpie," and Sheehan as "running off at both ends."

Sheehan filed an application for recusal — to have Eubanks removed from the case — because the judge's open-court statements, widely reported in the media, "brings the court's impartiality in this proceeding into question."

While Sheehan's secretary was typing the application, Gypsy Hogan, a *Daily Oklahoman* reporter, stopped by the office to see if she could get a story lead. She read the recusal draft, then asked Judge Eubanks if he knew anything about it. Eubanks said he didn't, but before Sheehan could file the application in the courthouse, he resigned from the Silkwood case, appointing Judge Luther Bohanon, who sat in the courtroom next door, to take his place.

Chapter 26

Shortly after Judge Luther Bohanon replaced Eubanks, Dan Sheehan was called to Washington to settle a running battle with the Nuclear Regulatory Commission, represented by the ubiquitous Glenn Whitaker, over the documents he had subpoenaed from the commission. A few days after Sheehan arrived in Washington, Judge Bohanon convened the court to rule on twenty-two motions left over from Judge Eubanks.

Son of a tenant cotton farmer, the seventy-five-year-old Judge Bohanon had worked his way through the University of Oklahoma, where he studied law, and then opened a practice in Seminole, Oklahoma, during the rowdy oil-boom days. Later, he moved to Oklahoma City, where he built a $100,000-a-year law practice, representing big business and big oil.

Appointed a federal judge in 1961, Bohanon selected as his motto "Do right and fear no man," and he carried a copy of the United States Constitution in his coat pocket. The media generally described him as a poor country boy at heart who had worked for everything he had and who believed everyone else should do the same; a conservative who hadn't changed a whit since he left school; and a man who wouldn't change his mind once convinced he was right.

It was clear from the moment Judge Bohanon rapped his gavel at the Silkwood hearing that he was not happy to be there. "After some very careful thinking about it," Bohanon told the court, "I told him [Eubanks] I would take it over. I find that I've

264

got a lot bigger job than I thought I had. The court would like to inquire — does anybody have any objection to me taking over?"

No one had.

The first order of business was Sheehan's motion to have the discovery period extended to March 1, 1978. Jim Ikard explained to Bohanon all the obstacles the Silkwood attorneys had faced. "We have run into attorney-client privileges, work-product privileges, state secrets, national security, First Amendment, newspaperperson's privilege, Fifth Amendment privilege against self-incrimination," Ikard pleaded. "We have run into the relevancy argument. In other words, denying us discovery on relevancy grounds . . .

"So every step of the way, whether it's a deposition, an interrogatory, a subpoena *duces tecum,* a request for production of documents, a request to examine the [Cimarron] plant . . . we have run into a stone wall. And that's why the extension is necessary."

"I'm afraid we are getting off into left field where all this discovery leads to nothing," Judge Bohanon said.

Ikard outlined the Silkwood complaint for the judge — the Kerr-McGee negligence in handling the plutonium that contaminated Karen Silkwood; the conspiracy to deprive her of her civil rights through wiretapping, bugging, and illegal surveillance; and the conspiracy to cover it up.

"Are you saying to the court," Bohanon asked, "that your proof will sustain that all these people named as defendants deliberately entered into a conspiracy to damage this plaintiff?"

"Yes."

"This *one* plaintiff?"

"Yes, that is what we have to prove."

"Mr. Ikard, how can you do that?"

"Well, the way that we will do that is —"

"That *this* type and caliber would enter into a conspiracy?" Bohanon cut in. "Well, you've got to *show* that they were a member of a conspiracy."

"Right."

"Nobody is a member of a conspiracy just because *you* say so."

"I understand that," Ikard said.

"They've got to do something —"

"That's right."

"That puts them in there."

Bohanon began to pepper Ikard with the names of those charged in the conspiracy. "What proof did you have when you filed it?" Bohanon asked with each name. "What proof do you have now?"

Ikard stumbled. He did not know the case as well as Sheehan did, and Sheehan was not there to jump in and help.

"Well, is it true that you have named a bunch of people here in this thing just because they were members of the board of directors of the Kerr-McGee Corporation?" the judge demanded. "Well, do you have any *proof* that they did anything?"

"I think that, yes, we do."

"The board of directors did something, passed a resolution?" Bohanon asked. "Did they take any physical action against Karen in any way?"

"At this time, I can't say that. No."

"We can't just go on in a state of dream," Judge Bohanon said. "There has got to be proof."

"But it can be inference," Ikard said.

"Where are your inferences?"

"The inference is that in this kind of program — coordinated, things going on," Ikard tried to explain, "the inference is clear that such a wide and sweeping conspiracy could not be done in a vacuum, that people have to know about it. Who were the people likely to know about it? Individuals that are directly in the line of communications."

Kerr-McGee's defense attorney Bill Paul took the floor. He argued that the Federal Bureau of Investigation, the Atomic Energy Commission, the Nuclear Regulatory Commission, the United States General Accounting Office, and the United States Congress had all investigated the Silkwood case and found nothing to support Sheehan's three allegations. "Judge, I am just trying to make *one* point," Paul said, "that the thing has been investigated to death, and this business about 'we need more time' has no merit to it at all. They've got all these investigations —"

"All right," Bohanon cut in. "Now, then, the court has read the order of Judge Eubanks in this case strictly limiting the period of discovery to November 30 or December 1, and the court will not disturb this order. And the motion for extension of time will be denied."

Bohanon also denied motions to compel several defendants who had already been deposed, such as Dean McGee, to answer questions they had either refused to answer or claimed they couldn't answer because they didn't have their records with them. Then Bohanon denied a motion to compel Kerr-McGee to release documents subpoenaed by Sheehan. K-M had argued they were not relevant to Silkwood and had refused to hand them over.

"I'm going to get this case ready for trial whether you want to or not," Bohanon threatened Ikard several times.

Sheehan had also filed motions to compel Oklahoma City Police Department intelligence officer William Vetter and Oklahoma State Bureau of Investigation officer Thomas Bunting to appear for a deposition. Bunting was an LEIU representative; Vetter had met with Reading frequently during 1973 and 1974, according to Fairy Godfather. But both policemen filed motions to quash their subpoenas, arguing that they knew nothing about Silkwood or wiretapping, and that a probe into the intelligence units of their respective organizations would cripple future intelligence work.

"Did you try to talk to them before you subpoenaed them to find out what they knew?" Bohanon asked.

"I don't think that's a predicate to serving a subpoena," Ikard argued.

"I never heard of anyone just subpoenaing somebody because they *thought* maybe they might know something, without having somebody go and talk to them," Bohanon said.

"That's not a ground for quashing a subpoena, Judge."

"Oh, I think so, unless you can prove that he is lying. He's got an affidavit here," Bohanon said. "I'm going to sustain the objection —"

"We are not to take their depositions at *all?*" Ikard asked.

"That's right. That's right."

"Judge, will the court entertain any *further* motion to compel?"

"No, no. You've had a year," Bohanon said. "You worked on this, you've done everything. You have even worked on it before you ever filed a lawsuit. You have sued a lot of people that . . . should never have been sued. You don't have any reason to sue them. It's a suit in the clouds in some regards."

"You are stripping us of an opportunity to prove our case," Ikard objected.

Bohanon wouldn't budge. But he did grant Sheehan one of his twenty-two motions. He allowed the Silkwood attorneys to visit the Cimarron plant, provided they promised to take only still photographs, not give them to the media, and return them to the court after the trial.

Ikard was stunned. "I got whipped," he told Sheehan on the phone that night. "We have to do something to get Bohanon out."

□   □   □

Fairy Godfather called Bill Taylor. He had read about Bohanon's appointment to the Silkwood case and Ikard's lashing in the courtroom. "Why don't you look at how Bohanon was appointed?" Fairy Godfather suggested.

Father Bill Davis learned that Luther Bohanon and Senator Bob Kerr had been good friends, and that it was Kerr who had pushed Bohanon onto the federal bench.

When Judge W. R. Wallace died in a car crash in 1960, Aubrey Kerr suggested that his brother, Senator Robert S. Kerr, propose Luther Bohanon to fill the vacancy. Bob Kerr agreed. "A man of humble origins who is now recognized as one of Oklahoma's most distinguished lawyers," Senator Kerr said of Bohanon when he submitted his name to President Kennedy.

When the Justice Department received a spate of letters from Oklahoma attorneys opposing Bohanon's appointment, Attorney General Robert Kennedy sent his assistant, John Seigenthaler, and Joseph Dolan, assistant to Deputy Attorney General Byron White, to Oklahoma to see what all the fuss was about. They found that most of the opposition to Bohanon either was political or stemmed from Bohanon's reputation as an establishment lawyer. Seigenthaler, however, became convinced that Bohanon was qualified for the judgeship, though not necessarily the most qualified, and reported favorably on him to Robert Kennedy.

The American Bar Association's eleven-member committee, made up of representatives of each circuit and the District of Columbia, investigated sixty candidates for judgeships that year, 1961. It found only two of them unqualified, and one of the two "unanimously unqualified." He was Luther Bohanon.

When the White House delayed for months the announcement of its appointee for the Oklahoma vacancy, Bob Kerr got angry and called the attorney general himself. Seigenthaler took the call for Kennedy, who was out of town. Kerr told Seigenthaler that Bohanon was his "personal friend," and that if Bohanon didn't get the judgeship, he, Kerr, would see to it that no one filled the vacancy.

Then one day President John F. Kennedy called Bobby Baker to the Oval Office to find out why his investment credit tax bill was stalled in Kerr's Senate Finance Committee. "Hell," the President said, according to Baker, "that goddamn bill ought to make Bob Kerr dance for joy. Wilbur Mills is going along in the House, but Kerr's dragging his feet. Why?"

Baker asked the President's permission to use his phone and called Kerr, who said, "Tell him to get his dumb fuckin' brother to quit opposing my friend . . . Bohanon for a federal judgeship in Oklahoma."

"Is that all?" Baker asked.

"Hell, ain't it enough?"

President Kennedy grinned, picked up the phone, and, attempting to imitate Kerr's flat Oklahoma drawl, told his "dumb fuckin' brother to quit opposing my friend . . . Bohanon."

On August 16, 1961, Senator Kerr and the President held negotiations in the Oval Office. Kerr later told his supporters that he gave President Kennedy three names for the Oklahoma judgeship — "Bohanon, Bohanon, Bohanon."

"If the President is unwilling to grant this courtesy," Kerr said, "Kennedy had better get himself another boy."

Kerr got Bohanon; Kennedy got Kerr.

Sheehan asked John Seigenthaler to sign an affidavit outlining the Justice Department–Kerr–Bohanon story. "There is no doubt that had it not been for Senator Kerr, Mr. Bohanon would not have been a federal district judge," Seigenthaler wrote.

Sheehan submitted the affidavit and the evidence he had collected from the public record to the Tenth Circuit Court of Appeals, requesting that Luther Bohanon be removed from the Silkwood case for conflict of interest, and that an out-of-state judge be appointed to insure a fair hearing.

□ □ □

Tony Mazzocchi called Dan Sheehan, who was still in Washington, fighting the Nuclear Regulatory Commission. "Can you come over?" he asked.

"Sure." Sheehan knew it must be important; Mazzocchi rarely called and was never mysterious.

As soon as Sheehan walked into the office where Silkwood, Tice, and Brewer had met three years earlier, Mazzocchi gave him a white OCAW envelope.

"I've been elected vice-president," Mazzocchi said. "We're moving to Denver. Sue took the clock off the kitchen wall and this device fell out on the floor. I picked up the pieces. Can your people tell me what it is?"

Sheehan controlled his excitement. "I'll find out for you."

Mazzocchi explained that the kitchen clock had not been moved for at least six months, and that it was his wife, not he, who immediately thought the device was a bug. "If it *is* a bug," he said, "it was in a good spot. With a young child at home, it's the only place Sue and I really talk at night."

Sheehan called David Waters, a former CIA electronics man who knew every kind of bug made, by whom, and for whom. "Can I talk to you?" Sheehan asked.

"Come on out," Waters said.

David Waters liked bugging so much, he had wired his whole house. With some pride he showed Sheehan how he could monitor a conversation in any room by dialing a special phone number, which would then activate a receiver hidden in the mouthpiece of the phone in that room.

Sheehan emptied the envelope on the table. "Can you identify it?" he asked.

Waters just smiled, and in less than a minute had the two-inch-long, three-quarter-inch-wide bug back together again.

"What are you doing that the NSA is so interested in?" Waters asked.

"NSA?" Sheehan was surprised. He thought the CIA or FBI might have planted the bug, but he hadn't given much thought to the National Security Agency, the most supersecret of all the United States intelligence corps, which specialized in satellite spying and intercepting coded foreign communications.

"I assume it must be the Silkwood stuff," Sheehan said.

"What Silkwood stuff?"

Sheehan told Waters the Silkwood story. The electronics expert had seen "The Reasoner Report," so he recalled the basic outline. "What are you doing on Silkwood that the NSA is so interested in?" Waters asked.

Sheehan told him about the Georgia Power Company, Art Benson, Jack Holcomb, NIA/AID, Boynton Beach, and said that Bill Taylor was checking them out. Sheehan was careful not to mention Andros Island, which both he and Taylor felt was too sensitive to reveal. But he reminded Waters that Silkwood had been an active member of the OCAW.

Waters looked nervous. "There's a guy you have to talk to," he said. "Let me call him."

Waters' friend, who worked at the White House, wasn't home, so Waters left a message. The phone rang an hour later. "There's someone here I'm *sure* you want to talk to," Waters said.

When the White House friend arrived, he asked Sheehan to tell him about the Silkwood investigation, and, once again, Sheehan repeated the story up to but excluding Andros Island.

"White House" became very concerned. "There's someone else you *must* talk to," he told Sheehan. "I'll try to get him this weekend. Do not talk about this on the phone. If you get a call to attend a beer and pretzels party, accept. You're beer, and he'll be pretzels. Come here to the house."

Early Sunday afternoon, Sheehan got a call. "Hello, Danny? Do you know who this is?"

"Yes."

"There's a beer and pretzels party I want you to attend."

White House had brought a co-worker with him, and the two visitors apologetically asked Waters to leave the room. As Sheehan told the story up to Andros Island for the third time, the two White House specialists kept exchanging glances and looking concerned. Sheehan knew Taylor must be on to something.

"You'll have to trust us," White House II said. "I have two questions for you. You don't have to answer them if you don't want to. First, did you run across anyone in your investigation with top-secret security clearance for Naval Intelligence?"

"Yes," Sheehan said. He told them about Jacque Srouji and her work for the Pentagon on Project Seafarer.

"Did you run across Wackenhut in your investigation?" White House II asked.

"Yes," Sheehan said.

The Wackenhut Security Agency was one of the three largest international uniformed-guard and spook companies in the world. It was headquartered near the NIA/AID building in Fort Lauderdale. Sheehan had learned that Wackenhut, strangely enough, also had a third-floor office in the building in Oklahoma City that housed the police department's intelligence unit. In fact, Wackenhut was the only outsider in the police building.

"I'll tell you something," White House II said. "We know that Wackenhut has a very large office in Venezuela, and that over half of its staff there are CIA. You're way over your head, Danny. You don't have any idea how sensitive this issue is. I know you don't have any reason to trust us, but you'd better contact your man in Florida and have him stand down."

"Why?" Sheehan asked.

"I'll tell you why. They'll kill him." White House II made it clear that "they" was not Wackenhut, but he did not indicate who "they" were. "And I promise you, no one will do anything about it."

Sheehan believed White House II, but so far he hadn't given Sheehan one piece of useful information or a single lead. Sheehan decided to bargain. "I'll direct my man to stand down for three days," he said. "Then I'll order him back if you can't convince me that he should stand down."

"Okay," White House II said. "I'll need to make a phone call first thing tomorrow. Can you meet for lunch?"

"Yes."

Sheehan got to a safe phone as soon as he could. "Listen," he told Taylor. "Promise to do what I tell you."

"If I can."

"No ifs."

Taylor promised.

"You're about to be killed. I've been contacted by the White House. They told me if I pull you off, they'll give us the information we need. Leave now and go to sanctuary. Call me from there. Go."

Sheehan got a call Monday morning. "Hello, Danny? You know who this is?"

"Yes."

"We'll have lunch tomorrow. I want you to go to the George Washington University cafeteria and get some lunch. Pick a table in the middle of the room. We'll meet you there at twelve-thirty."

The two White House people brought along a man whom Sheehan still refers to as "Roughneck" in order to protect his identity. Sitting in the cafeteria, Roughneck didn't say much; just listened. The four chatted, and Sheehan was beginning to suspect that they might be jerking him on a string. Eventually, Roughneck fed Sheehan some general information about private domestic surveillance that was so vague, it was useless. There were no facts, no empirical evidence, no names, no sources. Nothing.

"Danny, you said you had a meeting on the Hill," Roughneck lied. "Why don't I drive you?"

As soon as they got into the car, Roughneck began to pump Sheehan about Florida. "What was your man doing down there?" he finally asked.

"I can't really tell you that," Sheehan said.

"Can't you tell me anything?"

Sheehan baited him about Andros Island. "All I can tell you is my man called and said he needed to rent a boat. I told him, 'If you need it, rent it.' "

Roughneck slammed on the brakes and pulled over to the

curb. "Let's get out and walk," he said. Sheehan suspected the car was bugged.

"Danny, you have no idea how sensitive this is," Roughneck said.

"You're right. So far you've told me nothing. Unless I get something, I'm sending him back in tomorrow," Sheehan said, demanding information about NIA/AID as well as a meeting between Roughneck and Taylor.

"If I give you the meeting, will you hold your man?" Roughneck asked. Sheehan agreed.

Taylor and Roughneck met in Washington. NIA/AID, he told Taylor, is a top-secret operation with CIA connections. The CIA was in on it from the very beginning. Howard Osborn, the CIA's director of security, was present when Holcomb and Goodwin drew up the plan for the factory and school. The local Florida police have been told of the CIA connection and have agreed to cooperate in any way they can.

Roughneck confirmed Taylor's suspicion that potential CIA agents, double agents, and informants were recruited at the NIA and sent to Andros Island for more sophisticated training. The CIA also recruited domestic contacts from among the United States police who attended the bugging school, and the NIA was training students from Brazil, Uruguay, and Iran. The Iranians were members of Savak, the shah's secret police.

Roughneck also confirmed that both Art Benson and Jack Holcomb had worked with Howard Hunt on the Bay of Pigs invasion. He said Holcomb had been working for the CIA on and off for twenty years, but had never been on file as a full-time "company" employee. Currently, Roughneck said, Holcomb was doing some electronics work at AID for Naval Intelligence related to Seafarer.

"They'll never let you get Holcomb," Roughneck told Taylor. "They'll kill him first."

Taylor said he wanted some tangible evidence of a connection between NIA/AID and the CIA or the federal government. Some facts, a piece of paper, a document Sheehan could use in court.

"State Department covered everything down there," Roughneck said. "But I know what they missed. Go look at the fire inspector's record."

The Fort Lauderdale Fire Department told Taylor there was no fire inspection report for the NIA/AID building. But the Silkwood investigator was not about to give up. He went to the fire department substation that would have jurisdiction over NIA/AID and waited until noon, when everyone but one secretary had gone out to lunch. "I'm a fire-insurance investigator from Washington, D.C.," he told her. "I need a fire inspection report." He gave her NIA/AID's address, but the record was not in the file. Then she checked the chief fire inspector's desk and found the original record lying there as if it were just waiting for someone to ask for it.

The copy Taylor got of the one-page report showed that the building had been thoroughly inspected in 1970, and that the inspector had made fourteen recommendations for such things as fire extinguishers and no-smoking signs. After that inspection, the building had not been inspected again until January 1974, when the inspector wrote: "OK—Real Clean." But in September 1974, another inspector wrote in the report: "Refused admittance. Government surveyed. Need U.S. security clearance to enter. Mr. Martin assured all extinguishers are government inspected and maintained."

Almost two years later, Taylor would spot a car watching his house, which was more than 200 miles from Fort Lauderdale. He chased the car, got a good look at the driver, and took the tag number. The plate was issued to the Broward County Fire Department. (Fort Lauderdale is in Broward County.) The fire department told Taylor's associate that it had "loaned the car to a governmental agency."

Several weeks after he got the fire marshal's report, Taylor noticed a white Cadillac Coupe de Ville, with a foreign-looking man and woman inside, following him. The investigator was sure it was one of NIA/AID's roving patrols. Taylor thought he had shaken the car, but when he pulled into a Shell station near the bugging school to make a phone call, the white Cadillac passed by. Taylor couldn't tell whether the driver and his partner had seen him.

The investigator had been shadowed on previous visits to Fort Lauderdale. Once he even caught someone taking pictures of his license plate, and there had been a dozen mysterious calls at the motels where he stayed, even though he had used every trick to keep his whereabouts secret.

When Taylor returned to his motel, far from Fort Lauderdale, he spotted his door signal on the ground. Opening the door cautiously, he immediately sensed something was wrong. He had left the drapes open and the light on, as he always did when he was in the field. But now the drapes were drawn and the light was off. The room was black; there wasn't a sound.

As Taylor reached for the light switch, he brushed someone with his hand. Quickly he grabbed the visitor and began tossing him around the room, yelling and grunting like a Marine. The man was short and light. Taylor stooped close to the ground, listening, watching. He sensed two people in the room and knew he was in trouble.

One of his Smith-Wesson .38 special Model 12s was in his car, the other inside the mattress next to his pillow. There was no time to crawl to the bed; he caught the faint glimmer of a knife. The blade slashed in the dark, and Taylor heard someone yell. He lunged for the knife, grabbing the man's arm at the wrist. Then someone pushed him. As he toppled over a chair, a second man fell on him. Taylor felt blood ooze onto his chest and momentarily lost hold of the knife arm. But before the attacker could slash again, Taylor squeezed the knife blade close to the hilt and twisted the man's arm. The blade hit soft flesh — stomach or side, Taylor thought, as he pushed the blade in as far as it would go. Someone screamed, and blood gushed all over Taylor. For the first time, voices began to chatter in a foreign language. Then the door flew open and the attackers fled.

Taylor got up slowly and turned on the light. His clothes were strewn all over the room, the pockets of his coats and pants ripped, his suitcases emptied onto the floor. Taylor checked himself. He had cut his right hand when he grabbed the knife, but otherwise he seemed all right.

Taylor pulled his .38 from the mattress and ran onto the balcony. The attackers were gone. Using the lobby phone, Taylor called the police to protect himself. "I was just attacked." He gave the cops the motel address and his alias. "I'll meet you in the lobby."

When the police didn't squeal into the parking lot within a few minutes, Taylor got nervous. He moved his car so that he would have a clear view of the lobby and his room and could make a quick getaway if he needed to. Still no one came. Then, just as he was about to call a second time, a Plymouth pulled in. Two men in civilian clothes got out, walked to his room without stopping at the front desk, and knocked on his door. When no one answered, they jumped back into the Plymouth and left. How did they know which room he was in? Why didn't they look for him in the lobby? Taylor copied their license number and followed them, but lost the car in traffic.

Later, Taylor wrote down the key words he remembered his attackers using, then checked the local hospitals. "My friend was stabbed," he said. "I'm worried to death about him."

No hospital reported treating a knife wound that night. The key words were later identified as Iranian. The Plymouth was not registered to anyone.

Chapter 27

Bill Taylor's source inside NIA/AID told him that Leo Goodwin was not happy. The patron of the spook school, who was little more than a figurehead, was no longer satisfied with the snoop gadgets or the expertly forged New York Narcotics Investigator's I.D. card that NIA gave him to play with. Maybe the millionaire would open up to Taylor, the source suggested.

Taylor wanted to see the sixty-three-year-old Goodwin very badly, for he needed more evidence on the CIA's connection to the factory, school, and Holcomb. He wanted to know if the government was in any way financially supporting the twin operations, what kind of files Holcomb kept, and whether Goodwin would make a good witness for Dan Sheehan. So Taylor asked his source to set up an appointment for him to see Goodwin at his Fort Lauderdale home under the pretext that he wanted to buy some surveillance equipment. To prepare for the interview, Taylor cased Goodwin's mansion from the outside and watched the millionaire trim and dig in his garden.

Two days before the scheduled interview, Goodwin died and was cremated. Taylor asked the hospital for details and the chief medical examiner for a copy of the death certificate, but he got the brush-off. So Taylor and an associate developed a routine.

"I'm from the courthouse," Taylor told the clerk in the chief medical examiner's office. "Attorney Daniel Sheehan sent me over to get a copy of the death certificate of a Leo Goodwin — G-O-O-D-W-I-N."

278

Taylor's associate walked in. "Hi, haven't seen you for a while," she said. "You still clerking at the courthouse? How's Mr. Sheehan?"

The clerk checked the file for Goodwin's record without questioning Taylor. "I'm sorry," she said. "It's not here."

Taylor's associate turned to the clerk. "Having trouble? It's probably where the other one was. On the medical examiner's desk."

The clerk went into the chief medical examiner's private office and brought out Goodwin's record. While she was making a copy, Taylor noted some flashes of light coming from behind a panel. He grabbed the copy and left as quickly as he could. His associate met him out back as planned, and they sped off.

"Did you notice the flashes?" Taylor asked.

"Yes."

"What did it look like to you?"

"Someone taking pictures of us," she said.

Goodwin's death certificate noted that unnamed doctors had performed an autopsy. The primary cause of death was "congestive heart failure"; the secondary cause was cancer of the prostate. The document, signed by Dr. Richard Ferayorni, indicated that Ferayorni had not viewed the body after death. Taylor found that strange, but whenever he called Ferayorni's office, the physician's secretary said the doctor was not available.

FEBRUARY 1978

C. J. Royer II, assistant investigator in the Silkwood case working out of Oklahoma City, got a phone call from David Gallant, a local psychic who was having a recurring dream about a woman killed in a car crash. It finally dawned on him, Gallant told Royer, that the woman was Karen Silkwood.

Royer and Davis met with Gallant. Although highly skeptical, they were determined to leave no stone unturned in the Silkwood case. They knew that psychics have frequently helped solve crimes

279

or find missing persons, but all they expected from Gallant was a lead or a new hypothesis to check out. Something they had not thought of themselves. Gallant described with great detail and accuracy what Karen was wearing the night she was killed. Many of those details had never been published.

Royer and Davis drove Gallant to Crescent, hoping the accident site would give him some visions or dreams. They didn't tell him where Karen had crashed, but as they approached the culvert, Gallant said he began to feel strong vibrations that something terrible had happened nearby. With no prompting, Gallant picked out the exact spot where Karen died, but he could not get any clear psychic images there. He suggested Royer and Davis contact his mentor, Polly Estep, a well-known Oklahoma City psychic.

Royer gave Estep a jewelry box with three things taken from Silkwood's body the night she was killed — a Mickey Mouse watch, a pin in the shape of an owl, and a button. Estep reported to Royer a week later. She said:

Karen Silkwood was contaminated through her fingertips. Possibly someone had put plutonium in her nail polish, lipstick, or nasal spray. And some doctor or medical assistant at the plutonium plant had tampered with her medication.

Karen felt very threatened during the last few weeks of her life by three or four people, male and female. At least two of them had personally threatened her during an interrogation. Karen was also having sleepless nights. Two weeks before her death, she found a lot of information about specifications or rules of some kind. Part of that information disappeared from her hiding place, probably a closet, and this distressed Karen. The rest of her information — a substantial amount — had something to do with shipping or receiving or with bottles. The count did not balance and something was missing.

Furthermore, there was a smuggling ring at the plant and Karen had been involved in it inadvertently for about three months. When she learned what was going on, she wanted to get out. but was threatened. Not long before her death, she had a violent confrontation with a woman.

The night Karen Silkwood was killed, she was being followed by three cars — one dark blue, one white, and one black. The cars seemed to be in radio contact, and Karen seemed to recognize the man in the dark blue car directly behind her. Just before the crash, he flashed his bright lights at her as if to signal her to stop. This upset Karen.

The dark blue car smelled new, and the man in it seemed to be wearing some sort of hat. He had a check made out for a large sum of money, with which he was going to negotiate with Karen for her information. If she accepted, she would have been asked to quit the Cimarron plant, and Kerr-McGee would have recommended her for another job it already had lined up. If she refused to negotiate, he would have killed her.

Karen had been in Crescent, where she had attended a meeting and had eaten dinner about fifteen minutes before the crash. The man in the dark blue car seemed to have been there, acting as a spy or undercover agent, even though he seemed related to the plutonium plant. He appeared to be forty-three years old, tall, with a fair complexion, curly thin hair, and a receding hairline. He was wearing a concealed tape recorder and smoked a lot.

Just before Karen crashed, the man in the dark blue car swerved to avoid her, drove for approximately one and a half miles, turned around, and cruised slowly past the wreck. Then he sped off.

The black car, directly behind him, was driven by a man with a dark complexion and curly hair. The white car, possibly a Cadillac, was driven by a small man with white hair, brown eyes, burn scars on his leg, and a left shoe built up one quarter of an inch. The man in the black car stopped at the wreck, climbed down into the ditch, peered into the car at Karen, but did not touch the door. The man in the white car also stopped, but didn't get out. He called someone on his radio or car phone.

Dan Sheehan did not know what to think of Polly Estep's psychic scenario or how much she may have been influenced by the media, consciously or subconsciously. But one thing was certain — there was nothing in Estep's detailed account that contradicted the known facts. So he filed the psychic away with his other leads.

☐   ☐   ☐

Joe Royer was chasing a witness for Jim Ikard on a case unrelated to Karen Silkwood. He found her in the Don Quixote bar, but she was reluctant to talk and wanted to know whom he had worked for besides Ikard.

"For Dan Sheehan on the Silkwood case," Royer said.

"Silkwood?" she asked. "There's someone who used to work here you ought to talk to."

That someone was Trudy Preston; Royer found her in Daddy's Garage bar. She told him that in 1969 and 1970 she was a secretary in the police department intelligence unit, and that the officers were doing a lot of wiretapping and bugging at that time. She said they used to brag about it, and that she typed the transcripts of bugged or wiretapped conversations. She named Bill Vetter and Larry Upchurch as two of the electronics men. "They took pride in their ability to pick locks," she said, adding that they both carried lock-picking tools at all times. "They even gave me lock-picking lessons."

While Royer was chasing Trudy Preston's leads, Echo called Bill Taylor. At great risk, Echo said, he chanced a peek into the Bureau's June Mail file on Karen Silkwood. A random look, he stressed. He'd had less than a minute, so all he could do was read part of one page. Sometime in 1975, he told Taylor, the Oklahoma City Police Department had fired a gay intelligence unit officer. The officer was having an affair with a radio reporter in Oklahoma City who was getting some inside dope on Olson and the Bureau's Silkwood investigation. The radio station sacked the reporter soon after the department fired the officer.

It didn't take Joe Royer long to find the gay intelligence unit cop fired in 1975. Royer met him in Howard Johnson's, and the former intelligence officer confirmed that James Reading and Bill Vetter had kept close contact in 1973 and 1974 — something Reading had denied under oath. The man told Royer that the intelligence unit used nothing but the very best wiretapping and bugging equipment, bought from companies in California and in Fort Lauderdale (home of Audio Intelligence Devices). He said that Vetter and Larry Baker had been trained in Fort Lauderdale by the company that sold the equipment to the police department.

The former intelligence officer also told Royer about the illegal wiretapping of an Iranian student suspected of smuggling pistols to Iran through the Dallas–Fort Worth International Airport. He admitted being one of the policemen who had broken into the Iranian's home and tapped the phone, and said he had monitored the tape. (A discreet check in Dallas confirmed that United States Customs had asked the Oklahoma City police to check out an Iranian suspected of gun smuggling.)

The former cop identified Larry Baker as the "king of wiretapping" in the intelligence unit, but he clammed up on the Silkwood matter. He said if word got back to the department that he had been talking to Silkwood people, the intelligence unit would lean on the Private Investigation Committee to take his license away or somehow force him to get out of Oklahoma City. He told Royer he had been threatened by the department before because he knew too much.

The source confirmed everything Sheehan had learned so far about the department's wiretapping:

☐ Echo had told Taylor that the Oklahoma City Police Department had the best wiretapping and bugging equipment in the central United States.

☐ Taylor had found out that Jack Holcomb had an Oklahoma file, indicating that either AID sold equipment to the Oklahoma City Police Department, or the department sent some officers to NIA for training, or both.

☐ Trudy Preston said the intelligence unit was wiretapping in 1969 and 1970, and named Bill Vetter as one of the buggers.

Royer then learned that the secretary of the intelligence unit in 1974 was now secretary to the police chief. Sheehan called her for an interview, hoping she would turn out to be another Trudy Preston.

"I have to check with the chief first," she said. She called back shortly: "No, unless you talk to me in my office in my official capacity."

Sheehan and Royer were waiting when she arrived the next morning. As she was hanging up her coat, Sheehan introduced

283

himself. She dashed out the door and down the corridor, returning a few minutes later with Assistant Chief of Police Lloyd Grambling.

"What you boys want here?" Grambling asked.

"I'd like to talk about the places where your intelligence officers get their training," Sheehan said.

"What's this all about?"

Sheehan was direct. "I have some evidence that Karen Silkwood was wiretapped. I thought you'd be able to tell me where the surveillance schools are —"

"We don't know anything except what we read in the papers," Grambling said, adding that he had heard rumors that officers had been involved in the Silkwood case in an unofficial way.

"All I want to know is where do your people get their training?" Sheehan said.

"Wait." Grambling used the intercom. "We're not allowed to tell anyone where our boys get their intelligence training, are we? Nope? I didn't think so."

Sheehan reported to the press Grambling's statement about rumors of unofficial police involvement in the Silkwood case. "It's a lie and absolute falsehood," Grambling said after the story appeared in the *Daily Oklahoman*. "Nothing was ever mentioned to even insinuate that. If that's any gauge of the rest of their statements, then they must all be false."

Dan Sheehan felt that the police department would continue to deny that it had conducted illegal surveillance on Silkwood and others until he could prove it had the equipment to do so. And he felt certain that most, if not all, of the electronic gadgets came from Audio Intelligence Devices. Sheehan tried to smoke the information out of the department.

Dwayne Cox was a hungry Oklahoma City *Times* reporter who was following the Silkwood case. Sheehan baited him. Why not ask Police Chief Thomas Heggy for a list of the intelligence unit's surveillance equipment, Sheehan asked. The gear was bought with taxpayers' money, so Heggy couldn't keep it secret.

Chief Heggy had a different idea. "That's not public information," he told Cox. If Heggy wouldn't give him the list, Cox said, he'd write the story anyway, suggesting the police had something to hide. Heggy promised Cox the inventory in April.

□ □ □

Sheehan won another battle with the court. The Tenth Circuit Court of Appeals in Denver appointed Judge Frank G. Theis of Wichita, Kansas, to replace Luther Bohanon. The decision pumped new life into the Silkwood investigation.

Judge Theis viewed the case as important. He flew into Oklahoma City in February, soon after his appointment, and told the attorneys for both sides he fully intended to lay to rest the ghosts in the Silkwood case. "They'll either walk or I'll bury them," he said.

Judge Theis proved he meant business. He extended the discovery period to mid-June 1978; he ordered Bill Vetter, of the Oklahoma City Police Department, and Thomas Bunting, of the Oklahoma State Bureau of Investigation, to appear for the depositions they were trying to dodge; he informed Srouji, Reading, and Olson that they had to answer the questions they had refused to address in their previous depositions, and said that he, Theis, would listen in to make sure they did; and he ordered Kerr-McGee and the FBI to give the court all the documents Sheehan had subpoenaed (which they were holding back), and said that he, Theis, would decide what was secret and what was irrelevant.

Sheehan was delighted. For the first time he felt confident Karen Silkwood would get an impartial hearing in Oklahoma City. Sheehan was especially eager to depose Thomas Bunting, who frequently used the undercover name Tom Burke. Bunting, forty-four, had been with the OSBI since 1972, was supervisor of the Criminal Conspiracies Unit at the time of Karen Silkwood's death, and was the OSBI's representative in LEIU. Before he joined OSBI, Bunting had been with the United States Air Force Office of Special Investigations. His wife, Sue, had told Royer that Bunting was a wiretapping expert, taught photographic and electronic surveillance in the OSBI, and probably had wiretapping contacts outside Oklahoma.

Three days after the Oklahoma City papers reported that

Theis had refused to quash Bunting's subpoena, and that he'd have to testify after all, Bunting died. He had walked into his brother's home, passed out, was rushed to the hospital, and never regained consciousness. The doctor said he died of a heart attack, apparently induced by a cerebral hemorrhage. Bunting had not been ill and had had no heart problems. There was no autopsy.

"What good would it do?" Sue Bunting asked Royer.

Chapter 28

Dan Sheehan had one more chance to depose Jacque Srouji, and he needed to find out exactly who she was and what she was doing. Every time he had asked her in previous depositions whom she worked for *before* Karen Silkwood's death, Justice Department attorney Glenn Whitaker would say, "Not relevant," and every time he tried to find out what Srouji did *after* Karen's death, Whitaker would say, "I object. National security."

So Sheehan tried a legal maneuver to get around Whitaker's relevancy objection. He asked Judge Frank Theis for a precedent-setting federal court ruling that a journalist had an absolute First Amendment right to protect the confidentiality of news sources, but that a journalist forfeited that right by becoming an informant for any government agency. Then Sheehan argued for an evidentiary hearing to prove Srouji worked for the FBI, the CIA, or both. During that special hearing, Sheehan reasoned, Whitaker would not be able to object to his questions as irrelevant.

The Justice Department attorney slipped away like a bar of wet soap. Before Judge Theis ruled, Whitaker argued that Srouji was so tied up in national security tape that for Sheehan to unravel who she was and what she did for whom would be harmful to the nation. In order to prove his point, Whitaker asked Theis for a hearing *in camera* (in the judge's chambers) and for a gag to be placed on that discussion. Theis agreed.

After three hours with Whitaker, Olson, and Srouji, Judge Theis stripped the Nashville journalist of her newsperson's privi-

287

THE KILLING OF KAREN SILKWOOD

lege, but sustained Whitaker's national security objection. "I've never been so utterly convinced of anything I've had on the bench," Theis told Sheehan. "Nothing she knows, that has been divulged to me, has any remote relevance, and I mean remote to . . . this case. On the other hand, it involves things that are of more than confidentiality. I mean — that just shouldn't have exposure."

Sheehan was incredulous. "What about the specific question about who asked her to write the book?"

"It is all part of it. I can't divulge it," Theis said. "The implications are sinister."

Sheehan asked Jerry Spence to take a third deposition from Srouji, for he was toying with asking the Wyoming attorney to argue count one at the trial — the negligence charge against Kerr-McGee. Other attorneys would argue the conspiracy charges, because Spence didn't believe there were conspiracies and had tried to convince Sheehan to drop those charges. Sheehan felt if Spence deposed Srouji, he might come to believe there was more to the Silkwood case than K-M negligence.

By this time Sheehan suspected Srouji worked for the CIA. She wrote in *Critical Mass* that she had been in Vietnam, Cyprus, Israel, and the Dominican Republic during the wars and uprisings there. A confidential source told Sheehan that Srouji had gone to Russia for the government. When Father Davis asked Dominic de Lorenzo in 1977 where Jacque was, the editor said, "I got a card from her from Vienna." But when Sheehan asked Srouji during a deposition if she had been in Vienna, Whitaker popped up like a piece of toast. "I object," he said. "National security." Finally, Sierra had told Bill Taylor that he had seen information from Srouji under a CIA code.

Under Spence's questioning, Srouji sealed all the cracks in the story she told in her previous depositions. She denied that either Reading or Olson had given her documents; that Olson asked Reading to see her; that Olson introduced her to the Kerr-McGee security chief in the Hilton; that she ever told the FBI she would perjure herself before hurting Olson; and that she tried to influence Dingell's congressional investigation. She claimed she had erased all her tapes with Silkwood information. "Cassettes are expensive," she said. "I had to reuse them."

"Can you tell me," Spence asked, "how a memorandum by J. H. Reading 'to the File' relative to the Karen Silkwood case came into *your* possession?"

"No, sir," Srouji said.

"But a moment ago you were weeping, saying to His Honor that, you know, you have to protect the confidential source. Now, who were the confidential people you were protecting?"

"I don't remember who they were."

"Are you a member of the CIA?" Spence asked.

"No sir, I am not."

"Have you been?"

"No, sir."

"Are you being employed, or have you ever been employed, directly or indirectly, by the CIA?" Spence asked.

"No."

"Have you ever performed services for the CIA?"

"Not to my knowledge."

"Who was it that instructed you to write the book?"

"I object to that, Your Honor," Whitaker said.

"Sustained," Judge Theis ruled. "It is a matter that is privileged for certain, and directly connected with the [*in camera*] hearing I had."

Spence changed his tack. "Can you tell me, in a general way, what *kind* of a person it was who requested that you write the book?"

"Object to that," Whitaker repeated.

"Was that person somebody who you were operating with in your capacity as an informant with the FBI?"

"No, sir."

"Well, was the person who requested you to write the book connected directly or indirectly with Kerr-McGee?"

"No, sir."

Spence was getting nowhere with Srouji, so, on a hunch, Sheehan whispered to him to ask Srouji about her editor, Dominic de Lorenzo.

"Is he an informant for the United States Government?" Spence asked.

"I don't think you would call him an informant."

"Does he have a relationship with the United States Government?"

"Yes, sir."

"What is it?"

"You know — I can't — You would have to ask these questions of him," Srouji said.

"You know what it is. So *you* tell me!" Spence demanded.

"I know what he told me."

"What did he tell you?"

"I don't want to get somebody else in a lot of trouble," Srouji said.

"I know. But just tell me what he told you."

"You have to understand I didn't know all this when I signed the [book] contract, okay?"

"All right," Spence agreed. "But you just tell me what he told you."

"The book was out. Nobody put me up to it. And I didn't know what he was when I signed the contract."

"What did he tell you he was, *afterwards?*" Spence asked.

"He said he was with the CIA. I swear to God I didn't know that."

Dan Sheehan reported Jacque Srouji for perjury to the United States Attorney for Oklahoma City.

"You know," the attorney told Sheehan, "the *FBI* will have to investigate the perjury allegation."

"As long as it's not Larry Olson," Sheehan said. They both laughed.

Several weeks later, the Bureau called Sheehan to say he'd have to put the perjury allegation against Srouji in writing. Sheehan smelled a trap. If he wrote a report outlining Srouji's contradictions under oath, the Bureau could leak it to the press, reinforcing the image it was trying to create — that of Srouji as an unreliable woman. Besides, there was no law requiring a citizen who reports an alleged perjury to submit the charge in writing.

"I don't have to write it up," Sheehan told the Bureau.

Father Davis hand-delivered to the FBI Oklahoma City office the documents showing Srouji's contradictions under oath. To this date, the Justice Department has not pressed charges against Srouji, who had told the FBI she would perjure herself before hurting Larry Olson, Sr.

□ □ □

Father Davis had been struggling for months with an oblique reference in the subpoenaed FBI documents to "photographic" copies of sixty-six pages of notes that Bureau Internal Affairs had found on Olson's desk. The word *photographic* seemed odd.

One day, Davis was questioning Drew Stephens about the diary Drew had loaned to B. J. Phillips for a *Ms.* magazine article. Phillips had promised to return the personal notes, but never did. Drew told Davis in passing that a friend of his, Steve Campbell, had a copy. Steve was a photographer, Drew explained, and had taken pictures of each page in the diary, which Drew began after Karen's contamination to keep his head straight. The three Silkwood investigators — Bill Taylor, Bill Davis, and Joe Royer — began to dig into the Steve Campbell connection.

There wasn't much on Campbell — unemployed, a free-lance photographer without a photo lab, frequent visitor to police headquarters, and very knowledgeable about electronics. Campbell's friend Larry Andrews told Royer that when Steve had applied for a job with the National Security Agency, the NSA had sent him, Andrews, a confidential personal reference form. Andrews said he didn't know whether the top-secret intelligence agency had accepted Steve. Campbell later denied under oath that he had applied to the NSA.

Steve Campbell first met Karen and Drew in an Oklahoma City Pizza Hut on a Saturday night in 1974. Just before the Hut closed, Drew suggested that Campbell and Campbell's friend Bill Byler stop at Drew's home for a beer and a chat. Like Campbell, Bill Byler was a lensman — the chief photographer for the Oklahoma City police. He moonlighted as a uniformed security guard at the Pizza Hut.

Campbell left with Drew and Karen; Byler followed later. Drew stopped at an all-night grocery to buy a six-pack. Would

Officer Byler be concerned if they smoked grass at home, Drew or Karen asked Campbell. Steve told them it would be better if they didn't.

The four of them sat around until about six in the morning. Karen and Steve Campbell discussed nuclear power, the conditions at Kerr-McGee, and the union. Karen gave him a copy of a union book on hazards in the industrial environment. Drew and Bill Byler talked about auto racing, photography, and guns. Both Campbell and Byler saw "paraphernalia" used for smoking pot sitting around the house.

Campbell later tried to call Karen a few times, but he never caught her at home. When he read about her death in the papers, he went over to Drew's to offer his condolences. At some point, Campbell took pictures for Drew — the Kerr-McGee workers decontaminating Karen's apartment, Drew's diary, Karen's Honda in Bob Ivins' garage, and the accident site.

Sheehan and his investigators didn't know what to make of the Steve Campbell story until Judge Theis read Sheehan an entry from the list of the subpoenaed documents Kerr-McGee had reluctantly given the court. "There's something here about James Reading paying a Steve Campbell for some photographs," Theis said. The photographs were pictures of Stephens' diary. Sheehan couldn't wait to tell Bill Taylor.

"Do I have something for you," Taylor said before Sheehan could tell him about the new Kerr-McGee lead.

"I bet it's about Campbell and Byler," Sheehan said.

Taylor was amazed. "How did you know?"

They exchanged stories. Taylor told Sheehan that Drew and Karen had borrowed Bob Ivin's Volvo one Saturday night. In the trunk were two automatic rifles. Drew told Taylor that Bill Byler was interested in the guns, so he had showed them to the policeman. No wonder Byler was interested, Taylor told Stephens. It was against the law to have government-issued M-16s.

The new lead raised important questions: When did Byler see the M-16s? How did Reading find out about Drew's diary? When did he find out? There was only one way to get answers. Sheehan and Spence deposed Olson, Reading, Campbell, and Byler.

292

LAWRENCE OLSON, SR.: FBI Internal Affairs had interviewed Olson in June 1976 about the pictures of Stephens' diary they had found on the agent's desk. Olson had told the investigator that the pictures "were furnished to him by Drew Stephens, Silkwood's boyfriend." But during his deposition, Olson admitted he got the photos from James Reading. He said he had gone to see the Kerr-McGee security chief about another matter when Reading happened to mention he had Drew's notebook. When Olson asked to see it, Reading handed him a stack of 8-by-10 photographs.

"Did Mr. Reading tell you how he had gotten the notebook?" Spence asked.

"No, sir, he didn't."

"Did you ask him?"

"I exhibited some curiosity," Olson said. "It is my recollection he said he had some source that had access to it, and I was very curious about seeing it . . . to help resolve in my mind whether or not Drew Stephens was involved in the contamination incident."

"Did you inquire to determine whether it was illegally obtained?" Spence asked.

"No, I did not," Olson said. "I saw no indication that Kerr-McGee had any interest in union activities as such, conducted surveillance of union members, nonunion members, nuclear activists, or anything of that type before or after the contamination."

JAMES READING: Judge Theis had released to Sheehan an undated Reading memorandum explaining his dealings with Campbell. It was crucial to Sheehan's case to prove that Reading, Campbell, and Byler had collaborated *before* Karen's death. In previous depositions, Reading had claimed he never even heard Karen Silkwood's name before November 9, when Wayne Norwood asked him to investigate the alpha counter stolen from Karen's apartment during decontamination.

During Sheehan's questioning, the Kerr-McGee security director admitted he had paid Steve Campbell $180 for pictures of Drew's diary and for other "information" about his conversations with Drew and Karen. But Reading couldn't explain why his Campbell memo was undated. He swore the first time he met Campbell or Byler was around November 20.

Next, Sheehan tried to clarify Reading's relationship, as Kerr-McGee security director, to the Oklahoma City police.

"Where were you when you met Campbell?" Sheehan asked Reading.

"Bill Byler's home."

"You were at Bill *Byler's* home?" Sheehan emphasized. "Isn't Bill Byler a police officer in the Oklahoma City Police Department?"

"Yes, sir."

"What were you doing at Bill Byler's house?"

"Byler had called and advised that he had known the Silkwoods prior to Mrs. Silkwood's death — Mrs. Silkwood and Drew Stephens," Reading said.

"Now, where were you when Mr. Byler called you?"

"At my house."

"Is your phone number publicly listed?"

"No, sir."

"How did Mr. Byler get your number?" Sheehan asked.

"I think he got it from — I don't know what his title was then — I think it was, is, Captain Hicks or Lieutenant Hicks."

"Is it your testimony that Mr. Hicks, who is the head of the intelligence unit, is in possession of your private, *unlisted* phone number?" Sheehan asked.

"No, sir."

"Where did he get it?"

"I don't know."

"But it is your understanding that that is from whom Mr. Byler got it?"

"No, sir."

"No, sir?" Sheehan asked. "You just told me *yes,* sir."

"I did?"

"Yes!"

"I'm sorry," Reading said. "I'm not following your line of questioning."

"I'm sorry," Sheehan said. "I'm not following your line of

answering. You just got through telling me that Bill Byler got your phone number from Lieutenant Hicks."

"I must be daydreaming or something," Reading said. "It's been a long day. Let's see if I can clear this up."

"You might want to confer with your counsel on this point," Sheehan suggested.

"No, I don't," Reading said. "Lieutenant Hicks — boy, I don't recall. Maybe it's possible they called me at my office. I don't know. My telephone is unlisted."

Reading explained that when Officer Bill Byler called him about Drew's notebook, Reading was trying to find out everything he could about Karen Silkwood for Kerr-McGee. "Here was a golden opportunity of a source that was close to them," Reading testified. "I exploited it."

Reading went on to say he had not known Bill Byler personally before November 20, 1974, and that Byler had told him about the M-16s during their meeting on the twentieth.

"Can you tell me why you pursued the automatic rifle thing?" Sheehan asked.

"Yes, sir. I was concerned about if it was going to be used for some purpose that would hurt the company," Reading said. "We had incidents at the plant."

"What did you do?"

"I informed the Alcohol-Tobacco unit here," Reading said. "The federal people."

"Who did you notify?"

"I don't recall who I talked to up there."

"When?"

"Probably the next day."

"Did you talk to anyone in person or did you do it on the phone?"

"Just on the phone."

"Do you know whether the Alcohol-Tobacco unit undertook any investigation?"

"No, sir."

"Did you tell them who you are?"

"Yes, sir."

All Sheehan had to do to establish the exact date when Reading first heard about the M-16s was to ask the Bureau of Alcohol, Tobacco and Firearms for a copy of Reading's complaint. The bureau said there was no record of a complaint.

STEVE CAMPBELL: Drew Stephens had told the Silkwood investigators that Steve Campbell suggested Drew keep a copy of his diary in case someone stole the original, and that the photographer offered to take pictures of the pages from time to time. But Campbell swore Drew asked *him* to take the photographs.

Campbell said that Drew had showed Bill Byler the M-16s on the first night Stephens and Silkwood met the police photographer, sometime in October. Then, about a week after Karen's death, Campbell said, Byler invited him to his house because Lieutenant Hicks of the intelligence unit wanted to discuss "whatever information" Campbell had about Drew and Karen. Campbell said Byler had already told Hicks about the M-16s and whatever he knew about Karen's drug habits. Campbell also testified that he had already met with James Reading *before* November 20 (Campbell couldn't fix the date), and that he continued to take photos for Reading *after* the twentieth.

November 20 as the date when Reading saw Campbell's pictures was plausible, for Drew had begun his diary on the fifth, the day Karen was contaminated. But what about the Reading-Campbell meeting *before* the twentieth? Could that meeting have been before November 9, the day on which Reading swore he heard Karen Silkwood's name for the first time?

BILL BYLER: The police photographer contradicted Campbell on two important issues. First, Byler testified that the first time he met Karen and Drew was a few days before her death, just after she had returned from Los Alamos. "I got the impression the government had sent her . . . for some kind of extensive tests in their laboratories," Byler said.

It was physically impossible for Byler to have met Karen or Drew after they came back from Los Alamos because they had returned to Oklahoma City at 10:30 P.M. on November 12. Drew went right to bed; Karen stayed up for a while, chatting with Sherri Ellis. Karen was killed after work the next night.

Second, Byler testified that Drew showed him the M-16s *after* Karen's death. He also said that the day after he saw them he told a couple of intelligence officers during a coffee break that he knew Drew Stephens and Karen Silkwood, who had recently been killed.

"Later in the day, Lieutenant Hicks called me at the lab and asked if I could come over to his office for a visit, which I did," Byler testified. Byler said that at that time he told Hicks about the M-16s and whatever else he had learned about Karen and Drew. Hicks told him to report back if he learned anything else, and suggested he call Jim Reading. The next day, Byler testified, he met Reading for lunch to tell him about Stephens and Silkwood. Reading said he would like more information.

☐ ☐ ☐

Byler, Campbell, and Reading were absolutely clear about one date — each stressed that the police and Kerr-McGee became interested in Drew and Karen on or about November 20, a week after her death. There was one crucial inconsistency in their story, however. Campbell swore that Drew showed him the M-16s before Karen's death; Byler swore he saw the rifles after her accident.

Who was right?

Bob Ivins had been in the National Guard in 1974, so Bill Taylor reviewed the sequence of events with him and studied Ivins' National Guard records. The picture became absolutely clear.

> On Saturday, October 12, 1974, Drew Stephens and Karen Silkwood were visiting me at my home in the early hours of the night [Ivins said to a sworn affidavit]. At about nine P.M., Karen and Drew left my home . . . telling me that they were going . . . to the Pizza Hut . . .
>
> I loaned Drew Stephens and Karen Silkwood my 1969 Volvo which they brought home with them, instructing Drew to be sure to keep the trunk of the car locked because I had two M-16 National Guard automatic rifles [in it] . . .
>
> On Sunday, October 13, 1974, I went to Drew's house and retrieved my car and M-16s because I had to use the rifles that second day of my weekend National Guard drill . . . Drew specifically

told me that on the previous night . . . he had displayed the M-16 automatic rifles to Bill Byler and Steve Campbell . . .

I am absolutely certain that the events occurred on Saturday, October 12, and Sunday, October 13, 1974, because I used those M-16 rifles for the very last time that weekend.

Drew Stephens also submitted a sworn affidavit to the court denying he had asked Campbell to take pictures of his diary: "Campbell . . . undertook affirmative steps to persuade me to allow him to take photographs of the pages of this diary — promising to provide me with a copy of these photographs and to keep the negatives . . . in such a manner that it would not be possible for agents of the Kerr-McGee Corporation, the police, or the FBI to obtain copies — since I expressed a specific intent to him that none of these parties get to see the record I had been keeping."

Stephens also denied he gave Larry Olson any photographs, contact sheets, or negatives. "The only possible way that . . . Olson could have gotten copies of the photographs of my diary is by somehow learning that this Steve Campbell had taken photographs," Stephens swore.

Three things seemed apparent:

☐ Kerr-McGee and the police were collecting information about Drew and Karen without their consent.

☐ The information-gathering continued for at least two months after Karen's death, for Byler had testified that late in December 1974 he had taken pictures of some of Drew's friends for Reading.

☐ And Byler saw the M-16s in the wee hours of Sunday, October 13, just three days after Dean Abrahamson and Donald Geesaman had talked to the Kerr-McGee workers about the dangers of plutonium.

If Byler told the intelligence unit about the M-16s the day after he saw them (as he testified he did), then according to that testimony:

The very next day, October 14, Byler told intelligence commander Bob Hicks about the rifles and whatever else he had learned about Karen and Drew. Hicks told Byler to report back if he learned anything else, and suggested he call Jim Reading, whom Hicks identified as the Kerr-McGee security director. The next day, October 15, Byler met Reading for lunch to tell him about Stephens and Silkwood. Reading said he would like more information, and Byler continued to supply it.

Yet Reading had sworn more than once that he had never heard the name Karen Silkwood before November 9.

Chapter 29

Someone broke into Bill Taylor's house and took his Silkwood files and a gold pen of sentimental value. The investigator was angry, but not because the documents were gone — they were unimportant. He had sent word down the grapevine that his key Silkwood notes were stashed in a string of safe drops between Florida and Oklahoma, and not in his house. But Taylor concluded that the break-in was one more instance of childish harassment, and he wanted it to stop.

Taylor called his source, Sierra. "I don't know what the fuck is going on," he said. "So far, I have been very professional. I haven't done anything to anyone. I haven't bothered anyone's family. Someone took a pen that can't be replaced. One more fucking thing like this, I'm going to get even. You pass *that* along."

Taylor flew to Oklahoma City a few days later, checked into the Corpus Christi rectory, then left for a few hours. When he returned, the stolen pen was sitting on the commode, bent like a golden banana. The investigator got the message: "This can just as easily happen to you." He called Sierra again.

"You really do a good job," he said. "How about a list of your contacts?"

Sierra said no, he was scared to death.

Shortly after the gold pen incident, Jim Ikard got a break. George Sturm, a former Kerr-McGee laboratory technician, called to say he had some information about the Silkwood case. Sturm

300

had worked at the Kerr-McGee Research Center in Oklahoma City, and on November 7, 1974, had analyzed a Karen Silkwood fecal sample for plutonium contamination. It was very hot.

Sturm told Ikard: "If Karen Silkwood had ingested as much plutonium as her fecal sample indicated, she probably would not have lived more than a week at most. It would seem to me that the high level of contamination in the fecal sample would indicate that the sample was probably tampered with."

Sturm recalled that two days later, on Saturday night, November 9, the Research Center called him at home, saying it was very disturbed and asking him to come back to analyze more of Karen Silkwood's bio-assay samples.

Sturm said he told the center that he didn't want to do the analysis because Karen's previous samples had been so high that he was concerned about his personal safety. But he was told: "Don't worry about these samples, they are not anywhere nearly as hot as the others."

Sturm went back to the lab.

"There were four bottles of urine waiting in one of the laboratories, but they were not marked in any way," Sturm told Ikard. "I was told that these were urine specimens collected from Karen Silkwood."

Sturm found that the samples were not very hot. "I would have expected that since the fecal analysis of November seventh was so high," he said, "that samples collected two days later should have shown more plutonium contamination than they did."

"A very rapid drop-off?" Ikard asked.

"A very rapid drop-off, which indicates to me that either the first sample was deliberately tampered with, or that the urine sample, the second sample, was not actually . . . collected from Karen Silkwood . . . I would say categorically that the urine samples must have been tampered with." (The AEC had reported after Karen's death that someone had tampered with several of her samples.)

The next day, Sturm mentioned at the center that it was odd that the Silkwood urine specimens had not been identified in any way — no name, date, or Kerr-McGee identification badge number. "[Kerr-McGee] didn't seem to think that was very significant," Sturm told Ikard.

Then, two days later, Sturm saw three plastic bottles, labeled SILKWOOD URINES, in the trash barrel. Sturm told Ikard that no one else could have done a urine analysis on the contents of the bottles before throwing them away because he was the only technician in the Research Center trained to analyze bio-assay samples for plutonium.

Sturm said he mentioned the discarded bottles to Kerr-McGee. "I was told to forget it," Sturm told Ikard. "That I was trained to do a job and not ask questions, and if I wanted to keep my job, I'd just better keep quiet."

(The AEC had also reported that two of Karen's urine samples had been mislabeled. Kerr-McGee had blamed Karen for not labeling the plastic flasks properly when she handed them in.)

Kerr-McGee fired George Sturm three months later. "I think my problems . . . stemmed from the Silkwood incident," Sturm told Ikard.

□ □ □

Police Chief Thomas Heggy gave Oklahoma City *Times* reporter Dwayne Cox the list of police department surveillance equipment he had promised Cox in February. Besides the usual cameras and tape recorders, the list included transmitters and receivers for wiretapping, bugging, and tracking cars. The most sophisticated equipment, including a "de-bugger," came from two places: Fargo, Incorporated, in California, and Audio Intelligence Devices Corporation in Florida.

Lieutenant Ken Smith, who had replaced Hicks as commander of the department's intelligence unit (now called the organized crime unit), told the *Times,* "Most of this stuff was already purchased and in the inventory before I got here and I don't know, really, why it was bought because we cannot use it except in violation of the laws." (According to Oklahoma law, the police needed either a court order to wiretap or the permission of one of the persons being tapped.)

Sheehan subpoenaed Police Chief Thomas Heggy, Lieutenants Ken Smith and William Vetter, Sergeants Larry Upchurch and

302

Larry Baker, and Officer David McBride. He took depositions from them all.

The depositions, which took most of April and May, had a rhythm all their own:

☐ Nobody had heard of Karen Silkwood until after her death.

☐ Nobody was trained in wiretapping and bugging at Jack Holcomb's National Intelligence Academy or anywhere else.

☐ Nobody knew who purchased the AID wiretapping and bugging devices or when.

☐ Nobody knew either Jim Reading or Larry Olson very well.

☐ And nobody conducted illegal taps.

But Sheehan had learned from confidential sources that not only did Oklahoma City Police Department intelligence officers break into an Iranian student's home and wiretap his telephone; they had also sneaked into the Oklahoma City headquarters of the Black Muslims and photographed their mailing list, and they had bugged the hotel room of radical leader Abbie Hoffman. The sources suggested that there were more illegal taps.

Throughout the depositions, however, the department's attorney, Curtis Smith, objected every time Sheehan tried to probe a specific wiretapping or bugging allegation. "Not relevant," Smith would say.

LARRY BAKER: A former intelligence unit officer had described Larry Baker as the "king of wiretapping" of the Oklahoma City Police Department. But Sergeant Baker testified that he didn't even know how to wiretap with a transmitter and a receiver.

LARRY UPCHURCH: Trudy Preston, the former intelligence unit secretary, told Joe Royer that Upchurch was a lock-picking expert. Upchurch testified that he had lock-picking equipment, but denied he ever used it to enter illegally a home or private business. Sheehan asked him why he carried burglar's tools around.

"To enter my residence and desks within the police department," Upchurch said, adding that he picked locks for practice or to while away the time.

303

KEN SMITH: The commander of the department's organized crime unit said that the electronic surveillance equipment he found in the intelligence unit closet "did not work at all and it was junk."

Smith testified that some of the electronic gear had been purchased with Justice Department money under a Law Enforcement Assistance Administration (LEAA) grant, and that he had bought some body and "curtain" mikes from AID through its regional representative, J. W. Hand.

Finally, Smith admitted that the Oklahoma City Police Department was a Law Enforcement Intelligence Unit member, that he was the current LEIU representative, and that when he got his subpoena, he had notified LEIU's legal counsel about it.

BILL VETTER: Trudy Preston told Joe Royer that Bill Vetter, commander of the intelligence unit from 1973 until February 1974, was a lock-picker and wiretapper. Furthermore, a former intelligence officer told Royer that Vetter and Reading had maintained close contact in 1973 and 1974, and that Vetter and Baker had attended a wiretapping school in Florida.

But Vetter testified he did not meet with Reading in 1973 and 1974, and that he did not attend any wiretapping school. Sheehan handed Vetter a copy of the list of surveillance equipment prepared by Chief Heggy and asked Vetter if he had bought any of the gear. Vetter admitted he had purchased a few tape recorders, but denied he bought any of the sophisticated electronic gadgets. Sheehan then asked Vetter if any of the transmitters and receivers on the list were designed specifically for wiretapping.

"I don't know," Vetter said.

DAVID MCBRIDE: Officer McBride testified that he could bug a telephone, but said he didn't learn it at any school. He was self-taught, had fumbled his way by "trial and error" on the office intercom.

McBride admitted that in 1974 the intelligence unit had "drop-in" transmitters, which wiretappers place in the mouthpiece of a telephone, and that OCPD had bought three AID bugs that could be used either as room or body transmitters. McBride said that all the bugs and taps worked, that he knew how to operate them, but that he never used them illegally. Before Curtis Smith could cut him off, McBride testified that in at least one instance the intelligence unit had wiretapped without a court order or a one-party consent.

"Who was the representative of the Law Enforcement Intelligence Unit that you should contact in Oklahoma [in 1974] if you needed to talk with any representative in the state?" Sheehan asked McBride.

"Who the LEIU representative was *here?*"

"Yeah."

"I know who it was," McBride said.

"Can you tell me?"

"It was me."

THOMAS HEGGY: A former member of the intelligence unit, Police Chief Heggy headed department Internal Affairs between 1972 and 1975. Heggy testified that when he became chief, he ordered Lieutenant Smith to lock up all the wiretapping and bugging equipment. "I didn't want it used," he said.

"Hadn't it *been* used prior to that time?" Sheehan asked.

"No," Heggy said.

"Well, why did you issue that order if you didn't know that it was being used?"

"I wanted double assurance under my command that it was *not* used, period."

"Now in the light of the controversy that has arisen in connection with this equipment," Sheehan asked the police chief, "is it your intention to undertake any investigation to find out whether or not any officers of the Oklahoma City Police Department have ever been engaged in using that equipment unlawfully?"

"No, sir, not at this time."

"Can you tell me why you wouldn't try to do that?" Sheehan asked.

"Well, I wouldn't have an administrative investigation while there was a court trial, no, sir."

"Isn't it in fact *true,* Chief Heggy, that you are postponing any investigation into these matters so that incriminating evidence won't be obtained, that might be obtainable [by Sheehan] through a court order?"

"No, sir."

"Don't you, in fact, have access to the past records, as chief of the police department?"

"If they exist," Heggy said.

"You say that you have access to the information, and yet you're stating to us that you refuse to conduct such an investigation as long as there's a court inquiry going on. Is that correct?"

"That's *exactly* right," Heggy said.

Sheehan asked Heggy about Officer David McBride's allegation that the department had wiretapped without a court order and without one-party consent. "Have you undertaken any effort whatsoever to contact the officer to obtain information about that?" Sheehan asked.

"Not as yet, no."

"Do you plan to?"

"Possibly later."

"Possibly *later?*" Sheehan said. *"Do* you intend to do so?"

"Possibly later."

Sheehan then asked Chief Heggy whether the police had broken into the office of the Black Muslims and photographed their mailing list. But attorney Curtis Smith objected. "Irrelevant," he said.

"Are you refusing to answer my question?" Sheehan asked Heggy.

"On advice of counsel, yes."

"Now, are you maintaining that I have the *same* ability to get the information that *you* do?" Sheehan asked in frustration.

"Yes," Chief Heggy said.

Chapter 30

Dan Sheehan had been waiting for the chance to depose Roy King, the Kerr-McGee personnel director whom Oklahoma Highway Patrolman Rick Fagen had called for next of kin the night Karen crashed. But King was not eager to be interviewed by Sheehan. Someone had tried to kill him, and he was convinced that the murder attempt was somehow related to the Karen Silkwood incident. Now retired from Kerr-McGee, after more than thirty years of service, Roy King just wanted to be left alone.

Three months after Silkwood's death, someone had crept into the alley behind King's house in Guthrie, turned off the gas, waited until the flame was dead, then turned it back on again. It was February, and all the windows were closed. But fortunately for the Kings, every room but the garage was protected by a fail-safe system, which cut the gas off if the pilot went out.

In the middle of the night, Roy's wife nudged him and said, "Gosh, it's getting cold." King got up, noticed the flame was out, and concluded the gas company had shut off the fuel for repairs. To be certain, King checked the garage. He smelled gas.

The gas company checked its equipment, going so far as to take King's meter apart to see if the gas could have been turned off by a defective meter. "They said it absolutely could not," King later told Sheehan during his deposition. The Guthrie police guarded the Kings each night until they could move to Oklahoma City.

King had told Sheehan that he was still afraid someone might

307

try to kill him, and he pleaded not to be deposed. But Roy King was a defendant in the case, and Sheehan needed to question him about Kerr-McGee's alleged attempt to crush the union and about the night of the accident.

The OCAW suspected that Kerr-McGee was behind the attempt to get the Crescent local decertified. In the deposition, Sheehan asked King about that.

"Decertification was initiated by Don Hall," King testified.

"Can you tell me what you know about Don Hall?" Sheehan asked.

"I don't believe he belonged to the union," King said. "He asked for a little bit of guidance and who to contact regarding the necessary petition . . . that would be needed to make this decertification inquiry. And that was given to him. We made some contact for him, but that's as far as the company was involved in this thing."

"Did he ask *you* for this little bit of guidance?" Sheehan probed.

"He did. And I, in turn, counseled with our industrial labor relations director in Oklahoma City," King said, adding that he had no knowledge of Kerr-McGee's doing anything else in the decertification push.

Sheehan moved on to the night Karen Silkwood was killed. King explained that an Oklahoma Highway Patrolman (King couldn't remember his name) called to ask for Karen's next of kin. King drove to the Guthrie Police Station to see the patrolman, then to the hospital to identify Karen's body.

"I called the Silkwood people," King continued. "Then I came back down to the police station, and the highway patrolman made this comment. He said, 'There's a lot of things in her car that have Kerr-McGee identification insignia on them.' He said, 'I would like for you to join with me tomorrow and we will go down there and get those . . .'

"Well, about the time that next morning when he was supposed to be coming and getting me, I was thinking that maybe I ought to get in my car. And here comes the highway patrolman. And he said, 'Well, somebody has *got* those. So there's no need —

any point in us going down there to pick them up.' "

King took Sheehan by surprise. "Do you know who the Kerr-McGee people were that went out there to where the car was during the night and took the stuff out of the car?" he asked.

"I don't know anything about that."

"Did you ever —"

"I don't know if any of *our* people did take it out there," King corrected Sheehan. "I don't know that."

Sheehan later read Roy King's account to Oklahoma Highway Patrolman Rick Fagen. "I don't think that was me," Fagen testified. "I don't recall saying any of that."

But Fagen went on to admit that he was the only Oklahoma Highway Patrolman King had talked to at the Guthrie Police Station and the only patrolman to visit King at the Cimarron plant the next morning.

Exactly who did what in Ted Sebring's garage had always been fuzzy. Sebring had told the FBI that Kerr-McGee people checked the car with alpha counters; Rick Fagen had told Larry Olson that three AEC inspectors monitored the car for radiation after midnight. Were Sebring and Fagen confused?

After King's deposition, Bill Taylor and Bill Davis talked to Crescent policeman Joe McDonald, who was with Fagen the night the three men came to inspect Karen's Honda. McDonald confirmed Fagen's version of the story. Next, Davis and Taylor interviewed Jack Tice, who had been very helpful with leads so far and who had a good memory. Tice told the investigators that when he went to Sebring's to see Karen's car around ten o'clock the night Silkwood was killed, there were several Kerr-McGee men inside the garage. There were no AEC inspectors there, Tice said.

If McDonald's and Tice's memories were accurate, there were *two* visits to Ted Sebring's — one shortly after the accident by Kerr-McGee, one after midnight by the AEC. Spot Gentry, who helped Jim Reading investigate the Silkwood incident, later supported that conclusion.

"As I recall it," Gentry testified, "two groups searched it, I believe. Or else maybe the same group searched it once for nuclear materials and another for papers . . . I don't even have my own memorandum to go by now, but it seems to me that there was two searches made — one for the papers and one by the Geiger counters."

□ □ □

Dan Sheehan wanted to know whether the list of electronic surveillance equipment Chief Heggy gave the Oklahoma City *Times* was complete. The only way to be sure what gadgets Audio Intelligence Devices had sold to the Oklahoma City Police Department, who ordered them and when, was to get AID's records. So Sheehan subpoenaed AID president Jack Holcomb and his Oklahoma file.

When Bill Taylor went to Fort Lauderdale to serve Holcomb the subpoena — the law requires that an attorney see to it that the witness receives the subpoena — Holcomb was gone. Like Bill Lovin, Holcomb had left for West Germany. Taylor's source inside NIA/AID told him that Oklahoma City reporter Gypsy Hogan had tipped Holcomb off with a phone call asking why he was being ordered to testify in the Silkwood case.

As a poor second choice, Sheehan subpoenaed Jack Larsen, AID's director of marketing. Taylor was not about to be snookered a second time. He followed Larsen around until he became familiar with his car and home, work habits and driving patterns. Then Taylor told Sheehan to file.

As soon as Taylor got Larsen's subpoena, he rented an olive Dodge Charger with wide tires that would have traction in the Fort Lauderdale sugar sand. Then he drove down a power station road a quarter of a mile from NIA/AID and cut through the trees and underbrush to a spot he had picked across the road from the bugging factory where Larsen worked. From his stake-out, Taylor could see the front of the building and the parking lot without being spotted by the roving patrols or the TV cameras on the outside of the building. Taylor rolled the Dodge Charger back and forth to pat down the sand so that he wouldn't get stuck when it was time to move down the small hill onto the road.

Taylor waited in the breezeless Florida heat. Five o'clock came, but Larsen didn't. By 5:30, the parking lot was almost empty, but Larsen's Monte Carlo was still there. At six, Larsen's wife drove around to the back of the building. A few minutes later, she zipped out again and headed east, not west toward home. She was alone in the car.

Taylor waited. Why did she drive to the rear entrance? Why did she come back out of the building so quickly? Taylor dug out of the sand after Mrs. Larsen. When he caught up with her Dodge Dart, Jack Larsen was driving. Mrs. Larsen was at his side.

Larsen got off Highway 95 at the Pompano Beach exit. While Larsen's Dart was sandwiched between two cars at a red light, Taylor cut across the median strip and raced down the road against the oncoming traffic. Then he cut back across the median strip, blocked the intersection with his Charger, and walked over to the Dodge Dart.

"Hi, Jack," he said, standing to the left and the rear of Larsen. He had heard that Larsen packed a gun. "Here you go, Mr. Larsen. You are subpoenaed."

Taylor gave Larsen the court order and asked the insurance broker in the car behind Larsen's to be a witness. Taylor was as happy as a new quarterback who has just completed his first pass. He couldn't wait to tell Sheehan.

The investigator found an outside phone at a Shell station. Sweat dripped from his face, his socks were soaking wet from the perspiration that ran down his legs into his shoes, and he was exhausted from the strain. "Danny, you're not going to believe this," Taylor said after he finished telling Sheehan the whole story. "A bird just shit on my head."

□ □ □

Jack Larsen had been an Indiana State Policeman for twenty years and was a former intelligence unit captain. Like Thomas Bunting of the Oklahoma State Bureau of Investigation, and Bill Vetter, Ken Smith, and David McBride of the Oklahoma City Police Department, Larsen had also been an LEIU representative. When he retired in 1970, he sold bugging, wiretapping, and other surveillance equipment for Bell and Howell. Jack Holcomb lured Larsen to Florida, and since 1973, he had been AID's vice-president for sales.

Larsen denied that AID sold equipment to the Oklahoma City

police before 1975. But he did admit that in 1974 AID was selling drop-in taps for telephone mouthpieces, wall bugs, and bird-dog beepers to track cars. (Chief Heggy had listed those three kinds of devices on the inventory he gave Dwayne Cox.) Larsen also admitted that the Justice Department gave some AID customers money to buy surveillance equipment through LEAA grants.

But Larsen did not bring the files Sheehan had subpoenaed, claiming they were the personal property of Jack Holcomb. "Where is Holcomb?" Sheehan asked. The AID vice-president said he didn't know.

"Where are the files kept?" Sheehan asked.

"Which files?"

"Mr. Holcomb's," Sheehan said.

"Mr. Holcomb's office."

"Are they —"

"Purchasing files are kept in purchasing," Larsen clarified. "Our supervisor of the machine shop has files in the machine shop. The engineer has files in the engineering department."

"And you, as sales manager, have the files in your office relating to sales?" Sheehan asked.

"The sales files would be in the sales office."

"And you say they *all* belong to Mr. Holcomb?"

"That is true."

"What are the files that Mr. Holcomb has in his office if the various divisions have all the files relating to their division?"

"I don't know."

"But he has a whole bunch of files in his office, *too?*" Sheehan pressed.

"I would imagine. He has some file cabinets, so I imagine there are files."

"Have you seen them?"

"No."

"Can you tell me where are the files that relate to the sale of equipment to any law enforcement agency in Oklahoma?" Sheehan asked.

"Mr. Holcomb has them."

"How come you don't have them if they relate to sales?" Sheehan asked the vice-president of sales.

"Because Mr. Holcomb took them."

"Took them after the newspaper reporter told him that there was a subpoena for him?"

"He took them after he came down and stated he had a telephone call from a newspaper reporter in reference to something going on in Oklahoma, a lawsuit, whatever," Larsen said. "He came down and we reviewed the files, looked at the Oklahoma files, and he took the files."

"Do you have any idea where he brought them?"

"No, sir."

With Holcomb in West Germany, the Oklahoma file missing, and Jack Larsen not well informed, Sheehan took one last shot. He subpoenaed AID regional representative J. W. Hand.

□ □ □

Sheehan had sent Hand's subpoena by certified mail to the United States Marshal's Office in the Northern District of Texas, requesting that a marshal serve him as soon as possible. The request was routine. But when he called the district court a week later, Sheehan learned that the marshal had not even picked up the subpoena, even though Ikard had called and told the office it was coming. So Sheehan sent Joe Royer to Lewisville to serve Hand himself.

Royer saw Hand puttering in his yard on Sunday, June 4. When Royer returned on Monday with the subpoena, Hand was gone. Next, Royer went to the police station in Flower Mound — the village next to Lewisville — and told Sergeant William Junell that he was trying to find J. W. Hand. Royer showed Junell his driver's license and his private investigator's identification. After copying the information, Junell told him that Hand wouldn't be back until the weekend of June 9. Did Royer want to buy some electronic surveillance equipment from Hand, Junell asked. Royer didn't answer.

On Sunday, June 10, Sheehan flew into Dallas. Royer picked

him up at the airport and drove him out to Hand's home. As the deposition was scheduled for the next morning, they wanted to make sure the salesman received his subpoena.

Mrs. Hand answered the door. Her husband was not back yet, she said. Royer told her he had some papers for her husband. She asked him to wait, left the door open, and went into the living room. Royer heard her talking to someone.

"What kind of papers?" she asked when she returned.

"A federal subpoena for Mr. Hand," Royer said.

Mrs. Hand slammed the large wooden door in his face, shouting, "I didn't touch it. It isn't good service. We aren't accepting any service."

Royer told her he was placing the subpoena inside the screen door and that Mr. Hand was thereby served. On the way back to Dallas, Sheehan noticed a white, unmarked car with a tiny portable flashing red light following them.

"Uh-oh! I bet they are coming after us for serving the subpoena," he told Royer.

The investigator pulled over and he and Sheehan got out. Sergeant Junell, who was wearing a blue and white baseball cap, jeans, T-shirt, and tennis shoes, got tangled in the cord of the flashing red bubble as he tried to ease out of his white car. He finally jerked the light from the roof, tossed it into the car, and walked over to Royer and Sheehan.

"What seems to be the problem, Officer?" Royer asked.

"You're practicing as a private investigator in the State of Texas without a Texas license," he said.

"I wasn't," Royer said. "I was simply engaged in serving a federal subpoena."

"You were," Junell said. "You're in *real* trouble."

Junell turned to Sheehan. "And just who are you? Show me your identification."

"I don't believe that I have to," Sheehan said. "I wasn't driving the car."

"I told you to show me some identification," Junell threatened. "Now get out some I.D."

Sheehan refused.

"In the State of Texas," Junell said, "there is a law which

314

says that you have to display identification to a law officer any time he demands it."

"If there is," Sheehan said, "then it's not constitutional."

Junell ordered Sheehan back into Royer's car.

"I don't have to," Sheehan said.

"Well, then move away from [my] vehicle and stand behind yours."

Sheehan moved. Junell checked Royer's driver's license, used his car radio to call another policeman, then came back to Sheehan. "And now, show me *your* identification."

Sheehan refused.

"What's your name?"

Sheehan stood silent.

"Where do you come from?"

Sheehan stood silent.

"All right, get in." Junell pointed to his white car.

"Nor do I have to get into *your* car," Sheehan said.

"Oh yeah?" Junell said. He grabbed Sheehan's right wrist and arm, trying to hammerlock him. But Sheehan, who had studied martial arts in Green Beret officer's training school, squirmed out of the hold. "Officer, relax," he said. "There's no need for all this. Just relax. You're overreacting here."

Junell lunged at him again, trying another hammerlock. When Sheehan slipped away this time, Junell fell backward and slid into the shallow ditch along the roadside.

Junell came after Sheehan once more, squeezing his arm as if to force him into the car; but Sheehan held his arm straight and kept turning in a circle, talking to Junell, trying to calm him.

"You're resisting me," Junell said. "You're resisting arrest and that's a felony."

"I'm not under arrest," Sheehan said.

"You are too. For refusing to display identification."

"Well, if that's what you're saying, then I'll get into the car."

Junell released Sheehan.

"I know who you guys are and what you're doing down here," Junell said. "I spoke with the Oklahoma City police and they sure aren't happy with you up there. And we're not very happy with you down here either."

Junell let Royer go, saying, "Here's your license back. And here's your subpoena back. They don't accept service."

A police cruiser came with another officer, who ordered Sheehan out of Junell's car, made him spreadeagle, searched him, cuffed his hands behind his back, chained the cuffs to his belt, and hauled him off to the Denton County Jail. There, the police tried to interrogate Sheehan, but the attorney refused to talk, citing his rights.

The police fingerprinted him, took mug shots, and accused him of being an accessory to burglary. They took all his belongings — his wooden cross, belt, money, and a vitamin pill they said they were going to analyze for drugs. Then they told him that unless he came up with $1200 in cash as bail, they would lock him up for the night and haul him before the magistrate the next morning.

The Denton County magistrate released Sheehan the next morning on his own recognizance. Then, a few days later, he declared Sheehan's arrest and imprisonment illegal.

Sheehan left Texas in a hurry that night.

□ □ □

Dan Sheehan asked Judge Theis to extend the discovery period, which had ended while he was in the Denton County Jail. He still had subpoenas out for Jack Holcomb of NIA/AID, Edwin Youngblood of the National Labor Relations Board, and David Ellis, one of the first teachers at NIA. But before ruling on the extension request, Theis said he needed to catch his breath and study the motions before him. First, there were pleas to quash the subpoenas of:

□ Art Benson of the Georgia Power Company, who was now arguing that he didn't know anything about the Silkwood matter.

□ Paul Wormeli, a former Law Enforcement Assistance Administration official who had given LEAA block grants to police departments between 1970 and 1977, and who directed Project Search, which kept LEIU files on its central computer in East Lansing, Michigan.

□ Patricia Atthowe, president of Research West, a private investigation firm in California.

☐ Roy Leyrer, president of the Western Regional Organized Crime Training Institute.

Next, the Kerr-McGee defendants and the Kerr-McGee Corporation had filed a motion for summary dismissal of the case against them and for a summary judgment. Their attorneys argued that the plaintiffs had no evidence to prove the K-M defendants conspired "to violate the civil rights of Karen Silkwood or to prevent her or others from engaging in union activities or from making complaints to the Atomic Energy Commission." They pointed out to the court that each of the twenty-three Kerr-McGee defendants denied under oath that he had participated in any conspiracy.

Kerr-McGee attorneys emphasized that both plantiffs — Bill Meadows and Bill Silkwood — admitted under oath that they had no personal knowledge of the alleged conspiracy, and that what little they knew, they had learned from Daniel Sheehan, their attorney. The K-M attorneys pointed out to the court that when they deposed Sheehan, they learned that his charges were based on hearsay information to the fourth power: "Sheehan said (that's one) that either Howard Kohn or Peter Stockton or John Seigenthaler said (that's two) that Jacque Srouji said (that's three) that Jim Reading said (that's four times) that the telephone of Karen Silkwood was tapped."

The Kerr-McGee attorneys went on to remind the court that Reading had denied under oath the activity alleged in the complaint; that Reading denied under oath he had ever told Srouji that such activity occurred; that Srouji had denied under oath that she told Stockton or Kohn that Reading had said he had anything to do with tapping Silkwood's phone; and that even if Kohn, Stockton, and Seigenthaler were to testify what they alleged Srouji had told them, their testimony would be clearly inadmissible on the grounds that it was hearsay.

The Kerr-McGee attorneys also asked Judge Theis to dismiss the case against the K-M defendants and the corporation because there is no basis in the law for a civil suit against them. They argued:

☐ The Civil Rights Act of 1871 was passed to provide black persons, terrorized by such groups as the Ku Klux Klan, the means to get equal protection under the law. But Karen Silkwood was not black, and there was no alleged racial basis in the Silkwood case.

☐ The 1871 Civil Rights Act can also apply to "class-based" discrimination. But discrimination against individual members of a labor union cannot be defined as "class-based" discrimination.

☐ The Civil Rights Act does not allow dead persons or their estate to sue for damages.

☐ The Silkwood complaint fell into the category of unfair labor practices. But unfair labor practices belong to the jurisdiction of the National Labor Relations Board, not the federal court.

☐ If Karen Silkwood was injured by plutonium, she was protected by workers' compensation laws. Therefore, the negligence case against Kerr-McGee was outside the jurisdiction of a federal court.

☐ The Kerr-McGee Corporation was not liable for any alleged injury to Karen Silkwood because she worked for the Kerr-McGee Nuclear Corporation, a daughter company.

☐ And there was circumstantial evidence that Karen Silkwood "brought plutonium into her apartment herself, but no circumstantial evidence it came there by any act attributable to the Kerr-McGee Corporation." Therefore, the case should be dismissed.

Larry Olson and his former Bureau boss, Theodore Rosack, also filed motions for summary dismissal and judgment. Besides stating that Sheehan had no evidence to support his conspiracy case against them, they argued that as FBI agents they were immune from private lawsuits. Furthermore, they said, they were not liable for any alleged injury to Karen Silkwood because they were charged with joining the conspiracy after her death. How could they have injured her or deprived her of her civil rights if she was already dead?

Finally, Jacque Srouji filed a motion asking Judge Theis to disbar and censure Sheehan for unprofessional conduct. Acting as her own attorney, she accused Sheehan of using the Silkwood case as a "vehicle to pursue a monster intelligence scenario that has involved everyone from Walt Disney to Charlie Chaplin." She said that Sheehan's briefs "contain half truths, outright lies that smack of serious slander and deliberate character assassination," and that he had "managed to publicly disseminate his venomous attacks to a willing group of jugular journalists quite eager to print fiction rather than the truth."

After describing herself as an innocent, powerless woman without a legal team to reply or object to the "insane pleadings issued from Mr. Sheehan's garbage hopper," she went on to say: "To this day, he has not uncovered any evidence to prove his conspiracy theory nor to justify or explain defendant Srouji's presence in this lawsuit."

Srouji suggested that Sheehan talk to the Nashville *Banner* about her relationship to the FBI, arguing that her editors had instructed her to "cooperate fully and completely" with the Bureau. To prove that point, she submitted to the court a note written to her by Charlie Moss, the *Banner*'s late executive editor:

C. Moss — Fed

Name of Group — initials (nickname) appropriate membership

Purpose — leaders (connections, whether Red etc.)

Digest of thoughts expressed at meeting — conduct (sex etc.)

General rundown of trips to conference — who spoke (identify organization and where from. Conduct etc.)

Mississippi

Atlanta

Montgomery

Srouji described Moss's handwritten instructions to her as "genuine and leaves no doubt as to the *Banner*'s direction of Mrs. Srouji in the newspaper's involvement with the FBI."

Noting that much of Sheehan's behavior was immune from formal prosecution because he was an attorney, Srouji concluded that his personal vendetta against her, his slanderous scenarios, and his leaks to the press warranted that the court disbar and censure him for "conduct alleged to be in violation of the Standards of Ethics of the American Bar Association."

Judge Theis promised to study all these motions. In the meantime, he said, Sheehan could still depose Edwin Youngblood, whom Sheehan had failed to question on June 11 because he was still in jail.

Sheehan returned to Texas, where Youngblood lived. As the

319

director of NLRB's Region 3, which included Oklahoma, Young-blood had the responsibility to investigate Tony Mazzocchi's charge that Kerr-McGee had harassed Karen Silkwood, Jack Tice, Jerry Brewer, and other union members.

The FBI documents released in October 1977 revealed that the NLRB had investigated Kerr-McGee, written a report, sent a copy to the FBI, but filed no charges against the corporation. Months later, the FBI cleared K-M of the harassment allegation. The FBI documents raised a serious question. Why would the NLRB hand over its report to the FBI if the board had not concluded that Kerr-McGee committed a crime?

Sheehan had asked the NLRB for a copy of its Kerr-McGee investigation report, but the board refused, saying the document was not a public record. Next, Sheehan subpoenaed the report, but the NLRB argued that the document was sensitive and proprietary. Finally, Sheehan asked for the names of the persons who had investigated Kerr-McGee, but the board said it must protect the identity of its investigators. So, as a last resort, Sheehan subpoenaed Youngblood.

Edwin Youngblood testified that he had assigned two persons to go to Oklahoma City and Crescent to investigate the OCAW charges, and that each had concluded independently that Kerr-McGee had violated the National Labor Relations Act, a civil offense punishable by a fine. But even more important for Sheehan, Youngblood testified that the two NLRB investigators independently concluded that Kerr-McGee had also violated the Federal Nuclear Code by harassing whistle-blowers. Since a potential federal crime was more serious than a potential civil violation of the National Labor Relations Act, Youngblood testified, the NLRB gave the FBI a copy of its report and washed its hands of the case.

SEPTEMBER 1978

Judge Frank Theis dropped the conspiracy charges against Kerr-McGee and the FBI defendants. Theis argued that even if Sheehan could prove that Kerr-McGee conspired to, and actually did, deprive Karen Silkwood of her civil rights by wiretapping, bugging, and harassing her, there was no civil law under which the Silkwood estate could claim damages. And even if Sheehan could prove that the FBI conspired with Kerr-McGee to cover up those crimes, Theis argued, there was still no law under which the Silkwood estate could sue.

Sheehan had based his entire conspiracy case on Karen Silkwood's constitutional rights, spelled out to some degree in the Civil Rights Act of 1871, but Judge Theis argued that the act applied to blacks and other minorities, not to whites or members of a union as a class.

Theis ordered Kerr-McGee and the Silkwood estate to prepare for the trial on the negligence count, which charged Kerr-McGee with failing to exercise proper care in handling and safeguarding the plutonium (a hazardous substance) that contaminated Karen Silkwood in her apartment.

The Silkwood team — Davis, Nelson, Taylor — felt crushed. They couldn't charge Kerr-McGee or the FBI with a crime; the Justice Department had to do that with the investigative help of the FBI. And now they couldn't sue for a violation of Karen's civil rights, because Judge Theis had ruled that the Civil Rights Act did not cover her.

But Dan Sheehan was not discouraged. He had known all along that the application of the Civil Rights Act to Karen Silkwood was a controversial legal point, and he suspected that Judge Theis wanted to settle the issue *before,* rather than after, a long expensive trial. So far, lower-court decisions had been divided between a narrow and a broad interpretation of the act. Sheehan was a constitutional lawyer, and he instinctively felt his broad interpretation was right. So he appealed Judge Theis's decision to the Tenth Circuit Court of Appeals in Denver. If the appellate court wouldn't overturn Theis, the Supreme Court surely would.

In the meantime, Sheehan, his investigators, and his associate attorneys began to prepare their negligence case. And Sara Nelson began to beat the bushes for money.

Chapter 31

Arthur Angel joined the Silkwood legal team in Oklahoma in the fall of 1978. He was a Harvard Law School graduate and had recently embalmed the funeral industry for the Federal Trade Commission. Just thirty years old, short, with curly brown hair and an unruly mustache, he was friendly, charming, cynical, and tough.

Art Angel planned to help with the Silkwood negligence case against Kerr-McGee for a couple of months and then return to Washington. Like Jerry Spence, who had deposed Larry Olson, Sr., Angel wasn't convinced there were conspiracies, at least the kind that Dan Sheehan saw. And even if Sheehan could prove a Kerr-McGee plan to wiretap and bug Karen Silkwood, Angel reasoned, no Oklahoma jury would be incensed enough to do more than slap Kerr-McGee's corporate wrists. If the Silkwood estate wanted money, if Bill Silkwood and Dan Sheehan wanted to punish Kerr-McGee and warn the nuclear industry, and if Sara Nelson and Kitty Tucker wanted to educate the public about the dangers of nuclear power, they should kick Kerr-McGee in the pocketbook for gross negligence. Money was something Kerr-McGee, a jury, and the public could understand.

But the Washington attorney found the negligence case in legal shambles. The legal underpinnings for the charges had not been built into a case; Dan Sheehan had spent so much time chasing conspiracies, he had not yet pulled together the evidence and witnesses to prove negligence; Jim Ikard was so upset by Sheehan's focus on conspiracies and his bravado before the press that he was sorry he had ever got involved with Silkwood; and Judge Theis rumbled at Sheehan's bluntness. But Dan Sheehan wasn't flustered. The trial would begin in March, and five months was plenty of time to prepare for it.

Art Angel stayed in Oklahoma to mend fences. He drew Ikard back into the case, and the two of them crafted the negligence complaint in consultation with Sheehan. First, they argued that plutonium was an ultrahazardous metal; therefore, the Cimarron plutonium processing plant was an ultrahazardous operation. Under Oklahoma's strict-liability law, Kerr-McGee would be responsible for the ultrahazardous plutonium that had escaped from the plant and injured Karen Silkwood, even if K-M were not at fault.

Second, they argued that, over and above strict liability, Kerr-McGee was liable for Karen's injury because K-M was grossly and wantonly negligent in keeping its plutonium inside the chain-link fence.

Third, they argued that the richer Kerr-McGee Corporation was responsible for the damages rather than its poorer daughter, the Kerr-McGee Nuclear Corporation, because the parent tightly controlled the operations of the daughter.

In the amended complaint, the Karen Silkwood estate asked for $1.5 million as compensation for Karen's physical injury and emotional pain, and $10 million as punishment to Kerr-McGee. The money, after expenses, would go to Karen's three children — Beverly, Michael, and Dawn Meadows.

In order to win their total case, the Silkwood attorneys would have to prove to a jury that

☐ plutonium was ultrahazardous;

☐ Karen Silkwood was contaminated with *Kerr-McGee* plutonium;

☐ Karen did not deliberately contaminate herself;

☐ the contamination physically injured Karen between November 5 and November 13, when she died;

☐ Kerr-McGee was wantonly negligent in protecting workers and safeguarding its plutonium.

From October to March, Art Angel, Jim Ikard, and the Silkwood investigators began to interview former Kerr-McGee workers, collect and organize facts, and line up witnesses. Jim Ikard
prepared the experts chosen to testify about the dangers of plutonium. And Sara Nelson raised the money.

From September 1977 to the end of the trial, in May 1978,
Nelson collected $452,000 to pay for the investigation and trial.
About $100,000 of that came from a direct-mail campaign over
the signature of Representative Bella Abzug, who had volunteered
in 1975 to conduct a congressional investigation into the Silkwood matter (Kitty Tucker and Bob Alvarez had chosen Senator
Lee Metcalf instead.) Another $100,000 was in interest-free loans,
to be repaid out of the award if the Silkwood estate won. The rest
came from Bonnie Raitt and Jackson Browne rock concerts, the
Levinson Foundation, the North-Shore Unitarian Veatch Foundation, and small contributions. The Youth Project, a respected, independent, tax-free group, handled the grants, contributions, and
bookkeeping.

From the start, Kerr-McGee attorneys gave the Silkwood lawyers an advantage. By refusing to concede an undisputed point —
the plutonium that contaminated Karen Silkwood belonged to
Kerr-McGee — they made Judge Theis slightly sympathetic to the
Silkwood side. And by refusing to concede that plutonium was
ultrahazardous, and that the Kerr-McGee Nuclear Corporation
was a dependent daughter of the larger company, they gave the
Silkwood attorneys the right to probe areas of negligence and corporate power that Judge Theis otherwise would have ruled irrelevant.

With those givens in mind, the Silkwood team developed a
strategy:

□ To prove that she was not the kind of person who would contaminate herself, they would try to portray Karen Silkwood as a
union representative, aware of the dangers of plutonium, concerned
about health and safety, and candid with the Atomic Energy Commission and Kerr-McGee about the details of her contamination.

□ To prove that Kerr-McGee was directly or indirectly responsible for
Karen's contamination.

☐ To prove that $11.5 million was not too much to ask for, they would try to portray the Kerr-McGee Corporation as a spoiled, rich, callous company that must be punished — as a warning to other sloppy nuclear plants.

☐ To prove that plutonium was ultrahazardous, and that Karen had been physically injured, they would line up the best experts in the country to testify.

☐ To prevent Kerr-McGee from arguing that they were using Karen to destroy the nuclear industry, they would avoid appearing as antinuclear and would not call antinuclear witnesses.

Both Kerr-McGee and Silkwood attorneys faced the same problem. Neither could prove who actually had contaminated Karen Silkwood. At best, they could try to sway the jury with a convincing string of circumstantial evidence. And each side had one major weakness. Silkwood attorneys had a mound of convincing evidence that Kerr-McGee was negligent in protecting its workers against contamination with cancer-causing plutonium. That would soften and frighten any jury. But Kerr-McGee had evidence that Karen Silkwood was not injured according to the traditional definition of injury. That would make any jury skeptical. Without an injury, there was no case, no matter how great the negligence.

To win this case, therefore, each side would have to act as well as argue. The courtroom — the pit, as Judge Theis liked to call it — would be the stage; the jury, the audience; and Judge Theis, the stage manager.

Chapter 32

The Silkwood estate hired Jerry Spence, the Wyoming trial lawyer, for a flat $50,000 to crucify Kerr-McGee. Bright, cagey, and successful, Spence was a superb actor who had played to juries from the pit for more than twenty years, coaxing large personal injury awards from them. *Karen Silkwood* v. *The Kerr-McGee Corporation* — a challenging and precedent-setting case — had pumped him full of enthusiasm and adrenalin.

During the trial, Spence strode into the federal courtroom each day wearing a suede cowboy jacket and a ten-gallon hat, ringed with a band of silver buckles. He was husky and tall, with straight brown hair hanging over his ears, broad shoulders, and a middle-age paunch that rippled over his huge turquoise belt buckle.

Spence had the knack of explaining complex issues in graphic language, in a voice that ranged from a chortle to thunder. He paced in the pit as if he owned it, walking over to the Kerr-McGee bench to point out Wayne Norwood, striding from the witness stand to the blackboard, leaning over to look the jury straight in the eye, quoting Oklahoma's Will Rogers. He kept Kerr-McGee's chief attorney, Bill Paul, off guard, mouthing obscenities to Paul with his back to the jury, interrupting Paul's arguments and questions, and feigning sleep when Paul's witnesses rambled on.

Spence pushed everyone and everything as far as he thought they would go — Paul, Theis, the jury, the law — against the advice of his fellow counselors. He squeezed every drop of evidence

326

from his own witnesses, and tripped and twisted Kerr-McGee's. And he did it for ten weeks, during the longest trial in Oklahoma history.

Spence's supporting cast was also good. Art Angel prepared the direct examination of witnesses for Spence, continually passing him notes with new questions. Jim Ikard wrote most of the cross-examination questions, and was Spence's library on plutonium, cancer, and contamination. Angel and Ikard argued for him at the bench or in Judge Theis's chambers, where one quarter of the trial took place. Father Wally Kasuboski, a Franciscan priest, was Spence's document librarian, even sleeping with the files, lest one of Bill Taylor's "friends" drop in during the night. Alison Freeman, hired by Sara Nelson for the purpose, dealt with the media. Dan Sheehan watched, from the front row, offering suggestions like a drama coach. And each night, all of them fanned Spence's ego so that he'd have the confidence and energy to roar at Kerr-McGee the next day.

"It is a case about money," Spence told the jury in his opening argument. "That sounds kind of gross and crass, but that is what it is. It is true that the case is also about pain, and about death, and fear, and terror, and panic. The case is about that. And the case ultimately may be about the future survival of the American people. And it's about damages, whether or not damages are sufficient in size to make Kerr-McGee — and other defendants and other manufacturers and other companies like them — to perform their duties in the future in a way that the American people can live and survive healthfully. And it is about preventing . . . the dastardly escape of plutonium particles *ever again*."

Spence had Karen's children driven in just for the opening argument. "If you will look over to your right," he told the jury, "in the front row — you will see three little children with Mr. Silkwood, and they are his grandchildren and the persons he represents."

Spence introduced Dawn, Beverly, and Michael. "Would you stand up so we can see you?

"Now, this case is also principally about a human being," Spence continued. "It is about a woman, a young woman, who, you will be relieved to discover, wasn't perfect, and who, you may be shocked to discover, didn't view things the way many human beings do, who had perhaps different life styles than many. But the bottom line about Karen Silkwood was that she was a very ordinary woman. And by 'ordinary' I don't mean common. I mean she was a plain, ordinary human being like you and me. And the case is about her suffering, and the facts in the case will be about her terrors in the night for nine days before her death. And the case is about her daytime life as an employee at Kerr-McGee, where they attempted in those last nine days . . . to discredit her and to make her into something she was not."

Bill Paul objected in the middle of Spence's opening argument. The attorneys approached the bench. "Inflammatory, extraneous material, and prejudiced," Paul complained, asking Judge Theis to declare a mistrial. Theis overruled him.

On the advice of Art Angel, Spence began his case with expert witnesses who described the dangers of plutonium so that the jury would have a context in which to evaluate the charges against Kerr-McGee. Next, Spence examined a string of former K-M workers who spelled out the alleged Kerr-McGee negligence. And he ended with Bill and Merle Silkwood and Rose Mary Silkwood Porter, who talked about what Karen was like so that the jury could see the lab analyst as a person, rather than as an issue.

DR. JOHN GOFMAN: The nuclear chemist and medical doctor was Spence's star. Articulate, slightly professorial, and with a smell of history about him, Dr. Gofman told the jury he discovered how to isolate plutonium from uranium, and how he produced for Los Alamos the plutonium used in the atomic bombs that destroyed Hiroshima and Nagasaki. In 1941, he said, the world's stockpile of plutonium was so small, you couldn't see it even under a microscope. You knew it was there because its alpha particles left a radioactive trail.

Dr. Gofman spoke of plutonium with respect, almost awe. He told the jury that each nanocurie of plutonium — Dr. George Voelz said Karen had 8 nanocuries in her body — emits 2000 alpha particles every minute for 24,000 years. "To give an idea," he testified, "think of an M-16 rifle emitting bullets. Well, an alpha particle makes an M-16 rifle look trivial in terms of energy."

Dr. Gofman explained that there are between 100 million and 1 billion cells in every gram of tissue, and that each cell stores volumes of information.

"We would be lucky," he told the jury, "if the alpha particles would kill just a few cells . . . because a dead cell just gets absorbed and excreted. But it doesn't kill them. A certain number of the alpha particles rip into that cell and just hurt the instruction book. So you can say the page is torn out on how to be a team player, and that cell is cancer right then and there. It *is* cancer. We can't see that medically . . . What we later call cancer, five years, ten years, fifteen years later, is when this monstrous aberration has grown to millions, billions of descendants. But the cancer was formed the very instant that an alpha particle went in and destroyed the instruction book."

Dr. Gofman had studied the autopsy report on Karen and the Los Alamos full-body count numbers and analyses of Karen's tissues. "I am saying unequivocally that a person, like Karen Silkwood, exposed to that much plutonium, is married to lung cancer," he told the jury. "They are inseparable from that point on because the process is in motion. Unfortunately for us humans, we can't see it, we can't feel it, we don't sense it. But the diabolical damage is going on unknown to us, and it will manifest itself." He explained that the link between alpha particles and cancer had been known for more than fifty years.

Dr. Gofman went on to talk about how alpha particles cause genetic damage. Just before ovulation, he explained, each ovum divides, keeping half of its forty-six chromosomes so that the sperm can supply the other half. If the alpha particle hits the ovum before or after ovulation, it can kill the cell. "That would be lucky for everyone," he told the jury, explaining that the alpha particle can just as easily deform one of those chromosomes. "One of a hundred, actually thousand, varieties of defects that can lead to premature death, to mental retardation . . . to hemophilia, cystic fibrosis . . . to an increase in susceptibility to some diseases such as heart disease, arthritis, high blood pressure — all *that* can come from damage to these cells."

Dr. Gofman went on to attack AEC and Nuclear Regulatory Commission standards, pointing out that the numbers for the "safe amount" of radiation had been reduced steadily "as the corpses" had piled up.

"When they say forty nanocuries of plutonium in your body is the permissible dose, when they say that is safe, that is an unmitigated lie," Dr. Gofman told the jury. "It is an amount that the industry has been allowed to give, not because it was ever proven safe, but because that is the amount the industry can live with . . . The only problem is: How many lives is it all right to snuff out . . . ? How many cancers in Oklahoma City are all right for electric power in Los Angeles?"

Jerry Spence led Dr. Gofman to Kerr-McGee negligence. "Do you have an opinion, Doctor, as to whether or not a company who failed to adequately instruct its employees [about] the dangers of radiation from plutonium would be negligent?" he asked.

"I could only regard failure to come up with everything up front as willful and wanton and reckless dealing with people's lives," Dr. Gofman said.

Jim Ikard had asked Dr. Gofman to study the Kerr-McGee manual for health physics instructors and its safety booklet for workers.

"Do these documents adequately come up front and advise anybody who reads them as to the dangers of plutonium?" Spence asked.

"I do not believe they do," Dr. Gofman said. "These manuals talk about . . . how much plutonium is permissible. I saw nowhere in there an up-front statement stating: 'We must operate on the basis that with every increase in the amount of plutonium you are exposed to, you increase your risk of developing cancer' . . . In fact, I looked for the word 'cancer.' And, I found in two places the word 'malignancy,' and the word 'osteocarcinoma.' That is very interesting to me, not to be able to find the word 'cancer' . . . If I were writing that manual — a competent manual — I'd have the word 'cancer' on every page . . . and in *caps*."

Spence then read Dr. Gofman a sentence from the Kerr-McGee instructors' manual: " 'The benefits derived from . . . the use of radioactive materials far exceeds any risk that radiation workers are subjected to under normal conditions.' "

"That is sheer, absolute, false propaganda," Dr. Gofman told the jury.

"Would you consider forty hours of training of the employees, of the kind shown in that book, to be adequate?" Spence asked.

"Forty total hours of training?"

"Yes."

"I think that is exposing people to grave risk."

"What if you saw that they were trained maybe *eight* hours?" Spence asked.

"That would be a cruel joke. Gross, willful, and wanton negligence."

"What would you think about putting people to work in a plant like that without any training *whatsoever,* and to let them work in the plant until they get enough new people together to make a class?"

"You let them work before —" Dr. Gofman was shocked. "You just have no business putting a person into that kind of plant to work for one hour, or one day, without adequate training."

In his opening argument, Bill Paul had taken a piece of paper and shown the jury how easily Karen could have sprinkled some plutonium on it, folded it, put it in her pocket, and smuggled it out of the plant.

"What do you think of a security system that would permit plutonium to be taken from a plant in that fashion?" Spence asked Dr. Gofman.

"My answer is unequivocal on that," the doctor answered. "If that is the description of the plant, such a plant should never have been allowed to open, and should never have been allowed to operate for one single day — *ever.*"

Jerry Spence knew that Kerr-McGee would argue that it had done everything it could to care for Karen after her contamination, but that she wouldn't cooperate, even refusing to take the DPTA treatment to help flush soluble plutonium from her body.

"Would you state to the jury what your opinion is relative to that matter?" Spence asked Dr. Gofman.

331

"I think she was a wise woman to refuse to have that treatment. That is an experimental treatment. It isn't really cleared for human use."

Under Spence's probing, Dr. Gofman told the jury how the AEC had asked him to study radiation effects. The commission gave him 150 people and $3.5 million each year he reported favorably. "Then in the seventh year I said . . . that the risk of cancer from radiation was twenty times underestimated," Dr. Gofman told the jury. "Within two weeks I was labeled incompetent by the Atomic Energy Commission . . . That history has been repeated so often by the Atomic Energy Commission that it would be fair to say: 'If you want to last with funding from the Atomic Energy Commission and its successors, you just don't find radiation too harmful.' "

All during his examination of Dr. Gofman, Jerry Spence kept one eye on the jury. The doctor had them. Spence knew it, could feel it. Kerr-McGee would have to impugn Gofman's credibility in its cross-examination and defense, or the jury would accept Dr. Gofman's definition of injury — the assault of alpha particles on Karen's cells.

Bill Paul's co-counsel Elliott Fenton cross-examined Dr. Gofman.

"You understand that the plant is on standby basis, and would be available for inspection if you wished to go there?" Fenton asked Gofman.

"I didn't understand that. But knowing what I know of the plant, I have no great enthusiasm to visit."

"Do you think that there are great hazards and risks connected with the manufacturing of plutonium, even if done as safely as people can reasonably handle it?" Fenton continued.

"There are hazards, yes. But I think properly done, one can do plutonium-handling far better than the records convince [me] that Kerr-McGee has done."

"Now, Doctor, what do you recommend as a minimum permissible level of radiation exposure among the general population?" Fenton asked.

"I recommend that we do the one honest thing we should have done a long time ago — eliminate that term 'permissible.'

Let's square with our neighbors and our fellow Americans and say: 'Look, there are going to be cancers produced by radiation, and to talk about a permissible level is to talk about a legalized permit to murder.' "

Fenton let Dr. Gofman explain to the jury that radiation standards are set to "facilitate the industry," not to protect people. For example, the nuclear chemist said, the National Council on Radiation Protection set a standard in 1954 that — if it hadn't later been forced down by 180 times — would have caused 8 million cancer deaths per year. "A clear example of the irresponsibility of standard-setting bodies," Dr. Gofman argued.

Fenton then tried to make Gofman appear to be a disgruntled maverick whom other scientists did not take seriously.

"What about the International Commission on Radiation Protection?" Fenton asked. "They have never changed their standards, have they, Doctor?"

"Not yet."

"And you have made your views known to them, haven't you?"

"Yes."

"What about the Nuclear Regulatory Commission? Does it adhere to this standard?"

"It is right now having internal discussion," Dr. Gofman said.

"Dr. Gofman, I really believe I'm entitled to an answer to my question," Fenton demanded, "and that is: The Nuclear Regulatory Commission has not yet changed their standards, have they?"

"They have not yet."

"What about the Atomic Energy Commission? Does it adhere to this standard?"

"There is no Atomic Energy Commission."

"I mean when it did exist?"

"It did adhere, yes," Dr. Gofman conceded.

"Have you presented your views to that commission — Nuclear Regulatory Commission — to the effect that the standards should be changed and lowered?"

"On many occasions."

"Have they done so?"

"They are considering it," Dr. Gofman said.

333

"But they have not *yet* done so, have they?" Fenton pushed.

"No, sir. Things move very slowly in government bureaucracy."

"Dr. Gofman, you said . . . there are some awful fights in the scientific community," Fenton continued. "I'm sensing your fight with the Atomic Energy Commission is one of the awfulest. Is that a true statement?"

"I wouldn't think so. You see, we survived when the Atomic Energy Commission didn't," Dr. Gofman said. The jury smiled.

"How long have workers been involved in working with plutonium in this country?" Fenton asked.

"Since 1944." Sensing what Fenton was going to ask next, Dr. Gofman pointed out that researchers like himself could not get the health records of the companies who handled plutonium.

"Just answer this question, if you will, Doctor. Isn't it a fact that you do not know of a *single* death of a plutonium worker from lung cancer anywhere in the world since 1945?"

"I do not know how many have occurred," Dr. Gofman said.

"But you do not know of a single documented case, *do* you?"

"I do not know of any."

"Thank you, sir."

DR. KARL Z. MORGAN: The "father" of health physics, who had directed health and safety for thirty years at the AEC's Oak Ridge laboratory, supported everything Dr. Gofman had said, and more.

Dr. Morgan testified that he had reviewed hundreds of health and safety programs in universities, industry, and government. He blasted the Cimarron plant. "I felt it was one of the worst operations I have ever studied," he told the jury.

Jerry Spence asked him how dangerous it was to continue plutonium production when workers were wearing respirators. "One particle of plutonium — so small that you can't see it — is more than the permissible amount," Dr. Morgan told the jury. "So it is very easy for that to find its way into a mask . . . To allow operations to continue over a period of time — even one hour — using these safety masks, sometimes only partially safe masks, is inexcusable and irresponsible . . . callous, willful, wanton negligence."

334

Dr. Morgan emphasized the need for a certified health physicist to direct a plutonium plant's health and safety program. "It is like a hospital," he explained. "You wouldn't want to operate a hospital just with . . . students. You would want someone there with a medical degree."

Dr. Morgan went on to praise Karen Silkwood. "I think Karen — I never knew the young lady — had a terrific insight and realized that plutonium was an extremely hazardous material," he told the jury. "It was very much to her credit that she did all she could to bring this to the attention of the authorities, not only for her own protection but for her fellow employees."

In his cross-examination, Bill Paul cut right to the quick of the trial. He quoted to Dr. Morgan the Karen Silkwood contamination report of Dr. George Voelz: " 'It is my opinion that the exposure to plutonium was sufficiently low that there should have been no significant health hazard from exposure.'

"Now my question," Bill Paul said. "Would you agree or disagree with that opinion?"

"I would disagree."

"Tell me why, sir."

"The employee isn't going to fall over dead, or scream from anguish because of injury, but it does present to the employee an increased risk of cancer in the remaining years of her life."

JAMES V. SMITH: The former Kerr-McGee division manager told the jury that he had spotted design problems in the plutonium plant even before it was built — gaskets, ducts, valves, glove boxes — but that when he reported the defects, Kerr-McGee didn't correct them because it was "on a very tight schedule."

Jerry Spence pulled specifics out of Smith:

☐ Kerr-McGee used plastic gaskets on glove boxes instead of welding the seams tight, the way Rocky Flats, where Smith had worked previously, had done. The nitric acid in the glove boxes later dissolved the gaskets. "Now, if the gasket eats out, where does the plutonium go?" Smith asked the jury. He said K-M repaired the leaks with putty, but that no one knew how long that substance would last.

☐ Kerr-McGee used hundreds of feet of plastic ducts for the filter

335

system, instead of stainless steel pipes, as Rocky Flats had done. Plastic burns, and if it caught fire, it would release airborne plutonium.

☐ Kerr-McGee designed the wrong kind of pipe valves for the wall storage tanks. The plutonium mixture later dripped onto the floor.

☐ Kerr-McGee designed glove boxes with Plexiglas instead of the Homolite used at Rocky Flats. Homolite does not melt or burn as Plexiglas does, but it is two to three times more expensive. "With so much exposure to nitric acid," Smith told the jury, "[the Plexiglas] became so clouded, you couldn't see to operate."

Bill Paul had given the jury a set of pictures of the plutonium plant to demonstrate how neat and clean the Cimarron facility was. Spence handed the pictures to Smith. "There is not one picture in there of any kind of any of the production areas," Smith said. "These are all laboratory, front office, and so forth."

Smith told the jury about Kerr-McGee forcing workers to continue production in respirators for more than fifty hours while rooms were being decontaminated. "This shouldn't happen," Smith said.

"Why not?" Spence asked.

"Because there is always a certain amount of leakage in respirators . . . I've seen them [workers] lift them up and wipe their eyes off . . . without thinking."

Smith told the jury how Kerr-McGee sent him to Crescent to buy 100 gallons of paint to cover the contamination at the plant. The factory was built of cinder block, and the contamination was hidden in the holes. Paint sealed it in. "It was already peeling when I left," Smith told the jury.

"Supposing that somebody wants to clean that plant up — how are they going to handle the concrete walls?" Spence asked.

"I would say they would have to break it up and bury it."

Smith went on to tell the jury that plutonium waste was supposed to be solid, but that it constantly reverted to a liquid, eating through waste barrels and contaminating the hallways, loading dock, and trucks. Once, a truck had to be dismantled and the tires and floorboards buried because they were so hot. Another time, Kerr-McGee had contaminated the Cimarron River. "They had taken all the people down there that evening with shovels," Smith testified, "and were digging holes, trying to bury the fish to keep anybody from seeing them."

As he had in his deposition, Smith explained how poor Kerr-McGee security was; how facility manager Morgan Moore always seemed to know what was going on at the union meetings; and how the forty pounds of plutonium unaccounted for (MUF) could not have been in the Cimarron plant pipes.

After getting Smith to say that, at first, Kerr-McGee's training program was good, Spence asked, "What was the training of these people that came in later?"

"About one hour of health physics indoctrination," Smith said.

KENNETH PLOWMAN: A former Kerr-McGee health physics technician, Ken Plowman was now a machinist who raised cattle and a few horses. He had joined K-M in 1969, working first in the uranium plant and then in the plutonium plant, and quit in 1972.

"I felt the plutonium plant's program was going the same way the uranium plant's went," he explained to the jury.

"What was wrong?" Spence asked.

"Nothing was *right!*" Plowman said. "There was hardly any controls . . . The contamination was everywhere. The equipment leaked. There was no real — no real effort to control it, I don't believe. The supervisors didn't control it. The men didn't control it. It was just a battle that was lost."

Plowman was nervous, but Spence relaxed him. "Contamination everywhere, especially in the lunch room," Plowman told the jury. "Contamination out on the grounds. Bags of uranium taken to the burial ground and hid while the inspectors were there. Pellets being thrown around outside. Supervisors — who are supposed to be in charge — outside in the parking lots in their contaminated clothing."

Like Jim Smith, Plowman testified he knew days in advance when the AEC inspectors were coming. He said he didn't think the supervisors or the workers at the uranium plant really understood the hazards of radioactive materials, and that he had frequently complained about the health conditions.

"What happened?" Spence asked.

"Nothing that I know of."

Things weren't much better at the plutonium plant, where Plowman worked from 1970 to 1972. "I was trying to keep up with the controls in the plutonium plant," he told the jury. "And it seemed like things were going from . . . one emergency to another — and you didn't have time to do your routine monitoring and checking. It seemed like you had one problem after another — one leaky valve — one broken glove box — or one contamination incident — to where you ended up that that was all you were doing."

RONALD HAMMOCK: The Oklahoma Highway Patrolman had worked for Kerr-McGee from 1969 to 1972, a year and a half of that period in the plutonium plant. Hammock told the jury he received no specialized training and had attended no classes.

"Sometimes we would work twelve-hour shifts . . . in a respirator," he testified.

"Did that happen more than one day?" Spence asked.

"The year and a half I worked there," Hammock told the jury, "I would say fifty to sixty [days] I worked in respirators, if not more."

"Did anyone ever tell you . . . that contamination by plutonium could cause cancer?"

"No, sir," Hammock said. "They didn't."

"When did you discover that possibility for the first time?"

"Just a few months ago."

Spence asked Trooper Hammock whether workers got all the plutonium into the vault during tornado watches. "No way we could — all of it," Hammock told the jury. "All we could put in would be in a powder form. But the liquid would still remain in our system."

"Could you get all the powder in?"

"Not all the time," Hammock said. "It wouldn't be a great

amount — what I would call a great amount . . . several kilograms."

Hammock emphasized that there was no security at the plant while he worked there. "You could have taken [plutonium] out of there in any way you wanted to," he testified. "In your pockets . . . all you could carry."

Spence asked the former Kerr-McGee worker about the AEC inspections. "We didn't talk to anybody," Hammock said.

"Were you told not to?"

"That is correct."

"Who told you not to talk to the AEC people?"

"My supervisor," Hammock said.

During the cross-examination, Elliott Fenton tried to introduce documents from Ron Hammock's personnel file, proving the worker had received the Kerr-McGee safety booklet and had attended Kerr-McGee's five-day health physics course. Spence objected that the records were new evidence he hadn't seen. "Sandbagging," he complained to Judge Theis.

But while Spence was arguing in Judge Theis's chambers, Art Angel noticed something peculiar about the records. He told Spence to allow them to be entered as evidence. Spence showed them to Ron Hammock. "Do you recognize your signature on any of these sheets?" he asked.

"My name appears on this one," Hammock said. "However, it is not my handwriting . . . My last name is not even spelled correctly."

Hammock picked out three more signatures he claimed were not his. "I know the training sessions you are talking about if that is what you are trying to get at," Hammock told Spence. "I didn't attend them."

Art Angel and Jim Ikard had pleaded with Spence not to make a big issue of false signatures, arguing that one of Hammock's buddies might have signed the attendance record for him, or that maybe Hammock just couldn't recognize his own writing. But Spence smelled drama and took a risk.

"Would it be your opinion that these are — just to use the plain old English language — *forgeries?*" he asked.

"Yes, sir," Hammock said.

RANDY SNODGRASS: The Oklahoma City machinist went to work for Kerr-McGee when he was nineteen, loading raw plutonium into glove boxes, mixing it with uranium, turning it into cake, and then into powder.

"Of course you went to school, didn't you?" Spence asked Snodgrass.

"No, sir."

"Surely *somebody* told you something about the dangers of plutonium before you started?"

"Not that I recollect."

"Did they put you under the supervision of anybody?"

"No direct supervision," Snodgrass said. "Jim Smith took me on a tour of the plant. He explained to me what they did and how things started, where they ended up, and showed me where the health physics room was . . . and assigned me to a supervisor. He took me back and said: 'Stand here and watch this guy.' "

The next day, Randy Snodgrass was working in a glove box.

"Did anybody . . . ever *once* tell you that plutonium in the lungs could cause cancer?" Spence asked.

"No, sir."

"When did you learn that it might?"

"Last week."

"How did you find out for the first time?"

"Newspaper," Snodgrass said.

"The newspaper reporting the facts of this trial?"

"Yes, sir."

Snodgrass told the jury that a bristle from a hot wire brush poked through the lead-lined glove he wore and dug into his finger. Health physics technicians tried to wash the contamination off, but Snodgrass' finger kept reading hot. Then they tried to pinch the bristle out with tweezers, but it broke and went deeper. "I couldn't take the pain any longer," Snodgrass told the jury. "So they called the surgeon from Guthrie, and he came over and put my finger to sleep and sliced it out."

"Did you ever get to see any of your reports after that, relative to . . . contamination getting into your bloodstream?"

"No, sir."

Snodgrass told the jury how a filter blew out of a fifty-five gallon drum of radioactive waste one day, spraying contamination all over him "with the pressure of a garden hose." Snodgrass was wearing a respirator at the time, but the ammonia in the waste fogged it so badly that he couldn't see, and his beard made it almost impossible for him to breathe. He tore off the respirator to plug the leak.

"Did anyone take you to the doctor after decontamination?" Spence asked.

"No, sir. I went home."

Snodgrass went on to testify that he worked for ten straight days in a respirator. "We were really behind in production and we had to do it," he told the jury.

"How does that feel?" Spence asked.

"The respirator is terrible. It clings to your face and skin. It makes you sweat, and it is hard to breathe through it. It takes extra — it takes extra energy to suck air through the filter into your lungs. And you're breathing heavier. And the weight of the thing on your face — it made your neck sore after having it on thirty or forty minutes."

Snodgrass told the jury he would take the respirator off to wipe away the sweat on his face or the fog on the glass, trying to hold his breath until he could get it back on. Then Spence showed Randy Snodgrass a document that said: "I have completed the five-day Health and Training Program and I have received a personal copy of the Kerr-McGee Manual of Health and Safety."

"Is that your signature?"

"No, sir," Snodgrass said.

"Would it be accurate to say that that is a *forgery?*"

"Yes."

"Will you write the word 'forgery' under the signature in red."

Snodgrass wrote the word.

"Is *that* your signature?" Spence showed Snodgrass a second document.

"I don't believe it is."

"Now let's have you write on the bottom of that, the word 'forgery.' We'll call that forgery number two."

Spence gave the documents to the jury.

JAMES NOEL: The high school physical science teacher was a former Kerr-McGee lab analyst. Noel surprised attorneys for both sides by bringing a notebook-diary in which he had recorded some of the things that happened at Kerr-McGee during the five years he worked there. He told the jury the only training he received at the Cimarron plant was twelve and a half hours of lectures spread over five months, and that no one had told him plutonium was carcinogenic. Noel said workers were allowed to smoke cigarettes in the uranium plant.

"What did you think about that?" Spence asked.

"It seemed unwise to me."

"Why?"

"Well, you would be constantly moving your hands to and from your mouth," Noel said. "Possibly placing your cigarette someplace in a contaminated area."

Jim Noel had known Karen Silkwood, so Spence asked what she was trying to do toward the end of her life. "I know she was concerned about material unaccounted for, safety violations in the plant, and the concern that the workers and the public at large didn't know the hazards at the facility," Noel said.

"Could you be more specific?"

"She called me on the telephone the night of October 22, I believe it was, in 1974, to express concern about those three particular items and referring to the MUF. I remember her using the term seventeen kilograms [42.5 pounds]," Noel told the jury. "She was concerned that the employees were not taking her seriously about the nature of the possible danger they were in."

Noel told the jury that during the 1972 strike he had enrolled in college to finish his B.S. degree. After the strike, Kerr-McGee promised him the night shift so that he could attend day classes. Then, without warning, they switched him to the day shift. When he reported late two days in a row, Kerr-McGee fired him.

Spence pointed out to the jury that Kerr-McGee had consistently given Noel "outstanding" commendations, but that in January 1974, when the Oklahoma public school personnel director asked Kerr-McGee for a recommendation for James Noel, K-M supervisor Don Bristol wrote: "Mr. Noel consistently demonstrated lack of respect for authority while feeling free to express his views, in fact imposing his views on his fellow workers. In my opinion, Mr. Noel would have problems working under any restrictive work codes or rules."

The Oklahoma school system hired James Noel anyway.

WILLIAM APPERSON: The former Kerr-McGee worker told the jury that K-M had hired him as a maintenance man, but turned him into a welder, a welding supervisor, and a welding teacher. Apperson was twenty-three years old at the time and was not a "certified welder." In fact, he said, there weren't any certified welders on the whole maintenance staff, even though he and others had asked to go to school.

Apperson testified that he and the welding crew installed 30 percent of the pipes in the plutonium plant. "You can just imagine the importance of the welding over your head," he explained to the jury. "If it leaks, it could be . . . disastrous really."

Apperson said that he himself was contaminated twice a week on the average because the pipes were not welded properly. "Many times I've gone in and welded over a weld that leaked, that was poorly done, and you could see the holes before I ever started on the pipe."

Apperson told the jury that sometimes the welders were so badly contaminated, they couldn't get clean. "They would wash and scrub for a long time and pretty soon your hands would get sore, and they had people that wore gloves home, you know, rather than stay there all night."

RICHARD ZITTING: The president of the Kerr-McGee Nuclear Corporation from 1973 to 1976 told the jury that the K-M health and safety coordinator, Jerry Sinke, and the Cimarron health physics director, Wayne Norwood, were hand-picked by him and were "highly qualified." But under questioning from Spence, Zitting admitted he didn't know what kind of training and formal education each had had, adding, "I didn't believe, and I don't believe, that you would have to be a so-called certified health physicist to do their job."

343

"If employee after employee . . . testified that production came first and health second, you would say that is wrong?" Spence asked.

"Not true. Yes, sir."

Zitting told the jury that when he was president, no one ever told him workers continued production in respirators for days. "If you had known that, what would you say?" Spence asked.

"It depends on the circumstances."

"What if you were advised that there were . . . fifty days at a time where production was carried on in respirators?"

"I would think that would be a very long period of time," Zitting said.

"You think that would be evidence that production came before *safety?*"

"No, sir." Zitting explained that working in respirators indicated only that the plant had a problem. "Contamination doesn't necessarily mean an area is dangerous or hazardous to any substantial extent."

Spence asked Zitting three times what plutonium workers should be told about the dangers of radiation. Zitting failed to mention cancer. Finally, under Spence's prodding, Zitting admitted workers should be told that "some scientists" believe there is a link between plutonium and cancer. "I think it would be proper to tell them that nobody had ever gotten cancer from working in a plutonium plant, and *that* is the truth," Zitting said.

Zitting stressed that Kerr-McGee had a good safety record at the Cimarron plant. "We operated over two years without a lost-time accident."

STEVE WODKA: The OCAW leader told the jury how Kerr-McGee sent Karen lawyers, not doctors, on November 7, 1974. "She was quite scared. She was quite concerned for her health. She was quite upset. She was quite worried," Wodka testified. "She had no idea whether she was going to live another day."

Wodka described how Karen looked on November 8, when she met him at the Oklahoma City airport: "She was wearing big sunglasses. When she took the sunglasses off, there were big, deep, dark circles under her eyes, and big bags. She looked quite bad, and she was very depressed. And she didn't talk too much."

From the airport, Wodka and Karen went to the Northwest Holiday Inn to meet with the AEC inspectors. "Throughout the interview, she broke down and cried several times," Wodka told the jury. "She said in fact — and this is in quotation marks in my notes — 'I believe I'm going to die . . .' And then she cried again and she said — and I took this down because I thought it was, I just couldn't believe she was saying this — 'The tears burned the skin on my face from the salt.' Her skin was so raw."

Wodka testified that Karen told the AEC she was worried that someone might have put plutonium on the sandwich she ate in the plant on November 5. She complained that Kerr-McGee didn't even offer her a motel room after it quarantined her apartment.

Wodka told the jury that during the AEC interview, an inspector checked her for radiation. "They would take a wand and they would go over her arms and legs, and the thing would click away," Wodka said. "It was like she couldn't escape this contamination. I mean, it was still with her."

Spence asked Wodka why he hadn't pressed Karen to show him her documents on November 8 and 9, before he left Oklahoma. "Because I trusted her up to that point," Wodka told the jury. "I didn't have any reason to distrust her . . . I was more concerned about her health than the documents."

Wodka pointed out that in their last conversation on November 12, the night before she was killed, he told Karen, "If you can't put it together, I won't bring him [David Burnham] down."

Wodka told the jury she said she was ready. "Let's do it," she said.

Spence played for the jury a tape of Karen's conversation with Steve Wodka a month before she died. "In the laboratory, we've got eighteen- and nineteen-year-old boys . . ." Karen said in her soft drawl. "And they didn't have any schooling, so they don't understand what radiation is. They don't understand, Steve. They don't understand."

WANDA JEAN JUNG: Karen's friend Jean Jung left Oklahoma after the Cimarron plant closed. According to court rules for civil suits, witnesses need not be asked to travel over 200 miles to testify. Spence used the law to his advantage. Jung was an important witness because she saw Karen's documents just minutes before Karen was killed. Judge Theis forbade Spence to probe the accident itself, since the negligence case dealt with Karen's injury before her death. Jung was the closest Spence could get to the edge of Theis's ruling. The problem was that Jean Jung was not articulate, and Spence wasn't sure what Bill Paul would do to her under cross-examination or whether the jury would find her credible. So he hired Deana Cooper, head of the drama department at Oscar Rose Junior College, to read selected portions of Jung's deposition, which Sheehan had taken before the trial.

Jung described in great detail the folder and notebook Karen had with her in the Hub Cafe just minutes before she was killed.

"What did Karen Silkwood say to you?" Spence read, playing Dan Sheehan's role.

"That was when she told me about her having been contaminated so much that it could eventually kill her," Jung said. "Tears came in her eyes and she said: 'I can't believe who would do such a thing like that.' And she said: 'It has got to be somebody that works for Kerr-McGee that can get it out.' "

"Did Karen Silkwood tell you anything, or did she indicate that she had any knowledge about how her apartment came to be contaminated?" Spence read from the deposition.

"No! She asked me, she said: 'I can't understand how it came to be contaminated . . . And then I told her, I said: 'I have got three kids at home. I have to go on home.' And she said, 'I have got all of my proof ready that I have been working on for quite some time.' "

"Is that the first time you had ever heard about her gathering evidence?"

"Oh, no. She told me she was doing that, I believe it was before she got contaminated."

"What did she say her proof was?"

"She said she had proof of the falsification of records."

"When did she say that?"

"That night."

Chapter 33

The Silkwood attorneys were confident. They had owned the courtroom for a month, and they sensed there was little love for Kerr-McGee in the jury box. But they knew Kerr-McGee could still win easily, if its attorneys could only convince the jury that:

☐ Karen Silkwood was not injured, or that

☐ she had contaminated herself, or that

☐ the plutonium that contaminated her did not belong to Kerr-McGee, or that

☐ she was contaminated while doing the "business" of the company, in which case Oklahoma workers' compensation laws protected Kerr-McGee against personal injury suits.

Kerr-McGee attorneys chose to argue that Karen Silkwood was not injured and that, even if she were, the contamination was her own fault because she had deliberately spiked her urine samples.

Kerr-McGee's basic defense against negligence — which would influence the size of the award if the jury decided Karen was deliberately injured by someone outside the plant — was to argue that Kerr-McGee had carefully followed the regulations of the Atomic Energy Commission. Thus, to attack Kerr-McGee was to attack the AEC.

Bill Paul headed the Kerr-McGee legal team, supported by Elliott Fenton, L. E. Stringer, John Griffin, Jr., Larry D. Ottoway, and Bill J. Zimmerman. Paul was a bright lawyer, respected and

347

successful by Oklahoma City standards, and the president of the Oklahoma Bar Association. But from the minute he nervously began to shuffle papers in the pit, it was clear he was no match for Jerry Spence.

In his gray suit and dull tie, with short hair neatly parted, Paul looked like a law professor. Hugging the lectern in the pit, he played with his glasses, rarely looked at the jury, and only tentatively ventured out to approach a witness or write on the blackboard in flawless penmanship. He was a perfect Oklahoma gentleman who rarely displayed emotions, almost as if he were intentionally trying to make Spence look crass. The Wyoming lawyer ate him alive.

Bill Paul outlined Kerr-McGee's case in a clear, colorless opening argument, characterizing the Wyoming cowboy-attorney as a reckless man who attacked everyone and everything — Kerr-McGee, the Atomic Energy Commission, the Nuclear Regulatory Commission, the Kerr-McGee attorneys, the majority of nuclear scientists, and the standards they set to protect workers and the public.

Paul told the jury that the highly respected International Commission on Radiation Protection set 40 nanocuries as the permissible dose for plutonium workers, and that Karen Silkwood had less than 10 nanocuries when she died. "This case is about the exposure of Karen Silkwood to plutonium," he argued. "It is about the injury, medical effects, if *any,* for the nine-day period of November 5 to 13."

Bill Paul stressed that Kerr-McGee was not negligent in safeguarding its plutonium, for only .00003 grams had escaped from the Cimarron plant and contaminated Karen's apartment. Once Karen was contaminated, Paul continued, Kerr-McGee did everything it could to help her, but she refused to cooperate. As soon as Kerr-McGee discovered that her apartment was contaminated, it made arrangements for her to take DPTA treatment in Guthrie, but Wayne Norwood could not find Karen. Then Dr. Charles Sternhagen arranged for DPTA at Baptist Hospital, but Karen refused the treatment. Hours after Kerr-McGee learned that her apartment was contaminated, it called the AEC, which sent experts immediately, but Steve Wodka wouldn't allow the AEC to talk to Karen until he got there. Finally, Kerr-McGee called in Dr. Neil Wald all the way from Pittsburgh, and Dr. Sternhagen all the way from Albuquerque.

"We think the question in this lawsuit is: How did the plutonium get into her apartment? You are going to decide that," Bill Paul emphasized. "Let me make this clear. I'm not here to discredit Karen Silkwood. We are here to bring you the facts. We are here to bring you evidence that bears on this question. If the facts discredit Karen Silkwood, then it just has to be."

Paul told the jury that there were no eyewitnesses to the contamination, but there was evidence that Karen had contaminated herself. "Here are some of the things that our proof will show," he said. "First: access. Who could get into that apartment? Who was there? Well, Karen, Sherri, who lived there, and Drew Stephens, who spent the night there Wednesday, November 6. So let's start with access: obviously, Karen could get into her apartment. I know of no proof that anybody at *Kerr-McGee* could . . .

"The proof will be that in the lab — of all places the lab where Karen worked — that it was a relatively simple matter (if you intended to do it) to remove a very small quantity of plutonium. Our proof will show that Karen *had* that opportunity. She worked odd shifts. She was working during this time a lot of the times from four o'clock until midnight. And there were very few people there . . .

"You will learn that it would be very simple to intentionally remove a small quantity . . . a few drops in the bottle, dust a little bit on Kleenex, clean wipes, put it in an envelope — and no problem.

"Could she have *unintentionally* removed it? Answer: it would be very hard, if she were doing what she was supposed to be doing, monitoring herself at every station."

Then Paul told the jury that some of Karen's samples had been spiked with plutonium. "You're going to see evidence that every [sample] that tested high was collected — *where?* At her apartment . . . Isn't it a little funny that those collected at home are the *only* high ones, and those that were collected at Los Alamos under supervision, or at the plant, were all normal? So look

349

for that urine sample proof. It speaks, and speaks, and speaks, and speaks."

What motive might Karen have had to remove plutonium intentionally from the laboratory to her apartment for the purpose of spiking her urine samples?

> ☐ Karen was furious with Kerr-McGee about the November 5 reprimand for taking medication without reporting it.
>
> ☐ Karen was spying on Kerr-McGee for the union, trying to collect documents to embarrass the company. When she couldn't get the evidence, she had to do *something*.
>
> ☐ The union had become an obsession for Karen. Under pressure and distressed, she was driven to contaminate herself.

But before Bill Paul could open his defense, Judge Frank Theis jerked the rug. Paul had intended to call witnesses to testify that Karen Silkwood had abandoned her three children, whom Jerry Spence had paraded in front of the jury, and that she popped Quaaludes, smoked grass, and attempted suicide. The Kerr-McGee attorney argued in the judge's chambers that he wanted to prove that Karen's emotional pain was not caused by her contamination; it had begun months before November 5, 1974.

But Judge Theis knew that Bill Paul also desperately wanted to portray Karen Silkwood as the kind of person who would contaminate herself — an unfeeling mother and a mentally unstable junkie. Theis ordered Paul to scratch those character witnesses.

Paul objected. Hadn't Judge Theis permitted the Silkwoods to bring a family album to the stand and to talk about Karen? And hadn't he allowed Spence to show the book to the jury? But Theis couldn't be swayed. Quoting the law to Paul about permissible evidence, Theis called the album "innocuous" and Paul's strategy "character assassination."

If Theis left a big hole in Kerr-McGee's defense, the corporation's attorneys themselves created another. They called three kinds of witnesses: experts to prove that Karen was not injured, Kerr-McGee officials to prove the company ran a tight ship, and AEC officials to state that Kerr-McGee's Cimarron operations were good. But they failed to call one single worker to testify that the plutonium plant was safe and healthful.

WAYNE NORWOOD: The Kerr-McGee health physics director had a good bedside manner. A kind-looking man, Norwood spoke to the jury in a soft, firm voice. He explained that, although he didn't have a Ph.D. in health physics, he had worked with radioactive materials for more than twenty years and had great experience in dealing with contamination problems. He described Kerr-McGee's five-day health physics training course and defended it as better than adequate. He explained how filters, monitors, and alpha counters protected the health of the workers.

"Mr. Norwood," Bill Paul asked, "from the time the plant started until today has there ever been *one* day to your knowledge that you permitted anybody to work in an unsafe area?"

"No, sir," Norwood said.

"Or under unsafe conditions?"

"Not to my knowledge."

"Was there ever a time, to your knowledge, when health and safety was sacrificed at the expense of production?"

"No, sir," Norwood said.

The Kerr-McGee health physics director defended painting walls and floors to cover contamination as an approved health measure. He told the jury he was never given advance notice about unannounced AEC inspections, and that the hour the AEC allowed Kerr-McGee to collect all the plutonium into the vault during tornado watches was more than adequate. "It can be done in half that," he testified.

Norwood suggested to the jury, as he had to the FBI, that Karen contaminated herself, emphasizing three facts, under Bill Paul's direct examination. Once Karen suspected her apartment was hot, she warned Sherri Ellis about going into the bathroom and kitchen, the hottest rooms in the two-bedroom apartment. How would she know that, if she hadn't contaminated the place herself? Then Karen suggested that the cheese and bologna might have been contaminated because she had spilled urine in the bathroom. How could she draw that conclusion unless she had spiked her own urine sample? Finally, Karen asked Norwood about the hot specimens that she had donated earlier, even though the lab had not yet reported on them. How would she know they were hot if she hadn't spiked them?

351

Jerry Spence began his cross-examination by trying to impugn Wayne Norwood's credibility. First, he established that Norwood's University of Oklahoma degree was in poultry science, not in radiation sciences. Next, he got Norwood to admit that, although he was health physics director, he had no degree in health physics, was not specifically trained in health physics, and was not a certified health physicist. Then Spence pointed out how long Wayne Norwood had worked for Kerr-McGee and how good Kerr-McGee had been to him, suggesting that the elderly man was protecting the mighty corporation out of gratitude.

Spence got Norwood to admit that once Karen gave her urine samples to Kerr-McGee, she never saw them again, suggesting that it was just as easy for the company to spike the samples as it was for Karen. He also got Norwood to admit that everyone in the plant had access to her kits, labeled and sitting in the main hallway.

Finally, Spence got Norwood to concede that workers were not supposed to be in respirators for long periods of time during production, that it was against AEC regulations for workers with beards to wear the masks, and that Kerr-McGee did not buy a full-body counter like Los Alamos' to protect workers because it "cost too much."

WILLIAM UTNAGE: The former Kerr-McGee engineer said he had designed the Cimarron plutonium plant with a hand-picked team of experts, stressing that he took special care in planning the plant's contamination control system — monitors, filters, buzzers. Utnage described the vault as tornado-proof with an explosion-proof door similar to those in munitions factories. If a tornado ever hit the vault, Utnage explained, the most the twister would do would be to make a hairline crack in the steel.

Utnage emphasized how he had designed the glove boxes with great care and tested them for leaks before the plant opened; how he hired only experienced people to construct the plant; how he selected a certified health physicist, Allen Valentine, to design the plant's health physics equipment and to prepare health guidelines.

Jerry Spence picked William Utnage apart during a tough

cross-examination. First, Spence established that the Kerr-McGee engineer had never worked in or designed a plutonium plant before he came to Crescent, and had hired a team of design experts with no experience in manufacturing plutonium fuel rods.

Next, Spence got Utnage to admit that he not only didn't have a degree in radiation sciences, but had never taken courses in the subject. Then, after establishing that Kerr-McGee had commended Utnage for keeping the plutonium plant's expenses 25 percent under the projected cost the first year, Spence read contamination report after report, blaming the accidents on design faults. Utnage finally admitted that fifty workers were contaminated while he was facility manager, before Morgan Moore; but he denied Jim Smith's allegations of faulty design, insisting he had planned a "safe plant."

Like other Kerr-McGee managers, Utnage stressed that no scientist could point to even one case of lung cancer caused by plutonium, and that the maximum permissible radiation dosage approved by the AEC was "safe."

"We would expect there would be no adverse effects whatsoever," Utnage told the jury.

ALLEN VALENTINE: The certified health physicist from the Los Alamos Scientific Laboratory had written Kerr-McGee's health physics program, had designed the air sampling, decontamination, and criticality alarm systems, and had selected the contamination survey instruments, full-face respirators, and protective clothing. "I tried to buy the best I could," he told the jury.

Valentine recruited the health physics staff, including Wayne Norwood, whom he had known at Hanford, Washington. "We had a competent staff," he told the jury. "I think that in the area of procedures and training, that the practices . . . either met or exceeded practices at other plutonium facilities."

Valentine also wrote Kerr-McGee's health physics training manual. "I am a firm believer that you should keep radiation exposure to a minimum," he explained. "And I'm also a firm believer that with the standards that exist today, that there is an insignificant effect."

Jerry Spence was waiting for Allen Valentine. First, Spence

established that Valentine, Kerr-McGee's only certified health physicist, worked out of the K-M headquarters in Oklahoma City, not at the plant in Crescent. Then, Spence established that Valentine was certified *after* he had designed the Cimarron health physics program, not before, as Bill Paul had led the jury to believe.

Valentine admitted that most of the health physics staff at the Cimarron plant had no specific training in treating radiation exposure, and that some were machine operators turned into health physics technicians, even though the University of Oklahoma, and other schools, were graduating health physicists every semester.

Next, Spence questioned Valentine about full-face respirators. Valentine admitted that, as a trained health physicist, he would never allow workers to wear them for as long as ten hours during production.

In Kerr-McGee's license application to the AEC, Valentine had promised that the health physics program would follow the guidelines of the American National Standards Institute (ANSI). Spence read Valentine one of those standards: " 'Workers must be evaluated by competent medical personnel to insure that they are physically and mentally able to wear respirators under simulated and actual working conditions.' "

Valentine admitted that the standard was sound and that Kerr-McGee should have followed it but didn't. Respirators create stress and fatigue, he said. They are dangerous if workers can't see out of them. And they should not be worn by anyone with a beard, because the hair prevents a tight fit.

The ANSI also called the yearly medical exams for workers who use respirators "indispensable." But Valentine admitted that Kerr-McGee had not followed that guideline either. He told the jury that the K-M license application failed to warn the AEC that Kerr-McGee would make workers use respirators during routine production. To the contrary, Kerr-McGee promised to have the workers wear respirators only for "minor emergencies and control actions."

Valentine had also designed Kerr-McGee's system to collect and analyze bio-assay samples of contaminated workers. He admitted to Spence that it normally took thirty days for test results (with special handling, two weeks) before workers could learn how badly they had been radiated.

Spence went on to explore Valentine's attitude about radiation. "A permissible dose doesn't mean the same thing as a *safe* dose . . . does it?" Spence asked.

"In my perspective, it does," Valentine said. "It doesn't mean *absolutely* safe."

"You mean like part pregnant?"

Art Angel and Jim Ikard had studied the health and safety booklet Allen Valentine wrote for Kerr-McGee workers in 1970. When they found out that Valentine had relied almost exclusively on one 1959 scientific source article, they laid a trap.

Spence read Valentine a sentence from the K-M booklet he had written. " 'Experiments on animals show also that some of the inhaled material may accumulate in the pulmonary lymph nodes from which the elimination rate is slow, resulting several days later in a higher concentration in the nodes than in the lungs proper.' "

"Now that all sounds pretty," Spence told Valentine. "That particular paragraph deals with the most dangerous part of plutonium in the lungs, and that is the *only* information you gave these people, isn't that true?"

"Yes."

"You think you were being fair with those workers out there?"

"Yes, sir, I was," Valentine said. "In the light of the fact that I was not aware of a cancer case as a result of plutonium in the lungs — even today."

"Yes, we know that game, too," Spence thundered. "That is because when somebody gets cancer ten to twenty years later, they can't prove where it came from, *can* they?"

Bill Paul jumped up. "Your Honor, I object to the characterization of 'game' and to the questioning as argumentative."

"Sustained," Judge Theis ruled.

"Isn't it true that you simply plagiarized some of the statements?" Spence pointed to Valentine's booklet.

"That is correct," Valentine said.

"Word for word?"

"I believe they are."

Spence read from the 1959 source article: " 'The high incidence of lung cancer among the workers in mining operations in the Schneeberg and Joachimsthal Districts of Southeastern Europe was noted over four hundred years ago.' "

"Did you put that in your article so that your workers would know that?" Spence asked.

"No, because I believe it is referring to uranium."

"Well, you know that cancer is caused from alpha particles, don't you?" Spence demanded. "And that the alpha particles in uranium and the alpha particles in plutonium are the same — you knew that, didn't you?"

"Yes, but the —"

"And, so did you call attention to your workers that it was known over four hundred years ago that the alpha particles from these sources caused cancer?"

"It is not in the manual," Valentine said. "No, sir."

"You know, you can't pick and choose from an article . . . only those things which don't make sense and which are confusing, and leave out those things that people could readily understand."

Spence read another passage from the 1959 article that Valentine had left out of his workers' booklet. Large amounts of plutonium in the bone and liver can produce acute and immediate effects, the article said. Smaller amounts may result many years later in bone cancer, chronic anemia, and other diseases.

"If you had really told the workers what you actually knew and what was actually in this article, from which you quoted, you couldn't have gotten anybody to work in that plant, could you?" Spence asked.

"The answer to that question would be subjective on my part," Valentine said.

GERALD PHILLIP: The AEC investigator told the jury that at 4:30 on November 7, the day Kerr-McGee found Karen's apartment to be contaminated, he and three others were called into the

AEC conference room in Chicago for a briefing. They left for Oklahoma City that same night. One of the team was a public relations specialist sent to deal with the media.

Phillip explained that he had called Drew Stephens early the next morning, November 8, but that Drew would not allow him to talk to Karen until Steve Wodka arrived. Later that day, Phillip told the jury, he debriefed Karen for four hours, and that — by way of exception — he allowed Wodka to be with her.

"Karen appeared to be upset, pale, and I think she was wearing little or no makeup," Phillip said, referring to the meticulous interview notes he had made in 1974. "She wept, cried, three or four times during the interview . . . was sincerely upset. She indicated that she felt she had breathed plutonium into her system as a result of the July occurrence and, combined with the current situation, current problem, that she was going to die.

"She indicated she was in a hurry that morning [November 7] — didn't want to be late for work — and in the process of getting a sample, she had spilled a portion of the contents of the container on the commode. [Then] she had removed a package of bologna from the refrigerator and had taken it out into the bathroom and placed it on the lid of the commode, and that after doing so she had recalled that she still had part of a lunch from Tuesday in her locker at work, and so she decided not to take a lunch that day and returned the package of bologna to the refrigerator."

Phillip said that on November 13 he interviewed Karen again. She seemed calmer, wore makeup, looked better, and cried only a few times. Phillip said she ended the interview at 5:30 because she had a union meeting at the Hub Cafe, but promised to continue the discussion the next morning. She was killed that night.

Phillip told the jury that Karen admitted being "miffed" at the reprimand her supervisor had given her on November 5. And under Bill Paul's direction, Phillip testified that Kerr-McGee had been very cooperative with the AEC investigation team, suggesting that the company had nothing to hide; that Karen had told him about a urine sample still in her locker; and that when K-M tested it, it was not hot, suggesting that she hadn't had time to spike it with plutonium. Throughout the questions and answers, Bill Paul tried to insinuate that Karen was so upset because she had spiked her samples and was afraid of getting caught.

"Now in October and November of 1974, did Karen Silkwood possess the means to remove plutonium from the lab?" Paul asked.

"Yes, sir," Phillip said. "A small quantity of that kind would not be detected by the monitor."

Since the AEC was on trial as much as Kerr-McGee, it was important for Jerry Spence to either destroy or cripple Phillip's credibility. First, the Silkwood attorney tried to suggest that Karen was upset because she was scared of what might happen to her health.

"If you had received such contamination or evidence of it — like 45,000 d/m in your nose — would that upset *you?*" he asked Phillip.

"Yes, sir."

Next, Spence got Phillip to admit that the plutonium used to spike Karen's urine samples was insoluble and that, therefore, it was biologically impossible for it to be eliminated in urine. Did Phillip ask Karen, in his more than six hours of interview, whether she knew the difference between soluble and insoluble plutonium? Spence asked. The AEC official said he did not.

Then Spence began to chip away at Phillip's credibility as an AEC investigator. Phillip admitted that many people had access to the plutonium in the lab, including John Carver, Karen's lab supervisor. "Did you write their names down?" Spence asked.

"I didn't try to determine specifically who had access."

Phillip went on to admit that Karen told him she frequently left her apartment door open and that he omitted the fact from his AEC contamination report; also, that he didn't check the inventory of "contaminated" things Kerr-McGee had taken from Karen's apartment. Under Spence's questioning, Phillip testified that he didn't know that the plutonium in Karen's apartment came from pellet lot 29, or that the lot had been sent to Hanford three months before her death, or that there were forty pounds of MUF at the plant.

"Wouldn't those three facts be something that you would

want to know about or investigate further in determining how Karen Silkwood got contaminated?" Spence asked.

"I did not try to pull those pieces of information together and make something out of them," Phillip said, adding that the first time he had even heard about them was during his cross-examination.

Phillip had told Bill Paul earlier that he had had a long interview with John Carver, the lab supervisor who reprimanded Karen on November 5. But when Jerry Spence asked the AEC investigator to review those interview notes with the jury, Phillip said that he hadn't taken any. He admitted that he knew Carver was a leader in the move to get the union decertified.

Phillip also told the jury that Karen volunteered the fact that there was an unlabeled urine sample in her locker, that she gave him the key, asked him to get it, and told him which woman she trusted enough to go into the women's locker room for it. Phillip then admitted that Karen did not want him to give the specimen to Kerr-McGee.

Finally, Phillip said Karen was intelligent, open, cooperative, and that he had no reason to doubt her account of the contamination. Spence asked him if he had found it unusual for Karen to take bologna from the refrigerator to the bathroom.

"I have done the same kind of things myself in running late," the AEC investigator said.

CHARLES SCOTT: The handwriting expert Kerr-McGee hired to analyze Randy Snodgrass' signature brought blow-ups of the "forgeries," as well as several signatures on checks.

"Did you form an opinion?" Bill Paul asked the expert.

"I did."

"What was your opinion, Mr. Scott?"

"The signature 'Randy Snodgrass' on what we call a completion certificate dated 5–8–75 is the same handwriting and was written by the same person as the signature 'Randy Snodgrass' on all the documents submitted to me as standards of comparison."

Jerry Spence had egg all over his face. In his cross-examination, he established that Kerr-McGee did not ask Charles Scott to analyze Ron Hammock's signatures. The former K-M worker had told the jury that his name had been misspelled once and that two other signatures were not his.

Next, Spence tried to repair Randy Snodgrass' credibility. "Do you think an honest person can look at his own signature and say it isn't his and be wrong, but honestly believe it isn't his?" Spence asked.

"Yes," Scott said. "I don't think the average person is capable of analyzing his own signature, frankly."

MARY CAVENER: Randy Snodgrass and Mary Cavener had joined Kerr-McGee at the same time. "During September 1974," Elliott Fenton asked her, "did you attend a health and training session or sessions, held for three days at the uranium plant?"

"Yes," Cavener said.

"Did you know Randy Snodgrass at that time?"

"Yes."

"Did Randy Snodgrass attend those sessions at the uranium plant?"

"Yes," she said.

"You may cross-examine," Fenton told Spence.

"Why don't you just step down," Spence said. "I don't have any questions."

JAMES KEPPLER: As Jerry Phillip's AEC boss, James Keppler had been responsible for the inspection and regulation of the Cimarron plants since 1973. Keppler told the jury that the purpose of AEC inspections was to protect workers and the public from plutonium, and that the AEC's four annual inspections of the Cimarron plant were thorough. He explained to the jury that the AEC had three categories of noncompliance with regulations:

☐ One: Infractions that have a direct impact on health and safety, such as overexposure to radiation.

☐ Two: Infractions that could lead to such a direct impact, like failure to conduct a radiation survey when one is warranted.

☐ Three: Infractions that have little impact on health and safety.

Keppler testified that since 1973, when he took over the AEC regional office, his inspectors had not found a single Category One infraction at the Cimarron plants. "It was our view that the Kerr-

McGee plant was being safely operated," Keppler told the jury. "If that were not the view of my office, I would have shut the facility down."

"Do you have any evidence to indicate that the workers out there were receiving exposures that would cause them to have cancer in twenty years?" Bill Paul asked.

"No," the AEC official said.

Under Jerry Spence's cross-examination, Keppler estimated that Kerr-McGee had seventy-five violations of AEC regulations in its file, but insisted that none was serious enough to warrant a $500 to $5000 fine. Keppler told the jury that most of the inspections of the Kerr-McGee plant were unannounced and that he had learned Kerr-McGee had been tipped off only after Karen's death. Spence asked Keppler if he had tried to find the leak. "No," Keppler said.

Spence pulled out a memo Keppler had written, summarizing a meeting with Dean McGee and other top Kerr-McGee officers. In the memo, Keppler said that K-M was not committed to ALAP (exposures as low as possible), the Cimarron equipment was archaic and worker training inadequate, there was a high personnel turnover and a lack of worker supervision, and Kerr-McGee was careless with the plutonium held up in its pipes and ducts. But Keppler defended himself by claiming he didn't actually write the memo — a staff member had composed it from his conference notes.

After pointing out to the jury that Keppler had warned Kerr-McGee in 1973 that the management of the Cimarron plutonium plant was suffering from the loss of experienced personnel, and that an AEC consultant had warned Kerr-McGee in 1973 that its bio-assay program was poor, Spence asked Keppler to define "safe."

"Safe is a condition or situation in which a person is not subjected to substantial levels of radiation," Keppler said.

"Define 'substantial,' " Spence asked.

Keppler gave Spence the equivalent of 40 nanocuries of plutonium for a lifetime full-body burden.

"Now you are saying to the ladies and gentlemen of the jury that under [40 nanocuries] is *safe,* is that right?"

"I'm saying a person's life is not in danger."

"You mean *immediately*."

"That is correct."

"I finally realize that you're saying that 'safe' is what the *regulations* say is 'safe,' " Spence said.

"That is correct."

"What you saw was that the people at the plant were *regulatorily* safe . . . as distinguished from a medical opinion, isn't that true?"

"That is correct," the AEC official said.

DWIGHT GARY LONGAKER: The former Kerr-McGee lab analyst had worked with Karen off and on from 1973 to her death. He described her to the jury: "Karen appeared to be a vindictive person. She would get angry at you if you disagreed with her . . . mad at you for two or three days . . . She just didn't accept criticism."

Longaker told the jury about the Tylenol No. 3 incident for which Karen was later reprimanded. "To me it appeared that Karen was drunk or high," he testified. "She was at a [glove] box sitting there, and Don [Gummow] was sitting beside her. And they were talking and laughing like they had been drinking . . . She got up from the box at one time to go to another box, and she staggered a little bit."

He was concerned, Longaker said, so he went into the office, closed the door, called his wife, and told her to phone John Carver and say people were high on drugs in the lab. Longaker warned his wife not to give Carver her name. Carver was at home, Longaker explained, and when he got to the lab, Karen was gone.

Longaker went on to characterize Karen as "sloppy" and not concerned about the dangers of plutonium. She wore her hair long and wouldn't tuck it under her cap, as she was supposed to. "She wasn't very careful and didn't monitor herself very carefully," he said.

Finally, Longaker told the jury that Karen's emotional state had slowly deteriorated. "When I first came to work, Karen's attitude wasn't very different from anybody else's," he testified.

362

"Toward the last six months, or something, Karen's attitude changed about Kerr-McGee. She became antagonistic . . . Karen lost weight, became almost gaunt in her face — I especially remember that in . . . the summer of 1974. She became more nervous."

Jerry Spence went after Gary Longaker like a chain saw. "Did you ever walk up to Karen Silkwood when you observed her in this condition and confront her as I am confronting you — face to face, eyeball to eyeball — and say to her, 'What's wrong with you, Karen?'"

"No, I didn't."

"The next thing is, of course, you had the courage to go call the supervisor *yourself*, didn't you?"

"No, sir, I didn't," Longaker said.

"And what you did was to call your wife and tell her to call the supervisor anonymously, and then hang up, *didn't* you?" Spence demanded.

"Yes," Longaker said.

"And the truth of the matter is that during all of this period of time, Mr. Longaker, you were against the union, weren't you?"

"Yes, sir."

"As a matter of fact, you fought with Karen about the union, didn't you?"

"Argued with her." Longaker explained that only two of the twenty-two lab analysts supported the union.

"And to the day of her death that was still the attitude of you and the other people in that lab, isn't that true?"

"Yes, sir."

"Now, I think you have characterized your own sense of your own activity here as being a rat and a squealer, isn't that true?"

"Yes, sir," Longaker said. "It would have been better for me to have gotten the supervisor of the plant. I've thought about that since, and I wish that would have been what I had done."

DR. RICHARD BOTTOMLEY: Kerr-McGee ended its defense as the Silkwood estate had begun — with expert witnesses to testify about radiation injury. Elliott Fenton examined Dr. Richard Bottomley, professor of medicine at the University of Oklahoma Medical School and a cancer researcher.

"Do you have an opinion as to whether or not [Karen] had cancer on November 13, 1974?" Fenton asked.

"There was no evidence of cancer."

"Do you have an opinion as to whether she had any acute injury of any kind from contamination?"

"There was no evidence of this," Dr. Bottomley said.

"Would you define the term 'acute,' please?"

"Of short duration, as opposed to something which is chronic."

"Now do you have an opinion, based on reasonable medical certainty, as to whether Karen Silkwood had sustained any radiation sickness of any kind?"

"There is no evidence of this," Dr. Bottomley said.

"Do you have an opinion as to whether or not, as a result of the contamination, she sustained any pain, or any sensation generally associated with physical injury or damage?"

"Based on the dose that she would have received by the time of her death," Dr. Bottomley told the jury, "there couldn't have been any systemic damages from the radiation."

Art Angel cross-examined Dr. Bottomley. First, the young attorney established that, although the physician was an expert in cancer research, most of his research had been in the use of radioactivity to cure cancer, and that he had little expertise on how much radiation *causes* cancer.

"Isn't it true that the insult, or the injury to a cell, that follows a radiation exposure is an immediate one?" Angel asked.

"That is true."

Dr. Bottomley then went on to define cancer as a disease characterized by cells replicating themselves without the normal control. He told the jury that the process begins before the physician sees the cells under a microscope, but that until the doctor does see them, he can't diagnose cancer.

"So, it is what you can *see,* rather than what is actually *there?*" Angel asked.

Dr. Bottomley agreed.

"Isn't it true that in every single case, when you spot and

diagnose a cancer, that the process of cancer was taking place, and had already taken hold *before* you saw it?"

"That is true." Dr. Bottomley admitted there can be damage even if the doctor can't see it.

"Do you know how much plutonium it takes to assure that the process of cancer will take hold? Do you?" Angel asked.

"No," Dr. Bottomley said.

DR. GEORGE VOELZ: The physician was Kerr-McGee's star witness. Like Dr. Gofman, he was kindly, professorial, and spoke with authority and conviction. Bill Paul built him into an international radiation expert who had written articles on plutonium for the best scientific journals, who sat on the most important scientific committees, and who was health director at the Los Alamos Scientific Laboratory, one of the most prestigious nuclear research centers in the world.

Dr. Voelz explained to the jury that the full-body counts on Karen Silkwood indicated she had only between .33 and .35 nanocuries of americium, a plutonium daughter. He said that, although scientists didn't know the exact plutonium-to-americium ratio, he was certain that Karen had less than one quarter of the full-body burden permissible under AEC standards.

Then Dr. Voelz told the jury that his opinion was confirmed by thorough analyses of some of Karen's bone, lung, liver, and lymph nodes, which he had brought back to Los Alamos after the autopsy in Guthrie. The AEC permitted 40 nanocuries, he said; the analysis showed Karen had 8.8 nanocuries. Dr. Voelz also told the jury that an analysis of some of Karen's cells showed "zero" chromosomal aberration and that, therefore, she had suffered no genetic damage.

Dr. Voelz defended the 40 nanocurie standard for the body and 16 for the lungs. "I believe they are appropriate," he said.

"Are you aware of any reliable data which indicate the invalidity of those standards?" Bill Paul asked.

"No."

To prove his point, Dr. Voelz testified that he had studied twenty-six Los Alamos workers who had received between 7 and 230 nanocuries of plutonium in the 1940s. Some of the workers had a long history of smoking. But he found no cases of lung cancer, none of cancer of other internal organs, and only two of skin cancer, which were unrelated to radiation.

Dr. Voelz explained he conducted an even larger study of 224 men who had more than 10 nanocuries of plutonium in their bodies. Mortality tables projected eleven cancer deaths; only seven had died of cancer. And the tables projected 3.4 deaths from lung cancer; only one had died of it.

Dr. Voelz went on to explain to the jury the careful model scientists use to arrive at their standards. He said the long-term probability that Karen Silkwood would have got cancer from the plutonium in her body was between zero and five out of 10,000.

"Now, would you have an opinion, Dr. Voelz, based on reasonable medical certainty, as to the acute or short-term health effects resulting from the exposure that Karen Silkwood had?" Bill Paul asked.

"Yes, I do."

"Would you state your opinion to the jury, please?"

"I would feel there would be no health effects," Dr. Voelz said.

Jerry Spence had to destroy Dr. George Voelz or he would lose the case, all the evidence about K-M negligence notwithstanding. He kept the Los Alamos physician in the witness chair for two days, pounded him with questions, confused him, and wore him almost to a frazzle.

First, Spence began to pick at the AEC. He reviewed Dr. Voelz's employment history, emphasizing that the physician had always worked for the AEC, implying that the AEC had him in its pocket, and suggesting that the AEC was using Voelz to protect itself. Next, Spence established that the full-body count and postmortem analyses on Karen Silkwood were done at an AEC laboratory, paid for by the AEC, and conducted by scientists on the AEC's payroll. Then he got Dr. Voelz to admit that Karen Silkwood went to the AEC for help because she had no choice.

Spence had a copy of Dr. Voelz's study of the twenty-six Los Alamos plutonium workers with more nanocuries in them than Karen. Dr. Voelz admitted that, although none of the twenty-six had died of lung cancer, they had illnesses that radiation is known to cause: blindness, high blood pressure, thyroid nodules, mouth tumors, enlarged hearts, early-age heart attacks, tooth loss, and respiratory problems.

Then Spence attacked the model used to set the 40 nanocurie standard for the body and 16 for the lungs by showing that Karen Silkwood did not fit the model. He got Dr. Voelz to admit that plutonium in the lungs is more dangerous for smokers like Karen, females, young people, the poor, and people with asthma like Karen.

"Now, could you tell me, Doctor, what *exact* factor was placed in the model that you used for smokers?" Spence asked.

"The actual data that were incorporated in that model did not just include data from nonsmokers — it included smokers' data as an average — not as a specific factor."

"That is all a numbers game, isn't it? That has *no* specific reference to Karen Silkwood as she exactly was, isn't that true?"

"In these kinds of numbers, using averages of population is the best you can do," Dr. Voelz said.

"I didn't ask you that," Spence shouted. "Read the question to the witness."

The court reporter read: " 'That is all a numbers game, isn't it?' "

"Yes," Dr. Voelz said.

"Thank you. Now, what was the average age of the model?" Spence asked.

"I don't remember that detail."

"Well, what was Karen Silkwood's age?"

"Twenty-eight."

"What was the average sex of the model?" Spence asked.

"It was a mixture of males and females."

"Well, what relative to the model was the average race?"

"That gets into a detail," Dr. Voelz said. "Unless I —"

"Do you know?" Spence shouted.

"No, I do not know."

"There is a difference in the longevity tables for race, isn't there?" Spence asked.

"Yes."

"And the model assumed an average kind of race, didn't it?"

"Yes."

"Poor people don't get as good care as wealthy people?" Spence asked.

"True."

"What was the average economic level of the model, if you know?"

"I don't know," Dr. Voelz said. "It was the average."

"Yes," Spence said to the jury. Turning again to Voelz: "What was the average taken for education? That has something also to do with medical care and longevity, doesn't it?"

"Yes."

"What was the average education in the model, do you know?"

"No, I do not," Dr. Voelz said.

"Do you know what the average weight was?"

"No, I do not remember all the —"

"Do you know what Karen Silkwood's weight was?"

"It was 100 pounds."

"You don't know what you compared her to in the model, do you — do you think the model weighed more or less than she did?"

"I didn't have that detail," Dr. Voelz said.

"When was the last time you read *anything* about that model that you plugged in against Karen Silkwood?"

"When I read the model? Oh, it's been some years ago," Dr. Voelz said.

Spence asked Dr. Voelz how many alpha particles bombarded Karen Silkwood between November 5 and November 13, based on his estimate of the less than 10 nanocuries in her body. Dr. Voelz got confused. Spence told him to relax, take his time. He even helped the scientist with the calculation.

"Twenty-seven, twenty-six million," Dr. Voelz said finally.

"How many of those alpha particles are necessary to cause cancer?" Spence asked.

"Well, I'm not really sure what that answer is, but —"

"Do you *know?*"

"No, I don't really think I know," Dr. Voelz said.

"Do you know anybody who *does* know?"

"No."

"Do you recognize that *one* alpha particle hitting *one* cell . . . can do damage to the cell — do you agree with that?"

"I would agree with that in some cases," Dr. Voelz said.

"Would you be able to predict *which* alpha particle hitting *which* cell would cause cancer?" Spence asked.

"Not specifically."

"Do you think that somebody that is given DPTA should be told specifically that . . . included in it is the risk of kidney tube damage?" Spence asked.

"We normally do that," Dr. Voelz said. "Yes."

"I didn't ask you what you normally *do,*" Spence shouted. "I asked you: Do you think they *should* be told that?"

"Well, yes."

"What does a 'threshold dose' mean?" Spence's voice was soft.

"A level up to which there would be no effects."

"Is it true, then, that it is still a question in science as to how little . . . plutonium you can get without [its] causing cancer?"

"That is true," Dr. Voelz said.

"Do you think that a worker in a plutonium plant should be told that exposure to radiation might cause cancer?"

"I think that is a reasonable thing to do," Dr. Voelz admitted.

"Why?"

"People should understand what they are working with."

"Do you think they should be told that there never has been established a *safe* level of exposure?"

"I do believe that," Dr. Voelz said.

"Don't you think it is just plain common decency to give everybody who exposes themselves to a risk the right to know what he is exposing himself to?"

"Yes."

"If . . . basic known information is hidden from a worker — so that he can't know — would you agree with me that that is extraordinary misconduct?" Spence asked.

"Well, 'extraordinary' is a tough word," Dr. Voelz said. "But I would say it is a poor policy."

"Not only is it a poor policy, but it cheats the worker out of a right to make a decision about his life, *doesn't* it?" Spence demanded.

"Yes," Dr. Voelz said.

Next, Spence got Dr. Voelz to admit that his estimate of the amount of plutonium in Karen's lungs, based on the chest count readings at Los Alamos, could be off by as much as 300 percent, and that if he were off by that much, Karen would have inhaled 24 nanocuries in just three days. The maximum permissible dosage for a lifetime is 16. Then Spence attacked Dr. Voelz's postmortem analyses of Karen's bones and organs, using repetition to make his point to the jury.

"Now, you wanted everything so you could test and weigh and analyze all the samples that you got, isn't that true?"

"Yes," Dr. Voelz admitted.

"Did you get . . . femur samples?"

"Yes, we did."

"Did you test it?"

"No."

"You got the brain. Did you test the brain?"

"No."

"You got the gonad?"

"That's right."

"What is the gonad?"

"The ovaries," Dr. Voelz said.

"Did you test them?"

"No, sir, we did not."

"You got the heart?"

"Correct."

"Did you test the heart?"

"No, sir."

"Did you test the kidneys?"

"No."

"Did you get the kidney?"

"Yes."

"Now, there are portions of the lymph nodes that you got and did not test. Is that correct?"

"That is correct," Dr. Voelz said.

"Did you get muscle samples from her body?"

"Yes, we did."

"Did you test it?"

"No, sir."

"The spleen, did you get it?"

"Yes."

"Did you test it?"

"No."

"Now, you still have those parts?" Spence asked.

"Yes, sir."

"How many lungs did you take?"

"I believe we took both of them."

"Did you take the test that you have given to the jury from both lungs or just one?"

"I believe we took them from both lungs."

Spence pulled out the Los Alamos contamination report. "It shows that the sample was taken from the anterior right superior lobe, doesn't it?"

"That's right." Dr. Voelz was confused.

"That's the *right* lung, isn't it?"

"Yes, sir."

"You still have the left lung."

"We have lung material."

"Have you received any written authority from the Silkwood estate to keep the parts of Karen Silkwood's body?"

"No."

"Did you ever make an attempt to divide the samples so that some independent agency could run a duplicate test to make sure that what you did was indeed fair and just?"

"No, sir."

"That is possible, isn't it?"

"It could have been done."

"Did you ever think of that?"

"We were only responsible —"

"Did you ever *think* of that?"

"No," Dr. Voelz whispered.

"We have to, throughout the entirety of this case, accept *your* figures and *your* calculations and *your* measurements, or the figures of Kerr-McGee and *their* calculations and measurements, isn't that true?" Spence asked.

"That is true," Dr. Voelz said.

"Now, I was interested in a term that you used yesterday . . . You talked about number crunchers. You remember that?"

"Yes, I did."

"After all of your number crunching, you don't know the safe level for radiation, do you?"

"I don't know the safe level completely, absolutely," Dr. Voelz said.

"What you have told everybody else is safe comes out of figures and calculations, doesn't it?"

"Yes."

"It comes out of number crunching?" Spence asked.

"Comes out of data behind the numbers," Voelz said.

"And so what is or isn't safe is a calculation and a numbers game, isn't it?"

"It is numbers based on data of research that has gone on for years," Dr. Voelz said.

"And so the nation's safety, the safety of this nation, the safety of the workers in the nuclear industry, the safety of the people on the street, the safety of *all* of us, is actually dependent on the accuracy of these models and these numbers, isn't that true?" Spence asked.

"And the data that is behind them; that is correct."

"And your opinions, that you have been giving to the jury, are not anything but *opinions,* isn't that true?"

"It is judgment and opinion," Dr. Voelz said.

"And everybody, with or without numbers, using or not using these numbers, are entitled to opinions, aren't they — we all are?"

"Yes, I believe we all are."

"And men who have used your numbers have come to opposite opinions, haven't they?"

"Yes."

"And it isn't your purpose, then, to say to the ladies and

gentlemen of the jury that the people who have come to an opposite opinion from you are *dishonest* men, is it?" Spence asked.

"I never said that."

"Thank you. And now, Doctor, it's time to close out the testimony of this case, and it is time to close out this numbers game, and to put it in perspective . . . There are those who are crunching numbers over here who are employed by the government, such as yourself; and there are people over here who are crunching numbers on behalf of industry, such as many of those who have testified to the ladies and gentlemen of the jury; and there are those on the other side who are crunching numbers on behalf of people. Now, Doctor, in an area that is so fraught with uncertainty, so fraught with the problems that we've seen, wouldn't you agree that reasonable men ought to come down ultimately and finally in a conservative way that gives the benefit of the doubt of all of that to people? Wouldn't you agree with that?"

"I believe we've done that, and I believe it is true," Dr. Voelz said. The jury could barely hear him.

Both Kerr-McGee and the Silkwood estate rested their cases.

Chapter 34

The most important trial action took place in Judge Theis's chambers. After both sides rested, Theis called them in to haggle over the instructions he would read to the jury. Sitting in shirt sleeves and suspenders like a country lawyer, his potbelly keeping him eight inches from the conference table, Judge Theis refereed Angel and Ikard, Stringer and Griffin. He loved it.

Aware that *Silkwood* v. *Kerr-McGee* was a precedent-setting case, Theis moved cautiously to avoid being overturned. The grounds for appeal, if any, would be found in the back-room compromises and rulings. "I don't mind cutting a few new paths in the law, if they are supported by logic," Theis told the attorneys. "If you see me falling in the dung, drag me out."

After they had caught Allen Valentine purposely keeping "cancer" out of his instruction book for Kerr-McGee workers, the Silkwood attorneys felt the jury would give them anything — even more than $11.5 million. So they told Judge Theis they wanted to amend their plea to $70 million, sensing that even if the jury wasn't prepared to be that generous, it would get so used to $70 million that $11.5 million would seem a trifle. Pointing out that the Oklahoma law places no ceiling on what the jury can award, Theis told the Silkwood attorneys they could ask for the $70 million in their closing argument.

Next, Arthur Angel asked the Kerr-McGee attorneys if they finally were willing to concede that Karen had been contaminated with Kerr-McGee plutonium.

They caucused. "Yes," they said.

"Wise," Judge Theis told them. "Otherwise an irresponsible position."

Then the Kerr-McGee attorneys asked Judge Theis to dismiss the charges against their client because the corporation was protected by workers' compensation laws. Karen was contaminated at work, doing the "business" of the company, they said. Therefore the Silkwood estate could not recover in a civil suit. But Angel and Ikard pointed out that Kerr-McGee had never argued to the jury that Karen was contaminated accidentally at work; rather, they had built their case around Karen's taking plutonium from the plant and contaminating herself. Judge Theis agreed.

The Kerr-McGee attorneys next argued that if Judge Theis wouldn't dismiss the case, he should at least allow the jury to vote on the workers' compensation question. If the jury decided Silkwood was covered by workers' compensation, the case would be over. But Judge Theis refused to do so because Kerr-McGee had never introduced the workers' compensation question during the trial itself.

The Kerr-McGee attorneys were still refusing to concede that the fuel-rod fabrication at the Cimarron plant was an ultrahazardous operation demanding commensurate care. They argued that the jury should decide that question. Judge Theis said no; he would define plutonium as ultrahazardous as a matter of law (a precedent).

Kerr-McGee wanted Judge Theis to instruct the jury that if it found that K-M had substantially complied with AEC regulations and standards, it could not find the corporation negligent. Judge Theis challenged the Kerr-McGee attorneys to come up with a definition of "substantial" in twenty-four hours. When they couldn't, he denied their request.

□ □ □

Each side had four hours of closing arguments. Jerry Spence was exhausted, and he rambled through an effective, but at times overdone, emotional appeal for justice and money. Bill Paul and Elliott Fenton teamed for Kerr-McGee. They were clear, logical, and argued convincingly from the evidence they had.

Then Judge Theis explained the law to the jury in twenty-nine carefully worded instructions that took over an hour to read. He told the jury that each side had to prove its case with a preponderance of the evidence, that is, "to prove that something is more likely so than not so." He told the jury there are two types of evidence: "One is direct evidence — such as the testimony of an eyewitness; the other is indirect or circumstantial evidence — the proof of a chain of circumstances pointing to the existence or nonexistence of certain facts."

If the evidence is equally balanced, Judge Theis stressed, then the side with the burden to establish the issue loses. The Silkwood estate had the burden to establish three issues, he pointed out:

□ that plutonium escaped from the Cimarron plant;

□ that Kerr-McGee plutonium caused Karen Silkwood actual injury to her person or property;

□ and the nature and extent of the injury.

"If you find these elements established," Theis told the jury, "then the burden of proof is on defendant Kerr-McGee . . . to establish that Karen Silkwood took the plutonium from work to her apartment where she was allegedly injured.

"You are instructed that the court finds as a matter of law that the operation of the Cimarron facility . . . constitutes an abnormally dangerous activity. Therefore, if you find that the damage to the person or property of Karen Silkwood resulted from the operation of this plant . . . Kerr-McGee . . . is liable for this damage."

Next, Judge Theis went on to define negligence as the "lack of that degree of care as would be exercised by a very careful, prudent, and competent person." He instructed the jury that for Kerr-McGee to be found negligent in the Karen Silkwood case, the negligence must have *caused* her contamination in a direct, unbroken sequence. Judge Theis added that because the Cimarron plutonium plant "constitutes an abnormally dangerous activity, a duty of utmost care is placed on Kerr-McGee."

Judge Theis reminded the jury that it had heard a lot of testimony about AEC regulations and standards. "You are instructed, however, that you are not bound by these standards," he said. "Compliance with such standards does not necessarily mean injury cannot occur for which liability may be imposed. Your duty is to determine, according to your own best judgment, in the light of all the evidence, the nature and extent of the actual injuries suffered by Karen Silkwood, if any, and the emotional suffering and anguish proximately caused thereby, if any."

Judge Theis explained that if the jury found that Kerr-McGee had injured Karen Silkwood, it could further punish Kerr-McGee. "In any action like the one before you, the jury may give damages for the sake of example and by way of punishment, if the jury finds the defendants have been guilty of oppression, fraud, or malice," he said. "If a defendant is grossly and wantonly reckless in exposing others to dangers, the law holds him to have intended the natural consequences of his acts, and treats him as guilty of a willful wrong."

Finally, Judge Theis instructed the jury about Spence's $70 million plea. "In final essence, amounts of damages must be determined upon the jury's fair and impartial assessment of the evidence in their individual and collective judgments," he said.

The jury was out for four days. There were six members: a housewife, a telephone repairman, a retired schoolteacher, an electrical engineer, a clerk-typist, and a utility-company foreman. Only once did they come to Judge Theis for help. "What does physical injury mean?" they wanted to know.

Kerr-McGee attorneys argued that Judge Theis should refuse to answer the question, for his instruction would only bias them. But Judge Theis said he had an obligation to assist the jury. His definition of injury in radiation cases was another precedent.

"Certainly physical injury can include a nonvisible or nondetectable injury, and may include injury to bone, tissue, or cells," Theis told the jury. "If a person has suffered physical injury . . . on the basis of expert medical opinion, it is only necessary that a person believe he or she has been physically injured as a basis for mental pain and suffering to occur."

The courtroom was packed and hushed when the jury filed into the box on May 18, 1979. In unemotional, flat tones, the clerk read aloud the five questions Judge Theis gave them and their answers.

"Do you find by a preponderance of the evidence that Karen Silkwood intentionally — that is, knowingly and consciously — carried from work to her apartment the plutonium that caused her contamination?" the clerk read.

"No."

The courtroom stirred.

"Do you find that Kerr-McGee Nuclear Corporation was negligent in its operation of the Cimarron facility so as to allow the escape of the plutonium from the facility and proximately cause the contamination of Karen Silkwood?" the clerk read.

"Yes."

The courtroom was silent and tense.

"Enter the amount of actual damages, as defined in the court's instructions, that you find Karen Silkwood suffered," the clerk read.

"Five hundred and five thousand dollars."

The courtroom gasped.

"Do you find by a preponderance of the evidence that Kerr-McGee Nuclear Corporation is a mere instrumentality of the parent corporation, Kerr-McGee Corporation?" the clerk read.

"Yes."

"If you find that an award of exemplary damages is appropriate, as defined in the court's instructions, enter the amount."

"Ten million dollars."

The courtroom broke into applause.

In the front row, Merle Silkwood cried, holding tightly to Bill, and Sara Nelson hugged Kitty Tucker. In the pit, Dan Sheehan smiled, and Jerry Spence looked proud and tired.

After the verdict, Karen's friends drove out to Crescent, to Highway 74, to a concrete wingwall barely visible from the road and less than a mile from Kerr-McGee. They planted a huge sign in the hard, red Oklahoma clay:

KAREN GAY SILKWOOD

BORN	February 19, 1946
DIED	November 13, 1974
VINDICATED	May 18, 1979

Chapter 35

Indeed, Karen Silkwood had been vindicated. She had accused Kerr-McGee of negligence in protecting the health and safety of its Cimarron workers. She had collected bits and pieces of evidence — a string of stories suggesting patterns of callousness and ignorance. Without the power to subpoena witnesses or files, she had interviewed workers who were afraid of losing their jobs, and had collected documents late at night. The AEC had not believed her, but an Oklahoma jury did.

Like a cardboard effigy, her reputation had swung in a breeze of innuendo for almost five years. She was so unreliable, the story went, so emotionally disturbed and hooked on drugs that she had contaminated herself to embarrass Kerr-McGee. She couldn't defend herself, so the jury did it for her. After examining the evidence, it said unanimously that she did not contaminate herself.

Though pleased with the verdict, Bill Taylor was not satisfied. Like Sheehan, Davis, Tucker, and Nelson, he was convinced that Karen's death was no accident. But Judge Theis had ruled her car crash off limits. It was not relevant to her contamination injury, he had told the attorneys. Murder was not the issue at the negligence trial.

Even if it were, Taylor could never have proven that someone pushed or forced Karen into the concrete wingwall on November 13, 1974. True, there was circumstantial evidence — fresh dents in the left rear of her Honda, A. O. Pipkin's accident analysis, missing documents, inconsistencies, contradictions. But there was

379

no hard evidence, and there were no real suspects. Just possibilities and theories.

Furthermore, if Karen had been murdered, there would have to be a conspiracy to cover it up. But all Taylor had were gut feelings and unanswered questions. Judge Theis had closed the door to the answers when he dismissed the conspiracy charges against Kerr-McGee and the FBI.

Taylor was not about to give up. He knew that somebody was still worried, for no sooner had he landed in Oklahoma City during the last weeks of the trial than someone began to tail him. While Sheehan, Davis, and Nelson were in the courtroom cheering Spence on, Taylor sifted through cold ashes for more leads and new evidence. He went out to the Ellis farm, to the accident site, to the Kerr-McGee plant on the knoll above the Cimarron River. The trial and publicity had loosened a few tongues. People seemed a little less afraid. But Taylor found nothing.

Then Echo called. He told Taylor that he had been following the trial in the papers and was impressed. He had never believed the Silkwood case would reach the courts, he said. Even the negligence case.

He took a risk, Echo told Taylor. Chanced another peek into June Mail. A random look again. This time he grabbed a long report and read about two thirds of page 3 before he had to slam the file drawer closed. Echo was more than nervous. He sounded scared.

Taylor jotted notes on a yellow legal pad as Echo told him what page 3 said: "Followed . . . lost . . . about a mile into dirt road . . . west . . . 200 yards north of death scene . . . 74 . . . turn around road to right . . . north . . . she passed on narrow oil road . . . Karen hit brakes . . . to left . . . then to right . . . slid into at end . . . auto on left bank . . . tried to flag down . . . racing . . ."

Taylor cradled the receiver gently. It was all in the file. In June Mail. Buried. Karen Silkwood was run off the road, and the FBI knows who did it to her.

According to the FBI summary report, page 3, a car followed Karen Silkwood from the Hub Cafe, but lost her when she turned west (right), approximately 200 yards from the death scene, down a narrow oil road (the road to Sherri Ellis' father's farm).

The chase car eventually turned down the same road, looking for Karen; it met her returning to Highway 74. The chase car tried to stop her on the oil road, but she sped around it. Then the chase car raced to a turnaround on the right and followed her.

When she reached 74, Karen hit her brakes. The chase car slid into the Honda, lightly tapping it.

Karen turned left as if to head back to the Hub Cafe, then swung right toward Oklahoma City, ending up on the grassy left shoulder. The chase car was right on her, running parallel, boxing her in. The driver of the chase car tried to flag Karen down.

She was racing along the ditch . . .

Epilogue

Kerr-McGee appealed. It asked the court to lower the award, arguing that a $505,000 compensation for pain and injury was too much, and a $10 million punishment "unconscionably excessive."

Kerr-McGee also asked the court to declare a mistrial because:

□ Judge Frank Theis had refused to instruct the jury on the workers' compensation issue.

□ Judge Theis had allowed the jury to ignore the federal standards and regulations.

□ Judge Theis had erred in defining plutonium as "ultrahazardous." And the plaintiff had failed to prove that any negligent act of Kerr-McGee had caused Karen Silkwood injury.

□ The verdict had been contrary to the manifest weight of the evidence.

□ The trial had not been fair — excessive pretrial publicity; misconduct of Gerald Spence; allowing the plaintiff to offer evidence only remotely related to the case, while denying Kerr-McGee the opportunity to offer its evidence; and unfairly instructing the jury on "physical injury."

The Tenth Circuit Court of Appeals has not yet ruled on Kerr-McGee's appeal; but a three-judge appeals panel upheld Judge Theis's dismissal of the conspiracy charges against Kerr-McGee, Jacque Srouji, FBI agent Larry Olson, Sr., and others. The Silkwood estate is continuing that appeal.

Congressional investigator Peter Stockton filed a complaint against Dean McGee, Jacque Srouji, Larry Olson, James Reading, the FBI, the Kerr-McGee Nuclear Corporation, and others. He alleges a conspiracy to prevent him from discharging his duties in an office of trust; deprive him of his right to privacy; and libel him "by causing to be published to third persons false and derogatory information." All the defendants have denied the charges.

Notes and Sources

I became fascinated with the Silkwood story in 1977 when, as Washington correspondent for the *National Catholic Reporter,* I did a roundup piece on the complaint filed by the Silkwood estate. Several months after the article appeared in the weekly, the Quixote Center asked me to write a 25,000-word pamphlet summarizing the Karen Silkwood story. I accepted the assignment, hoping the research would lead to a book.

I spent the first four months of 1978 working on "Karen Silkwood: Union Sister," published by the Quixote Center in tabloid form. In 1979, I attended the last part of the ten-week negligence trial in Oklahoma City. Even after the verdict, I continued to be piqued by the unanswered questions in the Silkwood case.

There has been so much emotional ink spilled over the memory of Karen Silkwood that the facts surrounding her contamination and death have been lost in speculation and specious arguments. Aware of that, I have tried to present a straightforward account. But, unfortunately, Olson and Srouji declined to be interviewed. And Dean McGee said, through a spokeswoman, that it would be inappropriate for Kerr-McGee defendants to grant interviews at this time.

This book is filled with dialogue, and I want to emphasize that I have "created" none of it. The dialogue came from my interviews and from the more than 25,000 pages of documents in the public record, principally:

☐ 11,000 pages of trial transcripts,
☐ approximately 6000 pages of pretrial depositions,

☐ 2000 pages of FBI documents released in October 1978,
☐ approximately 2000 pages of legal papers,
☐ 1600 pages of transcripts of congressional hearings,
☐ an undetermined number ⌐f Kerr-McGee, Atomic Energy Commission, and Nuclear Regulatory Commission papers.

There is no way I could have found and read all these documents while writing a timely book without the help of my research assistant, Aaron Weinstein. Nor could I have written a readable account of the Silkwood story without the help of Paula Kaufmann and the patient guidance of Frances Tenenbaum.

CHAPTER 1

The background material on Karen Silkwood came principally from the depositions of Bill and Merle Silkwood; the deposition of Karen's former husband, William Meadows; the author's interviews with Drew Stephens; and an article by B. J. Phillips, "The Case of Karen Silkwood," published in *Ms.* magazine in April 1975.

The author made no attempt to interview Bill or Merle Silkwood about their daughter because the Silkwood estate sold the movie rights for the life story of Karen. The author did not want to get caught in contractual problems.

The description of how the Cimarron plutonium plant operated is based on a lengthy description by James V. Smith, a former Kerr-McGee plant manager, in his September and October 1977 deposition. The analysis of the Crescent OCAW local is based on the author's interviews with Drew Stephens, Jack Tice, and the 1974 OCAW regional representative, Earl Campbell.

The descriptions of conditions at the plant are based on Kitty Tucker and Eleanor Walters' research booklet, *Plutonium and the Workplace,* Environmental Policy Institute, March 1979, as well as on Atomic Energy Commission incident reports.

Page 12: After Karen Silkwood's death, Drew Stephens kept a diary for about a month. In the notebook was a short, emotional profile of Karen called "Karen Gay Silkwood — A Personal View." The quotation comes from that profile.

Page 14: Kerr-McGee investigators wrote an account of Connie Edwards' experience with Karen Silkwood in a four-page, undated document, "Karen Silkwood's Drug Use." Jacque Srouji used a copy of this document during her testimony before the House Subcommittee on Energy and Environment. Kerr-McGee denies giving it to her. The document also presented hearsay evidence of an earlier suicide attempt. Kerr-McGee deposed Connie Edwards and was prepared to call her as a witness in the negligence trial in April 1979, but Judge Frank Theis refused to allow Kerr-McGee attorneys to introduce evidence about Silkwood's alleged suicide attempts, stating that it constituted "character assassination."

Page 17: It has never been clear how badly Karen Silkwood was contaminated at the end of July 1974. Some of her bio-assay samples were spiked; some were mislabeled; others were not labeled at all. Evidence has been destroyed, and it is impossible to reconstruct the facts.

CHAPTER 2

The account of the meeting Karen Silkwood, Jack Tice, and Jerry Brewer had with the OCAW in Washington and the AEC in Bethesda, Maryland, is based on the author's interviews with Jack Tice, Steve Wodka, and Anthony Mazzocchi, as well as the depositions of Jack Tice, Jerry Brewer, and Steve Wodka. The summary of allegations and issues that Silkwood, Tice, and Brewer brought to the attention of the Atomic Energy Commission is based on an AEC report of the September 1974 meeting.

Page 19: In a two-part article in *The Saturday Evening Post* ("Karen Silkwood Without Tears," December 1979 and January 1980), syndicated columnist Nick Thimmesch alleges that Jack Tice blew the whistle on Kerr-McGee because there was a drive to decertify the union at the Cimarron plant. But correspondence between Tice and OCAW vice-president Elwood Swisher indicates that both the local and the international decided to do something about health and safety *before* the decertification initiative.

CHAPTER 3

Most of Chapter Three is based on a transcript of an October 7, 1974, phone call from Karen Silkwood to Steve Wodka, and a transcript of the Abrahamson-Geesaman question-and-answer session with the Kerr-McGee workers on October 10, 1974. In an attempt to prove that the OCAW used Karen Silkwood, Nick Thimmesch says in his *Saturday Evening Post* articles that Wodka secretly taped the conversation with Silkwood. The transcripts, however, clearly show that Silkwood knew she was being taped and agreed to it.

CHAPTER 4

The profiles of the Kerr-McGee Corporation, Robert S. Kerr, and Dean A. McGee are based on: Ann Hodges Morgan, *Robert S. Kerr: The Senate Years,* University of Oklahoma Press, 1977; Martin Hauan, *He Buys Organs for Churches, Pianos for Bawdy Houses,* Midwest Political Publishers, 1976; Bobby Baker, *Wheeling and Dealing,* W. W. Norton and Company, 1978; John S. Ezell, *Innovations in Energy: The Story of Kerr-McGee,* University of Oklahoma Press, 1979; Marquis W. Childs, "The Big Boom From Oklahoma," *Saturday Evening Post,* April 9, 1949; "Kerr-McGee: Lock on Uranium Future," *Business Week,* March 18, 1968; Daniel Seligman, "Senator Bob Kerr: The Oklahoma Gusher," *Fortune,* March 1959; "Uranium: Boom with a Bang," *Time,* July 30, 1956.

The material about the uranium miners is based on: Joseph K. Wagoner, "Uranium: The United States Experience — A Lesson in History," Environmental Defense Fund, Washington, D.C., 1980; Tom Barry, "The Navajo Lung Cancer Widows," *Navajo Times,* August 23, 1978; Amanda Spake, "Navajo Miners and Lung Cancer," *Outlook, Washington Post,* June 9, 1974; "The Grants Strike: Uranium Workers Fight Historic Battle," *Union News,* November, 1973; Administration of the Federal Metal and Nonmetallic Mine Safety Act, *Annual Report to Congress,* U.S. Department of Interior, 1977.

Page 46: The author reviewed a wide variety of statistics on how many Navajo uranium miners died of cancer and how many are now dying of the disease. The figures used here are based on an unpublished research paper, "Lung Cancer Among Navajo Uranium Miners," by Leon Gottlieb, M.D., and Luverne A. Husen, M.D., as well as on an interview with Joseph Wagoner, who is considered to be an expert on uranium miners and cancer.

Page 48: Statistics on the frequency of tornados in Oklahoma come from the 1979 National Severe Storm Forecast Center Study, Will Rogers World Airport, Oklahoma City, *Weatherscan Report.* On July 29, 1975, Kerr-McGee described its tornado protection program in its responses to Atomic Energy Commission questions 23, 25, 26, 28, 29, 30, 36, 37, and 39.

CHAPTER 5

Descriptions of the Karen Silkwood contamination incidents are based on: AEC Region III Investigative Report 74–09; Alvin Wayne Norwood's Kerr-McGee report, dated December 13, 1974; Karen Gay Silkwood Contamination Statement, given to Kerr-McGee lawyers and dated November 7, 1974. Also on FBI interview reports on Sherri Ellis, December 10, 1974; Donald Gummow, January 28, 1975; John Carver, April 9, 1975; Alvin Wayne Norwood, March 31, 1975; and Dean Abrahamson, October 3, 1975. And on the depositions of Wayne Norwood, October 1977; Drew Stephens, February 1977; James V. Smith, September and October 1977; and James Frey, February 1979.

The allegations of lax security at the Kerr-McGee Cimarron plants are based on: NRC Region III Report 070–1193/75–13; the trial testimony of Ron Hammock and James V. Smith, March and April 1979; Kitty Tucker and Eleanor Walters, *Plutonium and the Workplace,* Environmental Policy Institute, March 1979; AEC memorandum, "Possible Theft of SNM — Kerr-McGee Corporation Cimarron Facility, Crescent, Oklahoma," June 5, 1974; Peter Stockton memorandum to Representative John Dingell, "Re: Preliminary Questions on Safeguards at Kerr-McGee — MUF and

Physical Security and the Regulatory Response by NRC," March 8, 1976.

Page 50: Karen Silkwood's October 31, 1974, accident was investigated after her death by Lieutenant Ronnie Johnson of the Oklahoma Highway Patrol (accident report #92, November 26, 1974).

Page 50: Nick Thimmesch says in *The Saturday Evening Post* that Dr. Shields "gave her [Silkwood] a lecture" for mixing drugs. But Dr. Shields told this author that he did not lecture Karen, did not suspect she was overlapping drugs on purpose, and that she was very cooperative. Dr. Shields also said that he never talked to Mr. Thimmesch and doesn't know where Thimmesch got his information.

There is some evidence to suggest that Karen Silkwood had taken more than Tylenol No. 3 on the job. She was very definitely high on something, but it is not clear what.

CHAPTERS 6 AND 7

In addition to the sources listed for Chapter 5, these chapters are based on: the depositions of Steve Wodka, June 1978, and Merle Silkwood, June 1978; FBI interviews with Steve Wodka, January 28, 1975, and with Dr. Neil Wald, November 28, 1974; the author's interviews with Steve Wodka and Tony Mazzocchi; trial testimony of Dr. George Voelz, April 1979; and a memorandum written by John Davis, "Notes to File, November 14, 1974."

Page 70: The author visited the Los Alamos Scientific Laboratory in April 1980. Dr. John Umbarger, who had supervised the Silkwood tests in 1974, explained the full-body and lung count procedures. The vault in which Karen Silkwood was tested is still there, but is no longer used.

CHAPTER 8

The Hub Cafe incident is based on: the depositions of Jack Tice, September 1977; Frank Murch, February 1979; Wanda Jean

Jung, September 1978; and Gerald Brewer, September 1977. Also on the FBI interview with Drew Stephens, February 4, 1975; the author's interview with Jack Tice; and Wanda Jean Jung's affidavit, January 11, 1975.

Karen Silkwood's accident and the events immediately after her death are based on: the Oklahoma Highway Patrol's Karen Silkwood accident report, November 15, 1974; the Oklahoma State medical examiner's Karen Silkwood Autopsy Report, ML 501–74; the depositions of Lieutenant Rick Fagen, February 1979; Drew Stephens, February 1977; and Steve Wodka, June 1978. Also, on the author's interviews with Jack Tice and Steve Wodka; FBI interviews with Lieutenant Rick Fagen, November 27, 1974; Drew Stephens, February 4, 1975; Ted Sebring, November 27, 1974; Harold Smith, December 2, 1974; Ken Hart, December 9, 1974; John Trindle, June 3, 1975; James Mullins, June 3, 1975; and Law Godwin, December 9, 1974; as well as on Kerr-McGee interviews with John Trindle, January 31, 1975; Fred Sullivan, December 13, 1974; Dalton Ervin, undated; and Harold Smith, November 26, 1974.

CHAPTER 9

This chapter is based on: the author's interviews with Steve Wodka, Tony Mazzocchi, A. O. Pipkin, and Dr. Ernest L. Martin; the depositions of Drew Stephens, February 1977, and Dr. A. Jay Chapman, February 1979; an article about Pipkin by Gregory Curtis, "This Man Loves Car Wrecks More Than Anyone in the World," *Texas Monthly,* May 1975; an essay on the Pinkerton Detective Agency by George O'Toole in *The Private Sector,* W. W. Norton and Company, 1978; Pinkerton Detective Agency Report (Dallas Office) for Kerr-McGee, November 21, 1974; James Reading memorandum, "James Reading to File," November 21, 1974; W. Spot Gentry memorandum, "W. Spot Gentry to Reading," November 21, 1974; A. O. Pipkin, "Accident Reconstruction Lab Report on Karen Silkwood's Automobile Accident," December 15, 1974; Dr. Gerald U. Greene report to A. O. Pipkin, "Re: Karen Silkwood," January 20, 1975; Dr. Ernest L. Martin report to A. O. Pipkin, "Examination of Section of Automobile Fender," February 19, 1975.

CHAPTER 10

This chapter is based on: the author's interviews with Steve Wodka, Tony Mazzocchi, and Jack Tice; the trial testimony of James V. Smith, Ken Plowman, Ron Hammock, and James Noel in March and April 1979; Gerald Brewer deposition, September 1977; FBI interviews with James Reading, December 16, 1975, and Dr. Neil Wald, November 28, 1975; the prepared statement of Deputy Assistant Attorney General John C. Keeney, Criminal Division, Department of Justice, before the House Subcommittee on Energy and Environment, May 7, 1976; Kerr-McGee list of "Relevant Questions" for polygraph testers, undated; letter of Anthony Mazzocchi to NRC's John Davis, January 21, 1975; letter of A. F. Grospiron to OCAW Local 5–283, January 6, 1975; Morgan Moore memorandum to all Cimarron facility employees, December 30, 1975; the undated opinion of Paul C. Dugan, "In the Matter of Arbitration Between Kerr-McGee Nuclear Corporation and OCAW 5–283, and Jerry Brewer, Grievant"; Special Agent in Charge, Oklahoma City, memorandum to FBI director, September 13, 1975; letter of FBI director Clarence Kelley to Anthony Mazzocchi, June 2, 1975; Battelle Pacific Northwest Laboratory Report to Kerr-McGee, April 15, 1975; report of FBI special agent Lawrence Olson, Sr., April 17, 1975; Lawrence Olson, Sr., "FBI Summary of the Karen Silkwood Contamination," December 15, 1975; AEC Region III Investigative Report 74–09; AEC Report 44–339, December 19, 1974; AEC Region III Investigative Reports 070–925/74–05, 070–1193/74–10, 040–7308/74–03; and Wanda Jean Jung's affidavit, January 11, 1975.

Page 107: Westinghouse had consistently supported the AEC and Kerr-McGee on the question of the quality of the fuel rods K-M sent to Hanford, Washington. The only evidence to support allegations that the K-M rods were defective is the statements of the workers who assembled and tested them. The question could easily be settled once and for all if Kerr-McGee, Westinghouse, and the NRC made public *all* the records and memoranda dealing with the fuel pins.

CHAPTER 11

This chapter is based on: the author's interviews with Peter Stockton, Jack Tice, Dr. Clarence Shields, Jr., and Tony Mazzocchi; FBI interview with Barbara Newman, January 18, 1975; the deposition of Jerry Brewer, September 1977; Barbara Newman, National Public Radio, transcripts, December 1974 and March 1975; *New York Times,* December 29, 1974; and transcripts of "The Reasoner Report," March 1, 1975.

Page 111: The description of Sherri Ellis' attack on the Kerr-McGee Cimarron plant with a .22 rifle is based on FBI report 117–53, August 15, 1975, and William Davis' notes from his interview with Ellis, December 12, 1978. The author has tried to interview Ellis, whom he met in Oklahoma twice and to whom he spoke on the phone several times. She declined to be interviewed, because, she said, she was writing her own book on Silkwood. She has also refused to sell the right to use her name and character in the proposed movie about Karen Silkwood. Ellis knows a lot about the Silkwood case. Kerr-McGee subpoenaed her before the March 1979 negligence trial, but she got wind of the subpoena and hid so that she could not be served. She had told Silkwood investigators that while she was in jail after attacking the K-M plant, FBI agent Larry Olson, Sr., and K-M security director James Reading tried to interview her. She refused to talk. The FBI report about the visit mentions nothing about James Reading. It is signed by Lawrence Olson, Sr., and Jo Don Baker, an Oklahoma City agent. Ellis claims Baker never came to see her.

Page 114: Dr. Shields told the author that the second doctor whom he suspected of prescribing Quaaludes to Karen Silkwood was a physician in Guthrie, Oklahoma, now deceased. The author has not been able to verify his name or the fact that he may have prescribed drugs. But the author has seen some evidence that, on occasion, Karen used such hallucinogens as mescaline and LSD, as well as cocaine.

Page 117: There have been allegations that the Oklahoma Highway Department resurfaced Highway 74 outside Crescent to destroy the evidence at the accident scene. However, there are no facts to support that charge. The Highway Department had decided to repave twenty miles of Highway 74 the year before Silkwood was killed, and had let two contracts for the work, one in September 1973, the other in July 1974. The work close to the accident site began about a month after Silkwood's death. Everyone had ample time to study the road — the Oklahoma Highway Patrol, the FBI, and A. O. Pipkin.

Page 118: Airlie House founder and director Dr. Murdock Head sued the Washington *Evening Star* for reporting the allegations of William Higgs, a radical Washington attorney. In a September 1967 press conference, Higgs said he had studied Airlie's tax returns, land records, and other documents, and failed to find any source of substantial income. Higgs concluded that Airlie House had been covertly supported by the Pentagon, the State Department, and the Central Intelligence Agency. He charged that the Airlie House had an elaborate system of electronic bugging. Dr. Head denied the allegations and opened his books to the *Star.* The newspaper retracted its story, and Head took the *Star* to court and won a $519,800 damage suit, which U.S. District Court Judge Oliver Gasch trimmed to $60,000. Ten years later, a federal jury convicted Dr. Head of conspiracy to evade taxes and/or to bribe Representative Daniel Flood, Representative Otto Passman, and an IRS agent. See: Washington *Evening Star,* September 14, 1967, September 15, 1968, September 18, 1968; Washington *Post,* October 14, 1971, January 22, 1972, October 13, 1979, March 31, 1978, April 22, 1978; and Richard Pollock, "The Mysterious Mountain," *Progressive,* March 1976.

CHAPTER 12

This chapter is based on: the author's interviews with Sara Nelson, Kitty Tucker, Karen DeCrow, Tony Mazzocchi, and Patricia Welch; NOW, "Labor Task Force Begins Here," August 1975;

NOW, "Chapter Action — SOS," July 1975; NOW, "NOW Protests Violence Against Women in National Action on August 26, 1975," undated; NOW press release, August 25, 1975; "NOW Seeks Justice in Silkwood Case," *Do It NOW,* September/October 1975; "Memo: To John Seigenthaler from Patricia Welch," May 10, 1976; *New York Times,* October 27, 1975, October 28, 1975, November 15, 1975; Philadelphia *Inquirer,* October 25, 1975, October 27, 1975; NOW, "Majority Caucus," undated; *Perspective: Newsletter for the NOW National Advisory Committee and Leadership,* vol. 1, no. 1, 1976; *Do It NOW,* May/June 1975 and September/October 1975.

CHAPTER 13

Besides the author's interviews with Peter Stockton, Sara Nelson, E. Winslow Turner, and Victor Reinemer, this chapter is based on a series of internal congressional documents: "Karen Silkwood Investigation: Communication with the Justice Department," undated; letter of Senator Lee Metcalf to Attorney General Edward Levi, November 17, 1975; Stockton to Dingell memorandum, "Re: Silkwood Investigation — Request for Documents from the Justice Department," November 18, 1975; Stockton to Dingell memorandum, "Re: Ribicoff Assures Full Backing in Silkwood Investigation — Subpoena Power, Additional Staff," November 18, 1975; Turner to Metcalf memorandum, "Re: Ribicoff Meeting with NOW and Other Groups — Karen Silkwood Matter," November 18, 1975; "Prepared Statement of Senator Lee Metcalf," undated; Turner to files memorandum, "Silkwood Investigation: Request for Access to Justice Department and FBI Files and Personnel," November 21, 1975; Justice Department Order No. 116–56, 1956; Turner to Metcalf memorandum, "Re: Silkwood/ Kerr-McGee Investigation — Refusal of Justice Department to Produce Files," November 24, 1975; Turner to files memorandum, "Re: Karen Silkwood Investigation," November 25, 1975; Peter Stockton, "Observations After Looking at Justice Department Files on the Silkwood Case," undated; "Goldstein Notes from FBI Summaries," undated; Peter Stockton, "Observations on the Adequacy of the Justice/FBI Investigation of Silkwood," December 8, 1975; Peter Stockton, "Basic Attitude of Justice," undated; Stockton to Dingell memorandum, "Subject: FBI Summary on Diversion of Plutonium and Silkwood Contamination," March 24, 1976; Gene Schmitt, General Accounting Office, handwritten notes about his January 28, 1975, phone conversation with Justice Department attorney Thomas Henderson.

The author also used: Peter Petkas, *Lee Metcalf: Democratic Senator from Montana*, Grossman Publishers, 1972; and Justice Department, "Karen Silkwood Death: Fact Memorandum," undated.

CHAPTER 14

This chapter is based on: the author's interviews with Win Turner and Peter Stockton; Peter Stockton's affidavit, "Notes on the Srouji Affair," date not legible; and Peter Stockton, "Random Notes of Meeting with Source," December 12, 1975.

Page 140: In her book *Critical Mass*, Jacque Srouji gives an entirely different account of her first meeting with E. Winslow Turner. Among other things, she alleges that the attorney threatened to subpoena her documents. Turner denied the charge during his interview with the author.

CHAPTER 15

This chapter is based on: the author's interviews with Win Turner, Peter Stockton, Victor Reinemer, and Peggy McLaughlin-Doran; the depositions of Jacque Srouji, June 1977, and Dean A. McGee, June 1977; Peter Stockton telephone interview notes for February 13 and 14, 1976; Stockton to Dingell memorandum, February 10, 1976, untitled; Turner to Ryter memorandum, "Kerr-McGee/Karen Silkwood Investigation — Field Trip to Chicago and Oklahoma," March 3, 1976; Turner to Bill Brock letter, March 4, 1976; Barbara Newman, National Public Radio, "All Things Considered," April 6, 1976; OCAW, "Statement of A. F. Grospiron,

President," April 7, 1976; Patrick Mahoney and Rolly Hochstein, *John Dingell: Democratic Representative from Michigan,* Grossman Publishers, August 1972.

The comments on the American Security Council are based on: Wes McCuen, "Facts About the American Security Council," Group Research, Inc., May 25, 1962; American Security Council, "1980 President's Report"; Joan M. Jensen, *The Price of Vigilance,* Rand-McNally, 1968; *Congressional Record,* September 7, 1979; "The Right Report," May 6, 1977; Washington *Post,* January 8, 1974, and October 26, 1970; *New York Times,* August 17, 1970; and Harold Relyea, "Hawk's Nest: American Security Council," *Nation,* January 24, 1972.

CHAPTER 16

This chapter is based on: the author's interviews with Sara Nelson, Michael Ward, and Peter Stockton; the depositions of Jacque Srouji, June 1977, and Drew Stephens, February 1977; FBI interview with Dominic de Lorenzo, date deleted; Peter Stockton's affidavit, "Notes on the Srouji Affair," date not legible; Kirk Loggins untitled memorandum to John Seigenthaler, May 10, 1976; William Davis to Mike Tate telephone conversation notes, March 29, 1978; Kerr-McGee, "Karen Silkwood's Drug Use," undated; *Problems in the Accounting for and Safeguards of Special Nuclear Materials: Hearings Before the Subcommittee on Energy and Environment of the Committee on Small Business,* U.S. House of Representatives, April 26, May 7, and May 20, 1976.

CHAPTER 17

This chapter is based entirely on the author's interviews with Peter Stockton and Michael Ward, as well as the transcripts of the congressional hearings cited above.

CHAPTER 18

The material on Jacque Srouji is as accurate as the author can possibly make it. It is pieced together from a variety of sources. In general, the author used those facts which have been verified by two independent sources or stated twice by Srouji herself.

Some of the most accurate information on Srouji, this author believes, comes from John Seigenthaler, Jr., son of the *Tennesseean*'s publisher. In an undated paper, young Seigenthaler said he overheard part of Jacque Srouji's discussions with his father. The author called John Seigenthaler, Sr., several times for an interview, but did not receive a return call from the publisher.

In addition to the sources cited above, this chapter is based on: the congressional hearing transcripts cited previously; the author's interviews with Bill Kovach and with Michael Ward and Paul Kritzer; the depositions of Jacque Srouji, February and June 1977; Ken Brannon's interviews with Jacque Srouji, "Why Was Srouji Fired?" *Nashville!*, October 1976, and "Was Jacque Srouji Really a Spy for the FBI?" *Nashville!*, November 1976; Jacque Srouji, *Critical Mass*, Aurora Publishers, 1976; memorandum from Srouji to de Lorenzo, "Re: New Lead," April 22, 1975.

CHAPTER 19

The report of the Peter Stockton smear is based on: the author's interviews with Peter Stockton; memorandum from Peter Stockton to John Dingell, "FBI Attempt to Intimidate Staff by Releasing an FBI Document Containing False and Derogatory Statements Concerning a Subcommittee Staff Member," undated; and FBI documents released October 14, 1977.

The report of the John Seigenthaler smear is based on: an unpublished paper by John Seigenthaler, Jr.; FBI telexes about John Seigenthaler, May 4, 1976, May 10, 1976, and May 6, 1976; a syndicated story by John Seigenthaler outlining his FBI problem; and the congressional hearing transcripts cited previously.

The report of the John Dingell smear is based on: a John Dingell letter to Edward Levi, June 16, 1976; Detroit *News,* June

20, 21, 22, and 30, 1976; and Jack Anderson, "FBI Smear Tactics in the Silkwood Case," syndicated column, February 4, 1980.

Page 182: Unfortunately, Representative Dingell declined, through a spokesman, to be interviewed by the author, saying that he had more pressing energy-related matters to take care of.

CHAPTERS 20 and 21

These chapters are based on the author's interviews with Daniel Sheehan and William Taylor, as well as the November 5, 1976, Silkwood estate complaint against Kerr-McGee, et al. The author uses code names for confidential sources to protect their identities and their personal safety.

CHAPTER 22

This chapter is based on: the author's interviews with Daniel Sheehan, William Davis, and William Taylor; the depositions of Jacque Srouji, February and June 1977; affidavits of Daniel Sheehan and William Davis, March 1, 1977; and Claude E. Love, "Brief in Support of Application for Protective Order by Defendant Srouji," February 1977.

CHAPTER 23

This chapter is based on the author's interviews with William Davis, William Taylor, and Daniel Sheehan; a series of attorneys' confidential reports that the author is not at liberty to list; and the deposition of James V. Smith, September and October 1977.

The information on the Georgia Power Company and *Information Digest* is based on, besides the sources listed above: "Congressional Aide Spies on Left," *Counter Spy,* Spring 1976; George O'Toole, *The Private Sector,* W. W. Norton and Company, 1978; Frank Donner, *The Age of Surveillance,* Alfred A. Knopf, 1980; Atlanta *Journal,* September 9 and October 5, 1977;

Richard Pollock, "The Shifty Eye of Reddy Kilowatt," *Mother Jones*, May 1978; sworn statement of John H. Taylor, September 23, 1977; Arthur Benson deposition, December 13, 1974.

The information on LEIU is based on: George O'Toole, "America's Secret Police Network," *Penthouse*, December 1976; "LEIU Disseminating Political Data," *Organizing Notes*, October 1978; Jack Anderson syndicated column, September 19, 1978; *LEIU*, a draft paper of the American Friends Service Committee, March 1978; Frank Donner, *The Age of Surveillance;* and George O'Toole, *The Private Sector.*

The information on Jack Holcomb, the National Intelligence Academy, and Audio Intelligence Devices Corporation is based on: Jim Hougan, *Spooks,* William Morrow and Company, 1978; George O'Toole, *The Private Sector;* "Bugging School," *Newsweek,* March 10, 1975; Fort Lauderdale *News,* February 20, 1975; James Ellison, "A Report from the Wiretap Subculture," *Washington Monthly,* December 1975; and the Miami *Herald,* February 23, 1975.

CHAPTER 24

This chapter is based on the author's interviews with Daniel Sheehan, William Davis, William Taylor, and Sara Nelson; 2000 pages of FBI documents released on October 17, 1977; and a memorandum from James Reading to Dean A. McGee, untitled, March 4, 1975.

The author has not given full reference information for the FBI documents for the following reasons: names and dates are deleted; it is frequently impossible to tell if the document is a telex, LHM, or 302 form; and there is often more than one version of the document, some versions with more deletions than others. The author has read all the documents, made notes by topic, and summarized his notes, with the exception of a few direct quotations. Daniel Sheehan has a complete set of the bound FBI papers.

CHAPTER 25

This chapter is based on: the author's interviews with Daniel Shee-han and William Taylor; the depositions of FBI special agent Law-rence Olson, Sr., October 1977 and April 1978; and the deposi-tions of James Reading, October 1977 and April 1978.

CHAPTER 26

The material on the hearing before Judge Luther Bohanon, No-vember 18, 1977, is based on the transcript of that hearing. The Judge Bohanon–Robert Kerr material is based on: Ann Hodges Morgan, *Robert S. Kerr: The Senate Years,* University of Okla-homa Press, 1977; Bobby Baker, *Wheeling and Dealing,* W. W. Norton and Company, 1978; *Daily Oklahoman,* August 20, 1961; Oklahoma City *Times,* September 11, 1969, July 26, 1961, July 21, 1961, June 27, 1961, August 9, 1961, August 18, 1961, September 30, 1961, July 16, 1961; and John Seigenthaler's affi-davit, July 23, 1978. Other sources used in this chapter are the author's interviews with William Davis, William Taylor, Daniel Sheehan, and Tony Mazzocchi.

CHAPTER 27

This chapter is based on the author's interviews with William Tay-lor, William Davis, and Daniel Sheehan; confidential attorneys' working papers, which the author is not at liberty to list; the dep-osition of Jack Larsen, June 8, 1978; and the death certificate of Leo Goodwin, January 18, 1978.

CHAPTER 28

Besides the author's interviews with Daniel Sheehan, William Davis, and William Taylor, this chapter is based on the following depositions: FBI agent Lawrence Olson, Sr., April 1978; Steve Campbell, April 1978; Robert Hicks, May 1978; William Byler, June 1978; and the sworn affidavit of Robert Ivins, July 15, 1978.

CHAPTER 29

This chapter is based on: the author's interviews with William Taylor; a taped interview of George Sturm by Jim Ikard, May 4, 1978; "Supplemental Inventory, Organized Crime Unit, Oklahoma City Police Department," March 31, 1978; and the depositions of Larry Baker, May 1978, Larry Upchurch, June 1978, Ken Smith, June 1978, Bill Vetter, April 1978, David McBride, May 1978, and Thomas Heggy, June 1978.

CHAPTER 30

Besides the author's interviews with Daniel Sheehan, Sara Nelson, William Taylor, and Arthur Angel, this chapter is based on Judge Theis's "Memorandum Opinion and Order," September 25, 1978; the amended complaint, "The Silkwood Estate v. Kerr-McGee et al.," and the depositions of Roy King, June 1978, Rick Fagen, February 1979, W. Spot Gentry, February 1979, and Jack Larsen, June 1978.

The full account of Dan Sheehan's arrest is contained in the brief "The Silkwood Estate v. Kerr-McGee et al.: The Law Governing the Potential Admissability of the Evidence Sought by the Plaintiffs for the Silkwood Witnesses," June 21, 1978.

CHAPTERS 31, 32, 33, and 34

The author attended part of the Kerr-McGee negligence trial in April and May 1979. He based these chapters on his notes, observations, and interviews during the trial, as well as on the complete trial transcripts.

The Silkwood estate presented nineteen witnesses. The author did not summarize the testimony of seven of those witnesses because they duplicated the testimony of others. Those seven were Ronald Fine, Edward Martell, Anthony Mazzocchi, Gerald Schreiber, Rose Mary Porter, Bill Silkwood, Merle Silkwood, and Linda Silkwood. Schreiber's testimony was by deposition.

Kerr-McGee presented twenty-four witnesses. The author summarized the testimony of only ten because most of the other witnesses were Kerr-McGee personnel who repeated the same theme: Kerr-McGee ran a safe and healthful Cimarron facility. The witnesses whose testimony the author did not summarize are: Marvin Barger, Martin Binstock, James Carr, Parker Dunn, Don Majors, Robert Marshall, Dean McGee, Morgan Moore, Frank Pittman, Gerald Schreiber (by deposition), William Shelley, Randy Snodgrass (re-called), Charles Sternhagen, and Fred Welch.